1·17 02

Evolutionary Patterns

D0839008

Evolutionary Patterns

Growth, Form, and Tempo in the Fossil Record

In Honor of Alan Cheetham

Edited by
JEREMY B. C. JACKSON, SCOTT LIDGARD,
AND FRANK K. MCKINNEY

The University of Chicago Press
Chicago and London

JEREMY B. C. JACKSON is director of the Center for Tropical Paleoecology and Archeology at the Smithsonian Tropical Research Institute and the William and Mary B. Ritter Professor of Oceanography at the Scripps Institute of Oceanography. He is the author or coauthor of more than one hundred articles, coeditor or coauthor of several books, including *Evolution and Environment,* published by the University of Chicago Press.

SCOTT LIDGARD is associate curator of fossil invertebrates in the Department of Geology at the Field Museum and lecturer in the Committee on Evolutionary Biology at the University of Chicago. He is the author of many articles in the area of paleobiology.

FRANK K. MCKINNEY is professor emeritus in the Department of Geology at Appalachian State University and Honorary Research fellow, Department of Paleontology, The Natural History Museum, London. His publications include many scholarly articles and two books: *Bryozoan Evolution* (coauthored with Jeremy Jackson) and *Exercises in Invertebrate Paleontology.*

The University of Chicago Press, Chicago 60637
The University of Chicago Press, Ltd., London
© 2001 by The University of Chicago
All rights reserved. Published 2001
Printed in the United States of America

10 09 08 07 06 05 04 03 02 01   1 2 3 4 5

ISBN: 0-226-38930-8 (cloth)
ISBN: 0-226-38931-6 (paper)

Library of Congress Cataloging-in-Publication Data

Evolutionary patterns : growth, form, and tempo in the fossil record in honor of Alan Cheetham / edited by Jeremy B. C. Jackson, Scott Lidgard, and Frank K. McKinney.
      p. cm.
   Includes bibliographical references and index.
   ISBN 0-226-38930-8 (cloth : alk. paper) — ISBN 0-226-38931-6 (paper : alk. paper)
   1. Evolutionary paleobiology. 2. Paleoecology. 3. Evolution. I. Cheetham, Alan H. II. Jackson, Jeremy B. C., 1942–   . III. Lidgard, Scott. IV. McKinney, Frank K. (Frank Kenneth)
QE721.2.E85 E965 2001
576.8—dc21
                                                          00-068008

Contents

Part 3. Macroevolutionary Patterns and Trends

Preface

One could well ask why even try to study evolutionary process with fossils after all of the extraordinary recent advances in evolutionary biology? Detailed field studies of Galápagos finches have revealed the strongest natural selection ever measured in the wild. Evolutionary experiments with the bacterium *Escherichia coli* have resulted in large, sudden shifts in gene frequencies and adaptation to their experimental environment. Breeding experiments with the fruit fly *Drosophila* demonstrate clear relationships between genetic distances among closely related species and the mode and extent of reproductive isolation—and thus by the assumptions of the molecular clock and the biological species concept the time required for speciation. Phylogenies based on molecular sequence data are helping to resolve patterns of relationships and the order of divergence at all taxonomic levels from the origins of populations and species to the universal tree of life. So what can fossils possibly teach us that living organisms cannot?

One answer, of course, is the relevant timescales of evolution. Deep time gives us the pattern, tempo, and mode of evolutionary transitions that take too long for biologists ever to observe. Generation times of organisms limit direct biological measurements of evolution. Shapes of finches' beaks selected for during strong El Niño events are reversed in intervening years with little or no net phenotypic change. With generation times of finches measured in years, no more than a few such cycles can be observed in a scientist's lifetime. Comparable numbers of observations for *Drosophila* are perhaps 1,000 generations and 100,000 generations for bacteria. However, most fossil species persist for a million generations or more without net morphological change.

Even for *Drosophila,* we study the genetics of population variability, isolation, or local adaptation, but not the genetics of the actual process of speciation that has not been observed. Indirect measurements based on molecular divergence necessarily ignore untold numbers of extinct species scattered across the phylogenetic landscape and their unknown relations to those species still alive. Molecular phylogenies also tell us little about the tempo of evolution without independent calibrations based on the fossil record. Moreover, the very direction of evolution can be turned upside down by homoplasy in morphologically based phylogenies drawn only from recent species.

Fossils also broaden understanding of what is possible in evolution and how long evolutionary changes take beyond the narrow experience of the present. This is doubly important because the present is a very strange time in earth history: an interglacial geologic instant in the tumultuous climatic oscillations of the Pleistocene. Few species alive today originated under circumstances like the present.

Charles Lyell taught us that "the present is the key to the past," and there is a long and rich tradition of using the basic tool kit of organismal biology to interpret changes in morphology and occurrence in the fossil record. Collectively, we expect biological explanations of the fossil record that are mechanistic in addition to historical. Variations that arise under domestication can be studied to understand better the patterns of morphogenesis that yield strikingly different adult forms. Yet how should we extrapolate these observations to morphological changes over geologic timescales, given that the fossil record is made up primarily of adult skeletons? Basic observations of life habits, behavior, biomechanics, physiology, development, growth, and life histories of living organisms provide an essential context for interpretation of fossils that has been exploited fruitfully ever since Cuvier. That is, except when they do not.

More and more we are learning that the past is also the key to the present. Morphological stasis of species over millions of years is a fact, whatever one thinks about the theory of punctuated equilibrium, yet morphological stasis was entirely unexpected by evolutionary biologists. And as John Maynard Smith has observed, the important realization that, once discovered, stasis is compatible with modern evolutionary theory is beside the point that no one seriously thought about the possibility before. Indeed, punctuated speciation and stasis make up just one example of the nonlinearity of virtually everything interesting in evolution as revealed by the fossil record. Mass extinctions and the new catastrophism in geology provide numerous other examples of the evolutionary importance of events unknowable in the present.

A Tribute to Alan Cheetham

Alan Cheetham has contributed fundamentally to all these questions through the extraordinary diversity of his research on cheilostome bryozoans. Rigorous and thorough empiricism is a hallmark of Alan's work. This empirical work on patterns has contributed importantly to understanding some of the main processes that have shaped the history of life.

Discerning evolutionary process from pattern is not simple. Insights into both process and pattern are necessary to develop an understanding of evolution in the fossil record, but neither component is adequate by itself. In our view, this perspective departs substantially from a qualitative comparative tradition that tends to be historically interpretive of individual cases of evolution. For example, it can be argued that patterns validated by empirical tests may constrain an otherwise viable range of inferred evolutionary processes, excluding some observed in living organisms that would seem reasonable historical interpretations of a particular evolutionary transition. In this sense, this empirical perspective and indeed this book are about discerning process from pattern in evolution and the fossil record. A tenet of this empirical perspective is that both components are capable of being verified or disproved by fittingly structured observation or experiment. But these facts are expensive, at least as required to substantiate hypotheses linking growth and form, evolutionary tempo and mode. Legions of detailed observations and comparisons are needed to satisfy the demands of Alan's analytical methods. We suggest that perhaps too few of us have been willing to make the expensive investment in empiricism. This is a perspective that bids us to set aside our preconceived notions when the empirical results deny our beliefs in how evolution works.

The colonial perspective on evolution that is present in much of this book derives from the hierarchical or modular organization of colonial animals such as bryozoans and corals. Unitary organisms are limited in the range of morphology expressed in the phenotype of any given individual. In contrast, colonial organisms are inherently hierarchical systems. Their polyps or zooids express varying degrees of polymorphism within a genetic individual, the colony. Moreover, colonies may acquire different forms (and different ranges of variation) resulting from different patterns of modular growth. Zooids within a colony may show enormous variability compared to characters used to describe unitary animals, yet this range of variability can also provide a useful index of phenotypic variability within and among genotypes. Through Alan's work, the genus *Metrarabdotos* (and, by extension, bryozoans) has become a virtual *Drosophila* of evolutionary paleobiology. Not only is the range of zooid variability within bryozoan colonies highly heritable, but patterns of character covariation can be shown to discriminate otherwise cryptic species, to correlate with genetic distances among species, to be applicable to fossils, and to be recoverable from densely sampled stratigraphic sequences as well! We know of no other major taxon, solitary or colonial, that meets these criteria with such compliance. But Alan's work also extends to other taxa and monographs of entire bryozoan faunas. Even in these more traditional works, Alan contributed basic new insights into the constructional morphology and extensive variation in size and shape of bryozoan zooids that are essential to recognizing bryozoan species, growth and development of entire colonies, and the ecological importance of colony form.

With so much to celebrate, we narrowed our scope to just three of the general areas of evolutionary pattern and process to which Alan has contributed the most. The first concerns modes of development, hierarchies of morphological organization, and the adaptive significance of colony form. The second general area concerns discrimination of species, the tempo of speciation and extinction, and phylogeny. The third and final theme concerns macroevolutionary patterns and trends.

Part 1: Modes of Development, Hierarchies of Morphological Organization, and the Adaptive Significance of Colony Form

Alan's most celebrated work in these areas includes pioneering studies with Richard Boardman on the sources of variation in bryozoan zooids and their importance to ontogeny, astogeny, and polymorphism that run amok in the diversity of bryozoan colony forms. Closely related are the elucidation of the concept of degrees of colony integration that are basic to understanding how bryozoans grow and interact with their neighbors and environment, and the growth forms they achieve in the process. His 1971 paper is among the first we know of that examines the integration of zooid and colony architecture in the context of ecological and biomechanical constraints on form. Finally, Alan's analyses of architecture and design of rigidly erect colony forms are a biomechanical tour de force. Branching treelike colonies have evolved many times in bryozoan evolution. Many of these taxa grow according to rules of bifurcation and branch thickening that minimize interference and thus suggest evolutionary achievement of an optimal space-filling design. Erect taxa through time decrease in material strength, as shown by experimental tests. Yet by applying his empirical models of branching colony growth, Alan showed that later designs become increasingly capable of resisting flow-induced breakage and minimizing interference.

In this spirit, Leo Buss explores the extraordinary complexity of rules of growth and development in hydroids and their potential numerical description and modeling. Modeling of growth by computer involves iterative self-assessment of morphology that determines the quantity (size) and quality (e.g., bifurcation or no bifurcation) of growth in the next growth increment. Buss suggests that clonal organisms do conduct such self-assessment and respond via clonewide conducting systems and pattern-forming genes that generate responses to local states dependent upon global rules of growth. Dan McShea examines the general pattern of decreasing complexity of modular parts compared to free-living organisms across hierarchical levels of complexity. He finds decreases in types of parts and overall functionality at an incorporated level, from prokaryotic cell to complex multicellular colonies. In the final chapter of the group, Beth Okamura, Jean-Georges Harmelin, and Jeremy Jackson demonstrate that availability of food rather than interactions with enemies is more likely the primary sculptor of colony form where food is scarce. They find that "spot" colonies, which reach a small size relative to colonies of phylogenetically related taxa, occur preferentially where food is scarce. This distribution pattern fits much more closely predictions of distribution based on hypotheses of food limitation than those based on hypotheses of competitive refugia.

Part 2: Recognition of Species, and the Tempo of Speciation and Extinction

Alan was the first to come to grips with the enormous variation in size and shape of the modules of colonial animals that had obscured taxonomic resolution since Linnaeus. Beginning with his Eocene Brackelsham fauna paper in 1966, Alan's multivariate, phenetic approach to bryozoan species-level taxonomy became a watershed for the study of character evolution. He did not back down from the challenge of continuously varying characters, but chose instead to document the coefficients of variability for zooidal characters, evaluating their utility for species discrimination and moving beyond the monothetic, typological approach to colonial animal taxonomy that preceded him. In his 1968 paper on *Metrarabdotos*, Alan laid the groundwork for the study of species morphological evolution within a statistically defined morphospace. His work anticipated current paleobiological studies of taxonomic diversity and morphological disparity by nearly 30 years. Stratophenetic analysis of *Metrarabdotos* was the basis of Alan's remarkable confirmation of punctuated bryozoan speciation and reversal of his own previously held view of gradual species transformation. His extension of Charlesworth's rigorous criteria for examination of morphological change within and across species boundaries stands as the paradigm for analysis of the tempo and mode of speciation for all fossil taxa. Similarly, Alan's application of quantitative genetic techniques to fossils is the first successful attempt to reconstruct past roles of natural selection and genetic drift based on empirically derived lineages of fossils.

In this vein, Nancy Knowlton and Ann Budd address problems of resolving species boundaries in reef corals using all available kinds of evidence. The accurate differentiation of tropical coral species is a conundrum that they argue requires observation of living corals, genetic studies, and the most sophisticated, detailed, and discriminatory morphological documentation and statistical analysis that can be mustered. The analytical rigor of defining species and their ranges in space and time are, of course, the essence of tempo and mode. John Pandolfi, Jeremy Jackson, and Jörn Geister demonstrate geologically sudden extinction of two coral species in the absence of human or natural catastrophes. This is yet another example of the fundamentally nonlinear tempo of evolutionary change as revealed by the fossil record. Ross Nehm dissects three independent episodes of developmental changes in marginellid gastropods, each of which led to similar morphologies that are strikingly yet similarly different from their ancestral forms. Interestingly, the three similar episodes of developmental changes occurred in different ecological and environmental contexts, which might be considered as evidence for developmental canalization. Last, Stephen Jay Gould reviews the status of punctuated equilibrium with a full appreciation of Alan Cheetham's fundamental contributions. Fifteen years after Alan surprised himself with an incontrovertible demonstration of rapid evolutionary origin followed by protracted stasis in species of *Metrarabdotos*, his initial and subsequent studies still remain *the* definitive test of rates of species evolution.

Part 3: Macroevolutionary Patterns and Trends

Alan's great contributions to macroevolution concern the evolution of modes and patterns of growth and diversity of colony forms throughout cheilostome evolutionary history. The idea that growth patterns play an important role in determining large-scale macroevolutionary change has seldom been tested in the fossil record. Most notable among these studies are the transition from uniserial to multiserial colony forms and the fine-tuning of branch thickening and bifurcation angles of the branches of rigidly erect bryozoans over 100 million years. Perhaps less well known are Alan's contributions to biostratigraphy and to the paleoecological significance of bryozoan growth form based on quantitative patterns of relative abundance.

Here Lee-Ann Hayek and Efstathia Bura provide an innovative approach to the taxon range problem. Their method estimates confidence limits on stratigraphic ranges, and relies upon an empirically derived distribution function rather than an assumed distribution. Their method creatively uses all available data on absence as well as presence and therefore can examine beginning and end of ranges independently of one another. Mike Foote examines the complexities of using taxon-age distributions of living and extinct genera to obtain independent estimates of origination and extinction rates. Ann Budd and Kenneth Johnson contrast the differing evolutionary histories of rare and abundant coral species. They find differences in time of origination, susceptibility to extinction, and duration between rare and abundant reef corals in the late Cenozoic of the Caribbean. Eckart Håkansson and Erik Thomsen document the macroevolutionary ups and downs of asexual propagation among different growth forms of bryozoans. They find that, within a lineage, the proportion of species that are largely dependent upon asexual propagation varies consistently through time, and they also find small variations in colonial architecture to be important in asexual reproduction. Lastly, Frank McKinney, Scott Lidgard, and Paul Taylor demonstrate that patterns of clade replacement measured by taxonomic diversity do not track those measured by abundance across the Cretaceous-Tertiary mass extinction. This result points out that diversity plots do not fully portray the texture of evolution of organisms or especially of ecosystems—raising again the specter of the importance of abundance data rarely gathered by paleontologists.

This book grew out of the Paleontological Society Symposium "Process from Pattern in the Fossil Record," held in honor of Alan H. Cheetham at the 1997 Geological Society of America annual meeting in Salt Lake City, Utah. We thank the Paleontological Society for sponsoring the symposium, the numerous reviewers of the chapters for their thoughtful criticisms and suggestions, and the University of Chicago Press for their customary encouragement and support.

Selected References

Part 1

1969. Boardman, R. S., and A. H. Cheetham. Skeletal growth, intracolony variation, and evolution in Bryozoa: A review. *Journal of Paleontology* 43:205–33.

1970. Boardman, R. S., A. H. Cheetham, and P. L. Cook. Intracolony variation and the genus concept in Bryozoa. In *Proceedings of the North American Paleontological Convention*, 294–320. Lawrence, KS: North American Paleontological Society.

1971. Cheetham, A. H. Functional morphology and biofacies distribution of cheilostome Bryozoa in the Danian stage (Paleocene) of southern Scandinavia.

Smithsonian Contributions to Paleobiology 6. Washington, DC: Smithsonian Institution Press.

1973. Boardman, R. S., and A. H. Cheetham. Degrees of colony dominance in stenolaemate and gymnolaemate Bryozoa. In *Animal colonies: Development and function through time,* ed. R. S. Boardman, A. H. Cheetham, and W. A. Oliver, Jr., 121–220. Stroudsburg, PA: Dowden, Hutchinson, and Ross.

1980. Cheetham, A. H., L. C. Hayek, and E. Thomsen. Branching structure in arborescent animals: Models of relative growth. *Journal of Theoretical Biology* 85:335–69.

1981. Cheetham, A. H., L. C. Hayek, and E. Thomsen. Growth models in fossil arborescent cheilostome bryozoans. *Paleobiology* 7:68–86.

1981. Cheetham, A. H., and E. Thomsen. Functional morphology of arborescent animals: Strength and design of cheilostome bryozoan skeletons. *Paleobiology* 7:355–83.

1983. Cheetham, A. H., and P. L. Cook. General features of the class Gymnolaemata. In *Treatise on invertebrate paleontology: Part G (revised),* 1, ed. R. A. Robison, 138–207. Boulder, CO: Geological Society of America; Lawrence: University of Kansas.

1983. Cheetham, A. H., and L. C. Hayek. Geometric consequences of branching growth in adeoniform Bryozoa. *Paleobiology* 9:240–60.

Part 2

1966. Cheetham, A. H. Cheilostomatous Polyzoa from the Upper Brackelsham beds (Eocene) of Sussex. *Bulletin of the British Museum (Natural History)* 13: 1–115.

1968. Cheetham, A. H. Evolution of zooecial asymmetry and origin of poricellariid cheilostomes. *Atti della Societa Italiana di Scienze Naturale e del Museo Civico di Storia Naturale di Milano* 106:185–94.

1968. Cheetham, A. H. Morphology and systematics of the bryozoan genus *Metrarabdotos. Smithsonian Miscellaneous Collections* 153. Washington, DC.

1973. Cheetham, A. H. Study of cheilostome polymorphism using principal components analysis. In *Living and fossil Bryozoa: Recent advances in research,* ed. G. P. Larwood, 385–409. New York: Academic Press.

1975. Cheetham, A. H. Taxonomic significance of autozooid size and shape in some early multiserial cheilostomes from the Gulf Coast of the U.S.A. In *Bryozoa 1974: Proceedings of the Third Conference, International Bryozoology Association,* ed. S. Pouyet, Documents des Laboratoires de Geologie de la Faculté des Sciences de Lyon, hors série 3, fasc. 2, pp. 547–64.

1976. Cheetham, A. H., and D. M. Lorenz. A vector approach to size and shape comparisons among zooids in cheilostome bryozoans. *Smithsonian Contributions to Paleobiology* 29. Washington, DC: Smithsonian Institution Press.

1986. Cheetham, A. H. Tempo of evolution in a Neogene bryozoan: Rates of morphologic change within and across species boundaries. *Paleobiology* 12: 190–202.

1987. Cheetham, A. H. Tempo of evolution in a Neogene bryozoan: Are trends in single morphologic characters misleading? *Paleobiology* 13:286–96.

1988. Cheetham, A. H., and L. C. Hayek. Phylogeny reconstruction in the Neogene bryozoan *Metrarabdotos:* A paleontologic evaluation of methodology. *Historical Biology* 1:65–83.

1990. Jackson, J. B. C., and A. H. Cheetham. Evolutionary significance of morphospecies: A test with cheilostome Bryozoa. *Science* 248:579–83.

1991. Jackson, J. B. C., and A. H. Cheetham. Bryozoan morphological and genetic correspondence: What does it prove? Reply. *Science* 251:318–19.

1993. Cheetham, A. H., J. B. C. Jackson, and L. C. Hayek. Quantitative genetics of bryozoan phenotypic evolution: 1, Rate tests for random change versus selection in differentiation of living species. *Evolution* 47:1526–38.

1994. Cheetham, A. H., J. B. C. Jackson, and L. C. Hayek. Quantitative genetics of bryozoan phenotypic evolution: 2, Analysis of selection and random change in fossil species using reconstructed genetic parameters. *Evolution* 48:360–75.

1994. Jackson, J. B. C., and A. H. Cheetham. On the importance of nothing doing: An exhaustive study of tiny bryozoans supports the idea of punctuated equilibrium. *Natural History* 103:56–59.

1994. Jackson, J. B. C., and A. H. Cheetham. Phylogeny reconstruction and the tempo of speciation in cheilostome Bryozoa. *Paleobiology* 20:407–23.

1995. Cheetham, A. H., and J. B. C. Jackson. Process from pattern: Tests for selection versus random change in punctuated bryozoan speciation. In *New approaches to speciation in the fossil record,* ed. D. H. Erwin and R. L. Anstey, 184–207. New York: Columbia University Press.

1995. Cheetham, A. H., J. B. C. Jackson, and L. C. Hayek. Quantitative genetics of bryozoan phenotypic evolution: 3, Phenotypic plasticity and the maintenance of genetic variation. *Evolution* 49:290–96.

1999. Jackson, J. B. C., and A. H. Cheetham. Tempo and mode of speciation in the sea. *Trends in Ecology and Evolution* 14:72–77.

Part 3

1963. Cheetham, A. H. *Late Eocene zoogeography of the eastern Gulf Coast region.* Geological Society of America, Memoir 91. New York: Geological Society of America.

1971. Cheetham, A. H. *Functional morphology and biofacies distribution of cheilostome Bryozoa in the Danian stage (Paleocene) of southern Scandinavia.* Smithsonian Contributions to Paleobiology 6. Washington, DC.

1986. Cheetham, A. H. Branching, biomechanics, and bryozoan evolution. *Proceedings of the Royal Society, London,* series B, 228:151–71.

1996. Cheetham, A. H., and J. B. C. Jackson. Speciation, extinction, and the decline of arborescent growth in Neogene and Quaternary Bryozoa of tropical America. In *Evolution and environment in tropical America,* ed. J. B. C. Jackson, A. F. Budd, and A. G. Coates, 205–33. Chicago: University of Chicago Press.

1999. Cheetham, A. H., J. B. C. Jackson, J. Sanner, and Y. Ventocilla. Neogene cheilostome Bryozoa of tropical America: Comparison and contrast between the Central American isthmus (Panama, Costa Rica) and the north-central Caribbean (Dominican Republic). *Bulletins of American Paleontology* 357: 159–92.

2000. Cheetham, A. H., and J. B. C. Jackson. Neogene history of cheilostome Bryozoa in tropical America. In *Proceedings 11th International Bryozoology Association Conference,* ed. A. Herrera Cubilla and J. B. C. Jackson, 1–16. Panama City: Smithsonian Tropical Research Institute.

Modes of Development, Hierarchies of Morphological Organization, and the Adaptive Significance of Colony Form

Growth by Intussusception in Hydractiniid Hydroids

1

Leo W. Buss

Forms that appear complex to the human eye can be generated by the iterative application of a small number of global rules (Ulam 1962, 1966). A large literature has developed in which rules are chosen and the morphology of plants, fungi, and colonial invertebrates mimicked (e.g., Braverman and Schrandt 1966; Leopold 1971; Cheetham et al. 1980; Waller and Steingraeber 1985; Bell 1986; Kaandorp 1994). The correspondence of simulations to actual organisms is sufficiently faithful that such techniques are often the tools of choice for representations of organisms in computer graphics (Prusinkiewicz and Lindenmayer 1990). I am aware, however, of no study that establishes any such rule in any living organism.

The rule schemata that generate such faithful computer images are often of this sort: "inspect the grid, identify all stolons that have grown by four units, place a branch at midpoint of new growth," "inspect the grid, identify all locations where three or more stolons have anastomosed, place a polyp at each such location," and so on. These are instructions for a computer; no sessile invertebrate possesses the capacity to assess its state and update development in this way.

Or do they? Clonal organisms typically possess one or more clonewide fluid-conducting systems (e.g., the phloem system of vascular plants, the system of cytoplasmic streaming within the hyphae of a fungal mycelium or the plasmodium of a myxomycete, the gastrovascular system of cnidarians, and the blood vascular system of certain colonial ascidians). The functioning of these systems will generate local patterns of pressure distribution, shear stress, and surface tension experienced by the tissues, cells, and/or membranes lining the conducting system. If such features can be detected and the signals transduced to effect expression of pattern-forming genes, global rules responsive to local state arise.

I will explore the suggestion that hydroid colony form arises from such a process of self-inspection. All hydroids share one colonywide physiological conducting system, the gastrovascular system. If the functioning of the gastrovascular system generates local hydrodynamic signals which, in turn, trigger specific morphogenetic pathways, then perturbations of gastrovascular circulation should have marked effects on colony morphology. I review evidence that this is the case. Hydroid colony form largely derives from patterns of stolon branching and placement of polyp buds atop existing stolons. If colonies self-inspect, features of gastrovascular circulation may trigger one or both of these events. I suggest specific biophysical features that may govern polyp placement and stolon branching and provide new observations, collected in the honor of Alan Cheetham, in support of one such suggestion. Finally, a model of growth by intussusception implies that colony-patterning genes exist, specifically those associated with the placement of polyps and the branching of stolons, and that their expression is regulated in accord with hypothesized biophysical signals. I review the progress toward identifying such genes.

Manipulation of Gastrovascular Circulation Alters Colony Form

Gastrovascular circulation and colony morphology are inevitably linked. Colony morphology is the disposition in space of a system of pumps (i.e., polyps) coupled to one another by a system of pipes (i.e., stolons). Changes in the relative sizes or spatial distribution of pumps and pipes will, as a matter of mechanical necessity, modify the pattern of circulation of fluids contained within them. What is at issue is whether such changes are a two-way street. Specifically, do changes in circulation elicit subsequent changes in the morphology? Experiments to this end have been performed in *Podocoryna carnea* (Sars 1846) and *Hydractinia symbiolongicarpus* (Buss and Yund 1989).

Podocoryna carnea is an athecate hydrozoan of the family Hydractiniidae, which develops a typical filiform colony (fig. 1.1). The colony is composed of feeding polyps connected to one another by stolons, which adhere to the substratum. Colonies grow by the elongation of stolon tips, the branching and anastomosis of tips (which establish and eliminate growth zones, respectively), and the placement of polyps atop existing stolons.

The gastrovascular system of *Podocoryna*, like that of all colonial hydroids, comprises the digestive lumens of the polyps together with the lumens of the stolons which couple polyps to one another. Fluid within this lumen is circulated to effect colonywide gas exchange and nutrient distribution. Muscles and nerves are limited to the polyps (Stokes 1974; Schierwater et al. 1992). Hence, exclusive of epithelial conductance, the

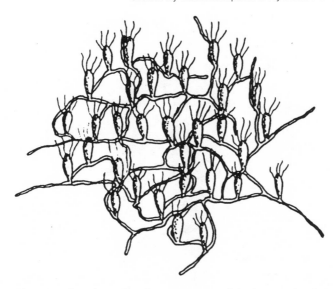

Fig. 1.1. Schematic of a colony of *Podocoryna carnea*. Colonies grow by elongation of stolons and budding of polyps atop existing stolons. New stolonal tips are generated by lateral branching and are eliminated by anastomosis.

gastrovascular system is the sole colonywide conducting system in this species.

Circulation of hydroplasm is effected by the periodic contraction of muscles in the polyps, which drives fluid from the digestive cavity of the polyp into the lumen of stolons (fig. 1.2), which, in turn, expand to accommodate the increased volume (Wagner et al. 1998; Dudgeon et al. 1999). Gastrovascular circulation is easily monitored in the peripheral stolons of a colony. Fluid flows into the blind tip of a stolon until the stolon is maximally distended, at which point the direction of flow is reversed and fluid flows away from the tip, collapsing the internal lumen (see, e.g., Wyttenbach 1973; Buss and Vaisnys 1993; Van Winkle and Blackstone 1997). Flow within stolons, thus, is sequentially bidirectional, and for peripheral stolons, the volume flux per cycle is proportional to the diameter of the stolon (Van Winkle and Blackstone 1997).

To assess whether perturbation of gastrovascular physiology induces alterations in subsequent colony morphology, a series of experiments has been performed wherein a colony was exposed to 2,4-dinitrophenol (DNP), gastrovascular flow assayed at stolon tips, and subsequent development monitored (Blackstone and Buss 1992, 1993; Blackstone 1997, 1998b). DNP is an uncoupler of oxidative phosphorylation; hence, application of DNP reduces the amount of ATP available to a colony. Since

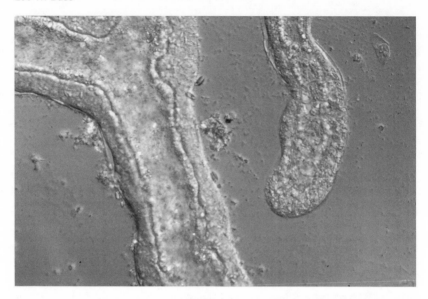

Fig. 1.2. Expanded stolonal lumen of *Hydractinia symbiolongicarpus*. Export of gastrovascular fluid from polyps into the stolon results in an expansion of the internal lumen of the stolon. Expansion of the lumen is not limited to the stolonal tip; rather, it occurs along the entire length of the stolon. Micrograph obtained under DIC illumination at 400×. Lumen diameter is ca. 30 μm.

the circulation of fluids within a colony is an active metabolic process, requiring muscular contraction, one might expect that application of DNP would disrupt gastrovascular circulation. This proved to be the case. Figure 1.3 shows the expansion and contraction cycles of a peripheral stolon of control and DNP-treated colonies. After 3 days of treatment with 1 μM DNP, the amplitude of stolonal expansion, and hence volume flux to the tip, is markedly depressed (fig. 1.3A). After 5 days of DNP treatment, stolonal oscillations have been effectively eliminated (fig. 1.3B).

The effect of DNP treatment on subsequent colony development was pronounced. DNP treatment increased the rate of stolonal branching and the rate of polyp bud formation (fig. 1.4). The relationship is that which one might expect. When the energy available to pump the fluid is decreased, the pipe must either be shortened or the number of pumps increased to maintain a given volume flux to the end of a pipe. Branching achieves the former, bud initiation the latter. Indeed, the colonies appear to be seeking to maintain a given volume flux to the tips; after 3 weeks of morphological response to DNP treatment, stolonal contractions in control and DNP-treated colonies display near-identical amplitudes (fig. 1.3C).

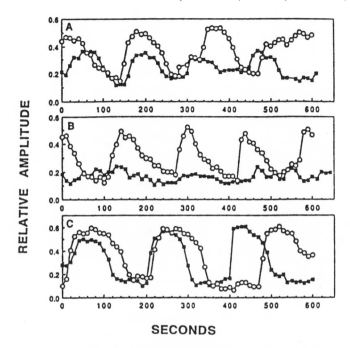

RELATIVE AMPLITUDE

SECONDS

Fig. 1.3. Effects of DNP treatment on gastrovascular circulation in *Podocoryna carnea*. Relative amplitude (y-axis) is a standardization that facilitates comparisons of stolons with differing diameters and is obtained by dividing the diameter of the stolon by the periderm-to-periderm diameter. Measures for several contraction cycles at a position ca. 250 μm from a peripheral stolonal tip for DNP-treated (■) and control (○) colonies. Stolon tip is full during peaks in relative amplitude and empty during troughs. *A*, Three days after initiation of treatment with 1 μM DNP. *B*, Five days after initiation. *C*, Three weeks after initiation. Note that application of DNP disrupts circulation (*A, B*) and that, following morphological changes (cf. fig. 1.4), normal circulation is restored (*C*). Figure redrawn from and methodological details available in Blackstone and Buss 1992.

The DNP experiments demonstrate that perturbations to gastrovascular circulation induce two pattern-forming systems: stolonal branching and bud initiation. DNP treatments transform relatively "runnerlike" colonies (sensu Buss 1979; Jackson 1979), characterized by longer stolons, fewer branches, and fewer polyps per unit stolon length into a "sheetlike" colony, characterized by shorter stolons, more branches, and more polyps per unit stolon length.

In these experiments, colonies of identical morphology were shown to develop differently if gastrovascular circulation was perturbed. Alternatively, one might ask whether colonies selected for differences in their morphology display corresponding differences in patterns of gastrovascular circulation. Blackstone (1998a) mated field-collected colonies of

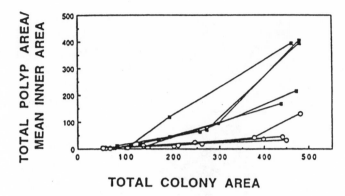

TOTAL COLONY AREA

Fig. 1.4. Effects of DNP treatment on colony morphology. Total polyp area/mean inner area (the y value) is an aggregate measure that increases with larger amounts of polyps (i.e., increased polyp area) and with larger amounts of stolons (i.e., decreased "inner area," substratum area bounded by the most peripheral stolons, but not covered by stolons). Plot shows ontogenetic trajectories for five DNP-treated (■) colonies and four control (○) colonies. *Lines* connect points measured through time for the same individual. Note that DNP-treated colonies produce more polyps/area and more stolons/area (i.e., are more sheetlike) than controls throughout ontogeny. Figure redrawn from and methodological details available in Blackstone and Buss 1992.

Podocoryna and established two genetic lines from the F_1. He brother-sister inbred both lines and, in each generation, selected for sheetlike morphologies in one line and runnerlike morphologies in the other. After six generations of inbreeding, he assayed flow to the stolonal tip. As expected on the basis of the DNP experiments, the line inbred for sheetlike morphologies displayed significantly lower volume fluxes to peripheral stolons than that of the line inbred for runnerlike morphologies.

The influence of colony morphology on patterns of gastrovascular circulation is particularly clearly illustrated in another hydractiniid hydroid, *Hydractinia symbiolongicarpus*. *Hydractinia* colonies differ from those of *Podocoryna* in one prominent feature. Early in colony ontogeny, the basal ectoderm of the body column of the polyp expands from the polyp base and overtops the stolonal nexus (Cartwright and Buss 1999). This tissue, the stolonal mat, expands peripherally as a sheet. Stolons within the mat are arranged as both radial and circumferential, or ring, canals (fig. 1.5A). Radial canals may or may not extend beyond the mat periphery as free stolons, and ring canals may or may not be continuous along the mat periphery (fig. 1.6A).

Dudgeon and Buss (1996) show that sheetlike *Hydractinia* colonies, that is, those with small ratios of free stolon length to stolonal mat area, display little variation in radial canal diameter, wider ring canal diameter, and higher frequencies of continuous ring canals than do runnerlike colo-

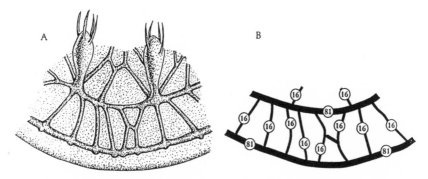

Fig. 1.5. Colony morphology in *Hydractinia symbiolongicarpus. A,* Schematic representation of the periphery of sheetlike colony. Unlike *Podocoryna carnea* (fig. 1.1), colonies display a uniform ectodermal plate, the stolonal mat. Note that endodermal canals within the mat are arranged either radially (i.e., extending from the center to the periphery) or in circumferential rings. *B,* Schematic representation of the relative volumetric distribution, indicated by relative line thickness, of gastrovascular flow among stolons depicted in *A. Circled numbers* are ratios of the diameter to the fourth power of a given stolon to the diameter to the fourth power of the smallest stolon. Figure redrawn from Dudgeon and Buss 1996.

nies (see Dudgeon and Buss 1996, tables 1 and 2). The greater disparity in stolon radii in runnerlike forms derives, in large measure, from the presence of free stolons in runnerlike forms, which bear larger radii and which frequently disrupt the continuity of ring canals.

Differences in radii of stolons is of considerable physical significance. The volume flux for flow through a pipe for the low Reynold's number, laminar flow regimes relevant here is approximated by Hagen-Poiseuille law, wherein volume flux is proportional to the fourth power of radius (see, e.g., Fung 1993). Thus, for a given pressure head and stolon length, the larger peripheral stolons of runnerlike forms will experience vastly greater volume fluxes to their tips than will the smaller radial canals within the stolonal mat (fig. 1.6B). If growth rate is proportional to volume flux, then free peripheral stolons may be expected to elongate at relatively faster rates. The pattern of gastrovascular circulation that is induced by the morphology of a runnerlike form may be expected, then, to generate faster growth of peripheral stolons than mat tissue, that is, to create the very conditions necessary to maintain a runnerlike form.

A converse argument applies to sheetlike colonies. Here the periphery is characterized by a wide circumferential ring canal into which smaller radial canals enter and out of which comparable-sized extensions of the radials exit (fig. 1.5B). Since all tips are of comparable size and all draw fluid from the same wide-bore ring canal, they may be expected to experience comparable volume fluxes and, hence, to elongate at comparable

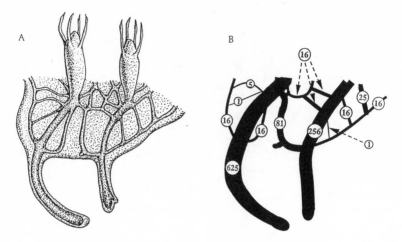

Fig. 1.6. Colony morphology in *Hydractinia symbiolongicarpus*. *A,* Schematic representation of the periphery of runnerlike colony. Runnerlike forms differ from sheetlike forms (fig. 1.5A) in three ways. First, in runnerlike forms radial canals frequently extend beyond the edge of the mat as free peripheral stolons. Second, the ring canals of runnerlike colonies are frequently not continuous. Finally, there is far greater variation in the radial canal diameter in runnerlike forms than in sheetlike forms, with those radial canals extending as free stolons having the largest diameter. In sheetlike forms the ring canals typically have the largest diameter, and radial canals exhibit.little variation in size (cf. fig. 1.7). *B,* Schematic representation of the relative volumetric distribution, indicated by relative line thickness, of gastrovascular flow among stolons depicted in *A*. *Circled numbers* are ratios of the diameter to the fourth power of a given stolon to the diameter to the fourth power of the smallest stolon. Note that stolon tips in sheetlike colonies (fig. 1.5B) experience vastly more equitable volume fluxes than those in runnerlike colonies. Figure redrawn from Dudgeon and Buss 1996.

rates. Equal circumferential growth rates result in sheetlike morphologies, just as unequal growth rates between free stolons and stolonal mat generate runnerlike forms.

If pattern-forming elements are responsive to signals produced by gastrovascular flow, then changing the pattern of circulation within a colony should be sufficient to change the morphology. Dudgeon and Buss (1996) have tested this claim. We attempted to transform a sheetlike colony of *Hydractinia* into a runnerlike colony and vice versa. The procedure was one of repeated microsurgery. The peripheral stolons of runnerlike colonies were removed two to three times each day. Severed stolons close "instantly" (Berrill 1953). Removal of peripheral stolons causes the fluid in these disproportionately large canals to flow into the smaller radial and ring canals of the stolonal mat, which correspondingly increased in diameter (fig. 1.7). After 17 days of such treatment, when runnerlike colonies had been surgically altered to adopt a gastrovascular configuration

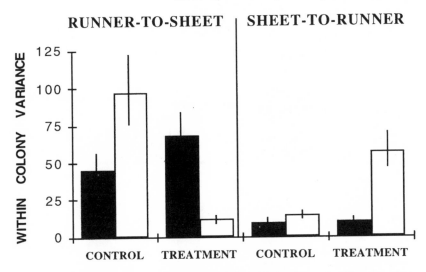

Fig. 1.7. Surgical manipulation of colony morphology. Variance in stolon diameters within replicate colonies for control and experimental treatments for both the runner-to-sheet transformation and the sheet-to-runner transformation. *Solid bars* represent initial, pre-surgical means with standard error, and the *open bars* the postsurgical values. Sample sizes were $N = 10$ for the runner-to-sheet experiments. Control and treatment sample size were $N = 8$ and $N = 3$, respectively, for the sheet-to-runner experiments. For the runner-to-sheet experiment, note the considerable variation in stolon sizes at the outset and that variation increased in unmanipulated colonies. In manipulated colonies, note that surgery reduced variation to levels comparable to that of sheetlike colonies. For the sheet-to-runner experiments, note that variation is low, and remains so, in control colonies, whereas surgery induces variation in stolon sizes comparable to that displayed by runnerlike forms. Figure redrawn from Dudgeon and Buss 1996.

akin to that of a sheetlike colony, the subsequent growth of the colony was monitored. Sheetlike growth resulted (fig. 1.8).

Sheetlike colonies were likewise transformed into runnerlike forms (Dudgeon and Buss 1996). Repeated severing of the ring canal at the mat periphery disrupts circumferential flow and increases variation in the volume flux to stolon tips. Severing the ring canal eight times per day resulted in variation in the radius of radial canals (fig. 1.7), that is, the condition that typifies runnerlike forms. After 10 days of such treatment, runnerlike morphologies resulted (fig. 1.8).

Gastrovascular circulation and morphology is a two-way street. Altering the physiology of circulation, as in the *Podocoryna* experiments, results in altered morphology. The altered morphology, in turn, restores a volume flux to the stolonal tips comparable to that of control colonies. In *Hydractinia*, surgical manipulations of morphology induced patterns of gastrovascular circulation comparable to that of different morpholog-

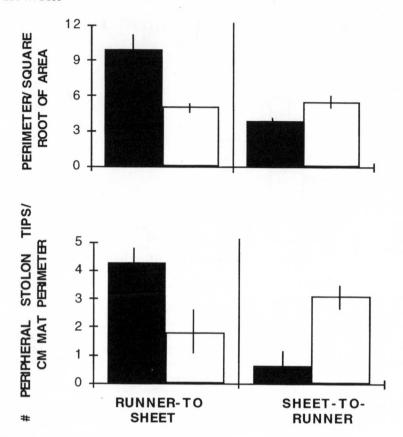

Fig. 1.8. Morphology induced by surgical manipulation of stolon sizes. Values are means and standard errors with sample sizes as given in legend of figure 1.7. *Solid* and *open bars* represent control and experimental colonies, respectively. *A,* Perimeter per unit square root of area is a measure of the gross morphology of a colony. Greater values of this index indicate a more runnerlike morphology (see Blackstone and Buss 1991 for discussion of this metric). Note that experimental colonies became more sheetlike in the runner-to-sheet treatment and that the converse occurred in the sheet-to-runner experiments. *B,* The number of free peripheral stolons per unit mat periphery is reduced in the runner-to-sheet treatment and increased in the sheet-to-runner experiment. Figure redrawn from Dudgeon and Buss 1996.

ies and subsequent growth was diagnostic of the morphology whose circulation pattern was mimicked. The simplest explanation is that colony form is regulated by aspects of circulation.

Polyp Placement and Stolon Branching

The experiments described above demonstrate that branching and bud formation are both elicited by perturbations to gastrovascular circulation, but provide little hint of what is being sensed to trigger either of these developmental progressions. Humoral elements (i.e., molecules being circulated in gastrovascular fluid), diffusible morphogens (Turing 1952; Meinhardt 1976; Müller and Plickert 1982; Müller et al. 1987; Plickert et al. 1987), metabolic gradients (Child 1941; Blackstone 1997, 1998a, 1998b), and/or hydromechanical signals are all conceivable devices. None of these mechanisms are necessarily mutually exclusive, and I make no attempt here to systematically review the alternatives. Rather, in light of the results discussed above, I concentrate exclusively on the heretofore largely unexamined possibility that internal hydromechanical signals play a role in triggering morphogenesis. Specifically, I ask what biophysical features might be sensed by cells, how such features might be expected to vary in colonies of simple geometry, and whether evidence exists supporting a role for them in inducing stolonal branching and polyp bud placement.

I first consider biophysical signals that might trigger polyp bud placement. Since polyps have a fixed upper size, a single polyp will only be capable of driving fluid a limited distance before continued extension of a stolon requires a new pump. Braverman (1963) reports that the ratio of polyp number to stolon length was a constant in a growing colony of *Podocoryna*, and Crowell (1957) notes that interpolyp stolon lengths are greater in colonies with wider stolonal radii in colonies of *Campanularia flexuosa* (Hincks 1868). The latter result, in particular, suggests that colonies detect features of fluid flow and use them to induce polyp bud placement at appropriate distances from existing polyps.

What biophysical features are most plausible as signals? Fluid expelled from a polyp is opposed by viscous forces imposed by the stolonal wall and by the fluid itself. The force balance equation for a flow in a pipe is shown in figure 1.9. The pressure difference (Δp) acting over the cross-sectional area of the pipe must equal the shear stress (τ) acting over the surface of the cylinder. Either shear stress or pressure differentials, then, might provide reliable sensors of gastrovascular circulation. Let us assume that the pressure differential is linearly related to polyp size and that colonies initiate new polyp buds at a length (L^*) corresponding to a particular value of shear stress. Then, for a one polyp:one stolon colony, the following relationship should hold:

$$PS = a(2L^*/r) + b,$$

where PS is polyp size, r is stolon radius, L^* is length to new polyp bud, and a and b are constants.

To test whether such a relationship holds, I established ten colonies of *Podocoryna* on glass coverslips using standard procedures (Blackstone and Buss 1991). After colonies had been established on the slides, one

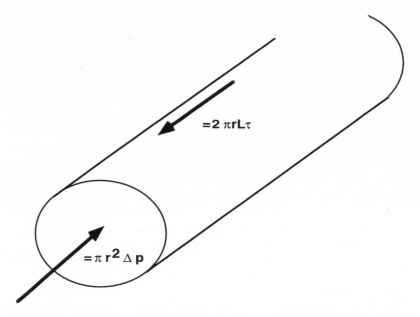

Fig. 1.9. Force balance equation for flow in a pipe. r is radius of the pipe, L is its length, Δp is the pressure differential, and τ is shear stress. Note that $\Delta p = (2L/r)\tau$.

polyp:one stolon systems were established by severing all connections between a given polyp and stolon and the colony. Initial stolon length was held constant, and polyp size was varied. Polyp size was measured by briefly exposing the colony to 0.5M KCl and measuring the contracted polyp diameter with an ocular micrometer. Colonies so established were maintained at 14°C in recirculating aquaria and fed 3–4-day-old *Artemia salina* (Linnaeus 1758) nauplii every third day. To maintain the original configuration of a single stolon with a single polyp positioned at the midpoint, colonies were inspected daily and stolon branches, if they occurred, were removed. When buds were detected, the length from the polyp to the bud was measured with an ocular micrometer at 50× and the diameter of the stolon measured. To obtain stolon measures, polyps were fed a single brine shrimp nauplius and the stolon visualized at 400× using a Zeiss Axiovert 35 under DIC illumination. Colonies were fed prior to measurement, as feeding elicits a stereotypic pattern of gastrovascular circulation involving maximal expansion of the stolonal lumen (Dudgeon et al. 1999). Stolon diameter was measured as the average of three measurements at maximum internal lumen amplitude at a point midway between the initial polyp and the new bud. Data on polyp size are the sizes measured when buds were first detected. The experiments were terminated when a new bud appeared on either stolon. In several colonies, buds were

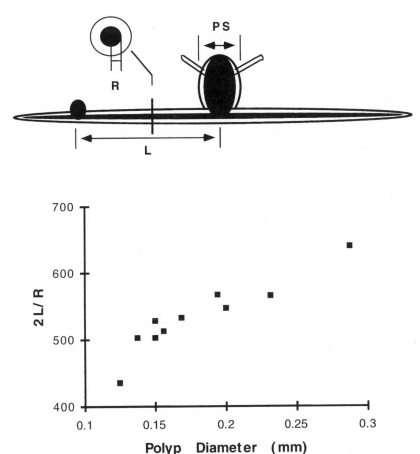

Fig. 1.10. Shear stress and polyp bud induction. Relationship between $2L/r$ and polyp diameter. As shown in the schematic, L is distance from polyp to bud, and r is the radius of the internal lumen of the stolon.

initiated on both stolons on the same day. In these cases, values for length and stolon radius were averaged.

Results are shown in figure 1.10. Excepting one observation at the smallest polyp size, the predicted linear relationship between polyp size and $2L/R$ was observed. These preliminary observations would seem to justify more extensive investigations using the same protocol to determine, for example, whether the value obtained at the smallest polyp size is repeatable and, if so, to determine if the relationship obtained is better explained by a logarithmic or hyperbolic fit. Two considerations, however, have led us to use an alternative technique. First, the force-balance equation on which this prediction was based assumes steady, unidirectional flow in an unbranched rigid pipe of infinite length (see, e.g., Fung 1993). All of these assumptions are violated in hydroid stolons, which leads me to suspect that increasing sample size may fail to increase precision. Moreover, the experiments reported above require one to establish and maintain a colony geometry that is unnatural. The one polyp:one stolon preparation typically leads to generation of stolonal tips. Branching may be expected to eliminate any simple relationship, and indeed does (unpublished data), suggesting that an experimental protocol that permits colonies to grow unimpeded would prove more revealing.

Steve Dudgeon and I have performed experiments in which we have allowed colonies to grow unimpeded in seawater differing in viscosity. Recall that shear stress is the product of viscosity and first differential of velocity with radius (see, e.g., Fung 1993). We varied viscosity from 1 to 2.2 centipoise by incubation in varying concentrations of dextran, and monitored the number of polyps and stolon branches produced by one polyp:one stolon colonies initiated with fixed polyp size and stolon length. Colonies were maintained in continuously rotating 10-ml tubes on glass coverslips and fed on days 1 and 3 of the 3-day-long experiment. Colonies were maintained at a constant 22°C in an incubator. Fresh solutions were provided twice a day. Colonies did not differ in their stolonal growth rate or the number of lateral stolon tips (i.e., branches) as a consequence of the treatment (Dudgeon, unpublished data). We found, however, that the number of polyps initiated per stolon length was linearly related to viscosity, with higher viscosity yielding colonies with more closely packed polyps (Dudgeon, unpublished data).

Together with the relationship shown in figure 1.10, these experiments are consistent with the hypothesis that shear stress acts as a biophysical signal regulating polyp bud placement. It bears emphasis, however, that shear-stress-based signaling is not the only interpretation of these results. In particular, similar findings would be generated under a hypothesis that colonies sense pressure differentials. Shear-stress-based signaling, though, is not without precedent. Control of vascular morphology by response to internal hydromechanical signals is increasingly well known in vertebrate vascular endothelium. Several genes are known to be differentially expressed upon perturbation of wall shear stress in a fashion that adjusts vessel radii to values which restore a systemwide constant shear stress (Bevan et al. 1995).

I turn now to the second component of colonywide patterning, that of stolonal branching. Recall that branching is unaffected by manipulation of viscosity (Dudgeon, unpublished data). An obvious candidate signal insensitive to changes in viscosity is the circumferential stress, or hoop stress, experienced in the stolonal endoderm. In a vessel with an elastic membrane under pulsate flow, hoop stress will vary with position along a vessel with each alteration of the direction of flow (see treatment by Fung 1993, 379–81).

As a polyp empties its contents into a stolon, the fluid is accommodated by an increase in stolon radius (Buss and Vaisnys 1993; Dudgeon et al. 1999). Change in stolon length is negligible (Wyttenbach 1973). Stolons, however, are not free to increase in radius without limit, as they are encased within a rigid peridermal sheath. Consider a polyp with a stolon of a length capable of accommodating the volume of the polyp without compression of the expanded stolon by the peridermal walls. The hoop stress experienced by the stolonal endoderm may be expected to vary along the length of a stolon and to oscillate in time. As the polyp grows, there will exist a polyp volume at which the stolon becomes compressed against the peridermal wall for some distance along the stolon. The relative capacity of the polyp to deliver a volume, and the stolon to receive it, may be expected to produce a pronounced gradient in hoop stress along the length of that stolon. Hoop-stress-mediated branching is a plausible signaling system. Production of a new tip will, with its subsequent growth, result in increased stolon volume. Hence, if circumferential stress-based branching occurs, an increase in the volume of the fluids being delivered to stolons will be accompanied by an increase in the volume of the stolon nexus into which that volume must be accommodated.

Experimental results exist that support this suggestion (Plickert 1980; Müller et al. 1987). These authors compressed stolons of *Hydractinia echinata* (Fleming 1828) and *Eirene viridula* (Péron and Lasueur 1809) along their margins with a blunt tungsten needle to a point at which the stolon lumen was narrowed to half its width. They found that mechanical compression of stolons anywhere along their length (exclusive of the specialized tip zone) for periods of 5 to 30 minutes was sufficient to generate stolonal branches at the point of compression, with highest frequencies of tip induction occurring with longer durations of compression. This effect was originally attributed to pressure differences (Plickert 1980; see also Petriconi and Plickert 1974) and later reinterpreted to reflect inhibition of stolon-tip production by a hypothetical morphogen (Müller and Plickert 1982; Müller et al. 1987; Plickert et al. 1987). The pressure distribution induced by compression of stolons in these experiments was likely equilibrated near instantaneously (just as pressure differentials are equilibrated instantly when a child presses her finger into an expanded balloon). In contrast, the differences in hoop stresses generated by the treatment will last as long as the stimulation is applied (and, potentially, far longer given the relative inelasticity of periderm).

While the observations outlined above suggest that shear and circumferential stresses are associated with polyp bud formation and stolonal branching, respectively, these suggestions must be regarded as preliminary. In particular, there is no reason to presume that biophysical signals are encoded as simple threshold functions or as functions sensitive to deviation from some set value. Shear and circumferential stresses will be time-varying, and signaling may be based on features of this variation (Blackstone 1998a). Calcium oscillations in cells, for example, trigger quite different cascades of subsequent events in response to deviation in frequency and amplitude components (Berridge 1997; Dolmetsch et al. 1997). Clearly assessing whether shear- or hoop-stress-associated signaling can account for colonywide perturbations, such as those reported in the DNP and microsurgery experiments, will depend on an understanding of how such signals are transduced. Moreover, even if this information was available, assessing whether a hypothesized signal generates buds and branches in predicted locations in a living colony will require a precise quantitative understanding of the time-varying volume fluxes within a living colony, which will, in turn, be highly sensitive to the precision with which the mechanical properties of polyp and stolonal walls are known. Indeed the complexity of establishing such an understanding suggests a more direct approach. If biophysical features play a role in inducing morphogenesis, perturbation of such features should elicit expression of genes known to be associated with early stages of branch and bud induction.

Expression of Pattern-Forming Genes

Perturbations to gastrovascular circulation alter subsequent colony ontogeny, and simple growth manipulations suggest putative hydromechanical signals linking such perturbations to stolon branching and bud placement. If colonies self-inspect, hydromechanical signals must be sensed at cell surfaces and the signals transduced to elicit differential expression of pattern-forming genes. While cnidarian developmental genetics is in its infancy, it is possible to identify genes that are expressed in early stages of the bud and tip patterning.

The first overt appearance of a polyp bud or stolon tip is characterized by the swelling of the ectoderm, either laterally in the case of stolon buds or apically in the case of polyp buds (Saint-Hilaire 1930; Berrill 1953; Müller 1964; Beloussov et al. 1972; Beloussov 1973; Beloussov and Dorfman 1974). The ectoderm in these regions is both cytologically and behaviorally distinct. Cytologically, ectodermal cells in both regions adopt distinctive orientations relative to the mesoglea and are distinguished by pronounced vacuoles in epitheliomuscular cells (Beloussov et al. 1989). These vacuoles are periodically filled and emptied, apparently under osmotic control (Kazakova et al. 1994), generating the distinctive growth pulsations of these tissues (Hale 1960, 1968; Wyttenbach 1968, 1974).

A gene whose expression maps to the first appearance of this specialized epithelium has recently been reported from *Hydractinia* (Cartwright and Buss 1999; Cartwright et al. 1999). *Cnox-2* is a homeobox-containing gene. The gene is not expressed in the free stolons of *Hydractinia*, but is expressed both in the tips of stolons and at the first appearance of polyp buds. Induction and elimination of this specialized stolonal epithelium results in concomitant changes in *Cnox-2* expression. *Cnox-2* protein, which is undetectable in proximal regions of the unmanipulated stolons, becomes expressed at high levels in the developing tip when cytologically and behaviorally distinct ectoderm appears (Cartwright 1997). Stolon anastomosis, which results in the elimination of the cytological and behavioral specializations (Müller et al. 1987; Lange et al. 1989), is likewise accompanied by a decline and eventual elimination of *Cnox-2* expression (Cartwright 1997).

Cnox-2, by virtue of its expression in patterning of both stolon tips and polyp buds, may serve as a useful tool. It is now conventional in studies of the vertebrate vascular endothelium to expose cultured cells to experimental manipulations of shear stress and monitor subsequent gene expression (e.g., Frangos et al. 1985; Davies et al. 1986). Hydroid cells have recently been established in continuous culture (Frank et al. 1994). The differential expression of *Cnox-2* at sites of morphogenetic activity in free stolons and its association with a specialized epithelium characteristic of early bud and tip formation may allow it to be used as a marker for analogous studies in hydroids.

Despite its potential utility as a marker, *Cnox-2* is unlikely to be directly regulated by hydromechanical signals. First, the gene is expressed in the ectoderm, whereas hydromechanical signals are endodermal in origin. Second, *Cnox-2* is up-regulated in both tips and buds, while these two morphogenetic responses are presumably signaled by different hydromechanical features. Finally, *Cnox-2* appears rather late in bud and tip formation, at a time when the distinct histological complement of these morphogenetic zones appears. More plausible candidates are those genes that are expressed earlier in the pattern formation, are expressed in endodermal tissues, and are expressed in either tip or bud production, but not both.

Plausible candidates are beginning to appear. The solitary hydroid, *Hydra*, produces buds off its body column, but lacks stolons. The first evidence of bud formation is the demarkation of a zone, or placode, of cells which ultimately migrate to form the bud (Otto and Campbell 1977). Two *Hydra* genes have recently been isolated and characterized, the expression of which corresponds to the endodermal extent of the bud placode. *Manacle* is a member of the paired/paired-like class of homeobox genes (D. Bridge, pers. comm.). *Manacle* is not expressed in the body column until periods just prior to bud formation, whereupon it is expressed in the regions of the endoderm corresponding to the bud placode, and its expression is subsequently down-regulated as the cells of the placode move into the now evident bud (D. Bridge, pers. comm.). *Budhead* is a *forkhead*-class gene (Martinez et al. 1997), members of which have been associated with embryonic "organizer" activities in a diversity of model systems. It is first expressed in a small group of endodermal cells at a position that corresponds to the point of future placement of the bud. Expression subsequently extends to the entire endodermal bud placode, and eventually localizes to cells in the head endoderm of the bud (Martinez et al. 1997). The isolation and characterization of these genes in a colonial hydroid would be of considerable interest.

Homologues of several genes that might be expected to be involved in polyp budding or stolon branching have not yet been sought in cnidarians. Among these are representatives of patterning genes involved in dorsal-ventral axis formation (e.g., *dpp/Bmp-4, sog/chordin*). The formation of a bud in the apical position and a branch in the lateral position requires such a coordinate system. Other attractive candidates include the genes associated with angiogenesis in the vertebrate vascular system (e.g., *VEGF*) and branching in insect respiratory systems. *Branchless* and *breathless*, for example, are fly fibroblast growth factor (FGF) and FGF-receptor homologues that have recently been shown to control the migration of tracheal cells to specify the location and the direction of branching (Klambt et al. 1992; Sutherland et al. 1996). Given the growing understanding of regulatory genes implicated in axis formation in triploblastic organisms, and the interest of several groups in the role of such genes in diploblasts, genes involved in colony-level patterning are likely to proliferate in the near future.

Why Intussusceptive Growth?

I have argued here that colony growth proceeds by an inspection of hydromechanical features sensitive to the state of a colonywide physiological system (i.e., the gastrovascular system) and by the triggering of a set of global rules (i.e., shear- and circumferential-stress-induced gene expression) whose local consequences (i.e., induction of bud and tip formation) alter the state of that physiological system. What has not been addressed is why a colony might be organized in this fashion.

In the case of a computer simulating the growth of a colony, the requirement for iteratively updated global rules is obvious. Real colonies, though, have no such requirement. A new bud in one part of a colony may be initiated by quite different rules than that in another. Under what conditions should a set of global rules evolve?

Global instructions responsive to the functioning of a colonywide physiological system should be found when natural selection favors the optimization of some overall performance feature of that system. The vertebrate vascular endothelium is instructive here. Murray (1926) reasoned that the vertebrate vascular design should be energy minimizing and then derived a series of static relationships defining the conditions under which such a minimization is realized. The optimum is one in which shear stress is maintained constant throughout (Sherman 1981; Zamir 1977; LaBarbera 1990), and, as mentioned earlier, compelling evidence exists that our vascular systems constantly readjust vessel radii to achieve precisely this condition (Bevan et al. 1995).

Gastrovascular circulation is an active metabolic process and very likely constitutes a considerable fraction of the overall energy budget of a colony (Blackstone 1998a, 1998b). Hydroid colonies may also seek energy-minimizing configurations of their gastrovascular system. Energy minimization, however, is only one possibility. If selection for growth rate or fecundity has priority, colonies may seek to maximize the volume flux to tips or gonozooids. Alternatively, global rules may seek to maximize mixing if colonywide nutritive support or gas exchange is primal. Designs that attempt multiple optimization (Cohen 1978) are also conceivable, as are sequential optimizations in different periods of ontogeny.

Exploration of optimal designs in hydroid gastrovascular systems will prove challenging. Murray (1926) studied vascular systems in which fluid flows unidirectionally through a dichotomously branching tree. Here the flow in a given vessel is a function only of the flow in the branch that preceded it. In cnidarians (and some colonial ascidians), flow is sequentially bidirectional, vessels anastomose to form a network rather than a tree, and fluid is typically propelled by a large number of pumps. The direction and velocity of fluid being circulated is, then, potentially dependent on the state of interconnected vessels and pumps over some spatial domain defined by the precise magnitude of the various material properties of the stolons and pressure heads of the pumps.

In the case of hydractiniid hydroids, it is known that the behavior of isolated polyps (Wagner et al. 1998) and of isolated stolons (Buss and Vaisnys 1993) can be accurately described as nonlinear oscillators and the colony, therefore, as a spatially distributed system of coupled oscillators. Spatiotemporal nonlinear dynamic systems are mathematically challenging and rarely amenable to analytical solution. Unlike many biological systems, however, the hydroid gastrovascular system is amenable to the gathering of high-resolution time-series data (Buss and Vaisnys 1993; Dudgeon et al. 1999), and it is therefore tractable to seek to develop rigorously parameterized models which can be explored numerically. While this goal has yet to be achieved, the arguments presented here define a clear research strategy. Simulations of flow within a colony under varying branching and production rules should permit numerical exploration of the energy requirements to drive fluid through such systems, the mixing generated by the observed circulation pattern, and the volume fluxes to specific growth zones. If the arguments presented here are viable, optimizing configurations would be expected to result from shear- and circumferential-stress-related production of pumps and pipes, respectively.

Acknowledgments

It seems appropriate to acknowledge the source of the word *intussusception*. I was first exposed to the term by its appearance in the title of a volume of essays published in honor of G. Evelyn Hutchinson (Deevey 1972). Among its definitions is "the deposition of new particles of formative material among those already embodied in a tissue or structure (as in the growth of a living organism)" *(Webster's Third New International)*. The term struck me as a particularly apt characterization of the perspective on colonial animal growth developed here (and, besides, I liked the idea of its appearance in a paper celebrating another scientist of exceptional distinction).

The ideas explored here have emerged from years of conversations with Neil Blackstone, Steve Dudgeon, and J. Rimas Vaisnys. They, Luis Cadavid, two anonymous reviewers, and the editors made helpful comments. Support was provided by the NRC Twinning Program and the National Science Foundation (MCB-9817380).

References

Bell, A. D. 1986. The simulation of branching patterns in modular organisms. In *The growth and form of modular organisms*, ed. J. L. Harper, B. R. Rosen, and J. White, 143–59. London: Royal Society.

Beloussov, L. V. 1973. Growth and morphogenesis of some marine Hydrozoa according to histological data and time-lapse studies. *Publications of the Seto Marine Biological Laboratory* 20:315–36.

Beloussov, L. V., L. A. Badenko, A. L. Katchurin, and L. F. Kurilo. 1972. Cell movements and morphogenesis of hydroid polyps. *Journal of Embryological and Experimental Morphology* 27:317–37.

Beloussov, L. V., and J. G. Dorfman. 1974. On the mechanics of growth and morphogenesis in hydroid polyps. *American Zoologist* 14:719–34.

Beloussov, L. V., J. A. Labas, N. I. Kazakova, and A. G. Zaraisky. 1989. Cytophysiology of growth pulsations in hydroid polyps. *Journal of Experimental Zoology* 249:258–70.

Berridge, M. J. 1997. The AM and FM of calcium signalling. *Nature* 386:759–60.

Berrill, N. J. 1953. Growth and form in gymnoblastic hydroids: 6, Polymorphism in the Hydractiniidae; 7, Growth and reproduction in *Syncoryne* and *Coryne*. *Journal of Morphology* 92:241–302.

Bevan, J. A., G. Kaley, and G. M. Rubanyi, eds. 1995. *Flow-dependent regulation of vascular function.* Oxford: Oxford University Press.

Blackstone, N. W. 1997. A dose-response relationship for experimental heterochrony in a colonial hydroid. *Biological Bulletin* 193:47–61.

———. 1998a. Morphological, physiological, and metabolic comparisons of runner-like and sheet-like inbred lines of a colonial hydroid. *Journal of Experimental Biology* 201:2821–31.

———. 1998b. Physiological and metabolic aspects of experimental heterochrony in colonial hydroids. *Journal of Evolutionary Biology* 11:421–38.

Blackstone, N. W., and L. W. Buss. 1991. Shape variation in hydractiniid hydroids. *Biological Bulletin* 180:394–405.

———. 1992. Treatment with 2,4-dinitrophenol mimics ontogenetic and phylogenetic changes in a hydractiniid hydroid. *Proceedings of the National Academy of Sciences, USA* 89:4057–61.

———. 1993. Experimental heterochrony in hydractiniid hydroids: Why mechanisms matter. *Journal of Evolutionary Biology* 6:307–27.

Braverman, M. H. 1963. Studies on hydroid differentiation: 2, Colony growth and the initiation of sexuality. *Journal of Embryological and Experimental Morphology* 11:239–53.

Braverman, M. H., and R. G. Schrandt. 1966. Colony development of a polymorphic hydroid as a problem in pattern formation. *Symposium of the Zoological Society of London* 16:169–98.

Buss, L. W. 1979. Habitat selection, directional growth, and spatial refuges: Why colonial animals have more hiding places. In *Biology and systematics of colonial organisms,* ed. G. Larwood and B. Rosen, 459–97. London: Academic Press.

Buss, L. W., and J. R. Vaisnys. 1993. Temperature stress induces dynamical chaos in a cnidarian gastrovascular system. *Proceedings of the Royal Society, London,* series B, 252:39–41.

Buss, L. W., and P. O. Yund. 1989. A sibling species group of *Hydractinia* in the northeastern United States. *Journal of the Marine Biological Association, UK* 69:857–75.

Cartwright, P. 1997. Characterization of a HOM/Hox homeobox gene, *Cnox-2,* and the evolution of coloniality in the Hydrozoa (phylum Cnidaria). Ph.D. dissertation. Yale University, New Haven.

Cartwright, P., J. Bowsher, and L. W. Buss. 1999. Expression of a Hox gene, *Cnox-2,* and the division of labor in a colonial hydroid. *Proceedings of the National Academy of Sciences, USA* 96:2183–86.

Cartwright, P., and L. W. Buss. 1999. Expression of Hox type homeobox gene, *Cnox-2,* and evolution of colony integration in the Hydrozoa. *Journal of Experimental Zoology.* 285:57–62.

Cheetham, A. H., L. C. Hayek, and E. Thomsen. 1980. Branching structure in arborescent animals: Models of relative growth. *Journal of Theoretical Biology* 85:335–69.

Child, C. M. 1941. *Patterns and problems in development.* Chicago: University of Chicago Press.

Cohen, J. L. 1978. *Multiobjective programming and planning.* New York: Academic Press.

Crowell, S. 1957. Differential responses of growth zones to nutritive level, age, and temperature in the colonial hydroid *Campanularia. Journal of Comparative Zoology* 134:63–90.

Davies, P. F., A. Remuzzi, E. J. Gordon, C. F. Dewey, and M. A. Gimbrone. 1986. Turbulent fluid shear stress induces vascular endothelial cell turnover in vitro. *Proceedings of the National Academy of Sciences, USA* 83: 2114–17.

Deevey, E. S., ed. 1972. *Growth by intussusception: Ecological essays in honor of G. Evelyn Hutchinson.* New Haven: Archon Books.

Dolmetsch, R. E., R. S. Lewis, C. C. Goodnow, and J. I. Healy. 1997. Differential activation of transcription factors induced by Ca^{2+} response amplitude and duration. *Nature* 386:855–58.

Dudgeon, S., and L. W. Buss. 1996. Growing with the flow: On the maintenance and malleability of colony form in the hydroid *Hydractinia. American Naturalist* 147:667–91.

Dudgeon, S., A. Wagner, J. R. Vaisnys, and L. W. Buss. 1999. Dynamics of gastrovascular circulation in the hydrozoan *Podocoryne carnea:* The 1 polyp case. *Biological Bulletin* 196:1–17.

Fleming, J. 1828. *A history of British animals.* Edinburgh.

Frangos, J. A., S. G. Eskin, L. V. McIntire, and C. L. Ives. 1985. Flow effects on prostacyclin production by cultured human endothelial cells. *Science* 227: 1477–79.

Frank, U., C. Rabinowitz, and B. Rinkevich. 1994. In vitro establishment of continuous cell cultures and cell lines from ten colonial cnidarians. *Marine Biology* 120:491–99.

Fung, Y. C. 1993. *Biomechanics: Mechanical properties of living tissues.* Berlin: Springer-Verlag.

Hale, L. I. 1960. Contractility and hydroplasmic movements in the hydroid *Clytia johnstoni. Quarterly Journal of Microscopic Science* 101:339–50.

———. 1968. Cell movements, cell divisions, and growth in the hydroid *Clytia johnstoni. Journal of Embryological and Experimental Morphology* 12:517–38.

Hincks, T. 1868. *A history of the British hydroid zoophytes.* London: John Van Voorst.

Jackson, J. B. C. 1979. Morphological strategies of sessile animals. In *Biology and systematics of colonial organisms,* ed. G. Larwood and B. Rosen, 499–555. London: Academic Press.

Kaandorp, J. A. 1994. *Fractal modelling: Growth and form in biology.* Berlin: Springer-Verlag.

Kazakova, N. I., K. Zierold, G. Plickert, J. A. Labas, and L. V. Beloussov. 1994. X-ray microanalysis of ion contents in vacuoles and cytoplasm of the growing tips of the hydroid polyp as related to osmotic and growth pulsations. *Tissue and Cell* 26:687–97.

Klambt, C., L. Glazer, and B. Z. Shilo. 1992. *Breathless*, a *Drosophila* FGF receptor homolog, is essential for migration of tracheal and specific midline glial cells. *Genes and Development* 6:1668–78.

LaBarbera, M. 1990. Principles of design of fluid transport systems in zoology. *Science* 249:992–1000.

Lange, R., G. Plickert, and W. A. Müller. 1989. Histoincompatibility in a low invertebrate, *Hydractinia echinata*: Analysis of the mechanism of rejection. *Journal of Experimental Zoology* 249:284–92.

Leopold, L. B. 1971. Trees and streams: The efficiency of branching patterns. *Journal of Theoretical Biology* 31:339–54.

Martinez, D. E., M.-L. Dirksen, P. M. Bode, M. Jamrich, R. E. Steele, and H. R. Bode. 1997. *Budhead*, a forkhead/HNF-3 homologue, is expressed during axis formation and head specification in *Hydra*. *Developmental Biology* 192:523–36.

Meinhardt, H. 1976. Morphogenesis of lines and nets. *Differentiation* 6:117–23.

Müller, W. A. 1964. Experimentelle Untersuchungen über Stockentwicklung, Polypendifferenzierung, und Sexualchimären bei *Hydractinia echinata*. *Wilhelm Roux's Archives* 155:181–268.

Müller, W. A., A. Hauch, and G. Plickert. 1987. Morphogenetic factors in hydroids: 1, Stolon tip activation and inhibition. *Journal of Experimental Zoology* 243:111–24.

Müller, W. A., and G. Plickert. 1982. Quantitative analysis of an inhibitory gradient field in the hydrozoan stolon. *Wilhelm Roux's Archives* 191:56–63.

Murray, C. D. 1926. The physiological principle of minimum work applied to the angle of branching of arteries. *Proceedings of the National Academy of Sciences, USA* 12:835–41.

Otto, J. J., and R. D. Campbell. 1977. Budding in *Hydra attenuata*: Bud stages and fate map. *Journal of Experimental Zoology* 200:417–28.

Péron, F., and C. A. Lasueur. 1809. Des caractères génériques et spécifiques de toutes les espèces de méduses connues jusqu'à ce jour. *Museum National d'Histoire Naturelle, Annuaire* 14:312–66.

Petriconi, V., and G. Plickert. 1974. Zum Einfluss der Strömung auf die Entwicklung von Hydrorhizastöcken. *Verhandlungsbericht der Deutschen Zoologischen Gesellschaft* 67:107–11.

Plickert, G. 1980. Mechanically induced stolon branching in *Eirene viridula* (Thecata: Campanulinidae). In *Developmental and cellular biology of coelenterates*, ed. P. Tardent and R. Tardent, 185–90. Amsterdam: Elsevier/North Holland.

Plickert, G., A. Heringer, and B. Hiller. 1987. Analysis of spacing in a periodic pattern. *Developmental Biology* 120:399–411.

Prusinkiewicz, P., and A. Lindenmayer. 1990. *The algorithmic beauty of plants*. Berlin: Springer-Verlag.

Saint-Hilaire, K. 1930. Morphogenetische Untersuchungen der nichtzellularen

Gebilde bei Tieren. *Zoologische Jahrbücher: Abteilung für Allgemeine Zoologie und Physiologie der Tiere* 47:512–633.

Sars, M. 1846. *Fauna littoralis Norvegiae: I Heft, Über die Fortpflanzungsweise der Polypen.* Christiania, Norway.

Schierwater, B., B. Piekos, and L. W. Buss. 1992. Hydroid stolonal contractions mediated by contractile vacuoles. *Journal of Experimental Biology* 162:1–21.

Sherman, T. F. 1981. On connecting large vessels to small. *Journal of General Physiology* 78:431–53.

Stokes, D. R. 1974. Physiological studies of conducting systems in the colonial hydroid *Hydractinia echinata*: 1, Polyp specialization. *Journal of Experimental Zoology* 190:1–18.

Sutherland, D., C. Samakovlis, and M. A. Krasnow. 1996. *Branchless*, encodes a *Drosophila* FGF homolog that controls tracheal cell migration and the pattern of branching. *Cell* 87:1091–1101.

Turing, A. 1952. The chemical basis of morphogenesis. *Philosophical Transactions of the Royal Society, London,* series B, 237:37–72.

Ulam, S. 1962. On some mathematical problems connected with patterns of growth figures. *Proceedings of the Symposium on Applied Mathematics* 14: 215–24.

———. 1966. Patterns of growth of figures: Mathematical aspects. In *Module, proportion, symmetry, rhythm*, ed. G. Kepes, 64–74. New York: Braziller.

Van Winkle, D. H., and N. W. Blackstone. 1997. Video microscopical measures of gastrovascular flow in colonial hydroids. *Invertebrate Biology* 116:6–16.

Wagner, A., S. Dudgeon, J. R. Vaisnys, and L. W. Buss. 1998. Non-linear oscillations in polyps of the colonial hydroid *Podocoryne carnea. Naturwissenschaften* 85:117–20.

Waller, D. M., and D. A. Steingraeber. 1985. Branching and modular growth: Theoretical models and empirical patterns. In *Population biology and evolution of clonal organisms*, ed. J. B. C. Jackson, L. W. Buss, and R. E. Cook, 225–57. New Haven: Yale University Press.

Wyttenbach, C. R. 1968. The dynamics of stolon elongation in the hydroid *Campanularia flexuosa. Journal of Experimental Zoology* 167:333–53.

———. 1973. The role of hydroplasmic pressure in stolonic growth movements in the hydroid, *Bougainvillia. Journal of Experimental Zoology* 186:79–90.

———. 1974. Cell movements associated with terminal growth in colonial hydroids. *American Zoologist* 14:699–717.

Zamir, M. 1977. Shear forces and blood vessel radii in the cardiovascular system. *Journal of General Physiology* 69:449–61.

Parts and Integration   **2**

Consequences of Hierarchy

DANIEL W. MCSHEA

Introduction

A curious pattern seems to characterize part-whole relationships at two very different hierarchical levels. At the cell level, metazoan and land-plant cells seem to contain fewer types of parts, fewer kinds of physical structures, than free-living eukaryotic cells. An extreme example would be the sieve cells of certain land plants, which function mainly as simple tubes and contain almost no macroscopic structures. In contrast, a free-living *Euglena* has a large number of macroscopic structures (e.g., nucleus, mitochondria, plastids, a flagellum, contractile vacuoles, and so on; fig. 2.1). These extreme cases aside, even the more typical metazoan cells seem to be less complex in this sense, on average, than typical free-living cells. This pattern has been observed a number of times; for example, Gerhart and Kirschner (1997, 242) remark in passing that "metazoans as multicellular communities are far more complex than any single-celled organism, even a ciliate protozoan, although each metazoan cell in itself is simpler than a protozoan."

At the colony level, in colonial marine invertebrates, the clonal units or modules (e.g., polyps in hydrozoans or zooids in bryozoans) in the more "integrated" species (those in which colony formation has progressed further) seem to be internally simpler than modules in the more solitary species. Wood et al. (1992, 132) observed that in the more colonial forms, "the units are often smaller and less complex than in solitary forms." Beklemishev (1969, 484) saw this too, noting the "simplification of structure of zooids as compared with members of related free-living species."

The possible common pattern at these two disparate levels is the smaller number of types of internal parts in organisms that are integrated into larger functional wholes relative to comparable organisms that live more independently. (Notice that it is number of *types* of parts that is at

Fig. 2.1. A drawing of a *Euglena* cell, showing its many parts.

issue, rather than absolute number of parts, although the two might be correlated.)

If the pattern is real, one possible explanation is selection for economy. First, consider the cell level: historically, as clones of eukaryotic cells became integrated to form the first multicellular entities and as these entities acquired functionality—the ability to feed, reproduce, defend themselves, and so on—the functional demands on the component cells would have been reduced. Then selection would have been expected to favor the loss of functionality in those cells and therefore a loss of part types, mainly to reduce energy and material costs in development, physiology, and so forth.

Obviously, a necessary assumption here is that functions tend to be localized in specialized parts, and therefore that loss of function entails loss of part types. I discuss this assumption later.

A similar mechanism may be at work at the colony level. In species in which the colony has become functional, selection for economy is expected to favor lower numbers of part types in the lower-level entities—polyps and zooids—that constitute the colony.

The argument is easily generalized, raising the possibility that the same part–whole relationship might be found at many hierarchical levels in biology, from prokaryotic cells to colonies, and perhaps at higher levels yet. For convenience, I will refer to this broader argument as the "hierarchy hypothesis." Other explanations for the common pattern can be imagined; one is discussed later.

A number of ideas have been offered about the relationship between the complexity of a system at a given level and its functionality at a higher level (see especially Bonner 1988). Most concern the role of selection in producing differentiation among lower-level entities, or polymorphism, within multicellular organisms or colonies (e.g., Bell and Mooers 1997; Bonner 1988; Gerhart and Kirschner 1997; Harvell 1994; Schopf 1973; Valentine et al. 1993; Wilson 1968). Notice, however, that selection for polymorphism and the hierarchy hypothesis target different lower levels. For example, if the higher level is a colony, selection for polymorphism is concerned with the complexity *of the colony* (understood as the number of types of entities that constitute it), while the hierarchy hypothesis is concerned with the complexity *of the entities that constitute it* (understood as the number of part types they contain). (See Karsai and Wenzel [1998] for a discussion of complexity at both levels in social insects.)

I emphasize that the pattern itself is still purely impressionistic, that is, undemonstrated. It might, for example, be an illusion produced by a perceptual bias, perhaps a tendency to focus on certain extreme cases. And the evolutionary explanation offered is entirely speculative. Despite its intuitive appeal, the proposed mechanism might be too weak to make a significant difference at every hierarchical level. Thus, the purpose of this paper is not to make any strong empirical or theoretical claim, but rather to raise possibilities, and to show why further investigation might be worthwhile.

This paper is also a contribution to the broader investigation of the origin and evolution of biological hierarchies, both in particular taxa (e.g., Beklemishev 1969; Boardman and Cheetham 1973; Coates and Oliver 1973; Coates and Jackson 1985; Lidgard 1986; Lidgard and Jackson 1989; McKinney and Jackson 1989) and in general (e.g., Simon 1962; Pattee 1970; Allen and Starr 1982; Leigh 1983; Eldredge and Salthe 1984; Salthe 1985, 1993; O'Neill et al. 1986; Buss 1987; Bonner 1988; Maynard Smith 1988; Maynard Smith and Szathmáry 1995; Michod 1999).

I begin with a discussion of terms—*level, integration,* and *part*—explaining how each is to be understood for present purposes, and how they are related to each other. (The understandings adopted are not new, but a somewhat detailed explanation is necessary for clarity in the later discussion.) I then show how the proposed pattern can be investigated at the cellular level, using preliminary data to compare numbers of part types in small samples of both metazoan cells and free-living eukaryotic cells. The preliminary data show that, as expected, the free-living cells have more part types. Finally, I explain the hierarchy hypothesis in more detail, and offer an alternative version of the hypothesis that is also consistent with the patterns observed.

Levels, Integrations, and Parts

Levels

Figure 2.2 shows a schematic of a hierarchy. The circles are entities of some kind, which interact to form larger entities, which in turn interact to form still larger entities. The arrows represent the interactions. In biology, the entities correspond to what have classically been called "levels of organization"; a standard example is the series . . . organelle, eukaryotic cell, tissue/organ, organ-system, multicellular organism, colony. . . .

Salthe (1985) argued that three levels need to be considered in most hierarchical analyses: the level of interest, or the "focal level," plus one adjacent level above and one below. Parts—as the term is used here (see below)—occupy the level just below the focal one; they both constitute the focal level and are contained within it. For a book, the parts might be the pages, binding, dust jacket, and so forth. The relationship with the next higher level—which for the book might be a row of books on a shelf—is precisely parallel: focal-level systems constitute and are contained within the system at the next higher level.

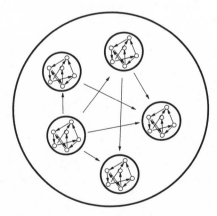

Fig. 2.2. Hierarchical levels. *Circles* are entities, and *arrows* show interactions among them.

Notice that the concern here is only with "object hierarchies," also called scalar or ecological hierarchies (Eldredge and Salthe 1984; Salthe 1985). "Process hierarchies," also called control or specification (Salthe 1993) hierarchies, are a separate domain. (See McShea 1996a, 1996b for further discussion and references.)

Notice also that levels do not correspond precisely with size ranges or scales. The level just below the cellular level, for example, may have parts ranging in size from mitochondrion to atom. The key criterion for occupancy of a given level is not absolute size but degree of nestedness. Thus, some genes are not parts of a cell, but rather are parts within parts, or subparts; or they might even be sub-subparts, depending on the number of levels thought to intervene. However, a sodium ion might be a cell part if it is appropriately nested (McShea and Venit 2001).

Figure 2.2 is obviously highly simplified in that real biological hierarchies are usually not so neatly separable into discrete levels. Thus, even in cases where entities at the focal level are well defined, as cells usually are, entities at adjacent levels may not be. For present purposes, this ambiguity creates difficulties only at the lower level, just below the focal level (cells), making identification of parts difficult; protocols for dealing with this are discussed later.

Integration

Integration is degree of connectedness. A connection is any relationship between two entities that results in a correlation between them in their behavior (Campbell 1958; Olson and Miller 1958), including "solid" links (e.g., chemical bonds), force fields of various kinds, signals transmitted, and so on. A highly integrated system then would be one in which all or most parts are connected in some way with all or most other parts.

Thus, a bar of iron is highly integrated in that the movement of each iron atom is tightly correlated with that of other iron atoms on account of the bonds among them. But a system without solid connections, such as a multiparty telephone connection, a "conference call," can also be highly integrated. Although the connections among the parts, the people, consist only of signals and the distances among them might be large, their behaviors during the conference call—specifically, the timing and content of what they say—are likely to be well correlated.

Integration, as I use the term, is independent of functionality. A system can be integrated and functional, as organisms are, or integrated and not functional, as iron bars are (in many contexts). My view therefore follows Boardman and Cheetham (1973) and Cheverud (1996), all of whom distinguished what they call "functional integration" from various other types. In what follows, I will not use this phrase, but will instead refer simply to the functionality of a system, or to its functional aspects, reserving the word *integration* for connectedness independent of function.

A robust literature on developmental integration has flowered in recent years (e.g., Olson and Miller 1958; Cheverud 1995, 1996; Zelditch 1996). Developmental integration refers to the connectedness among the entities involved in generating an organism. My concern is not with development, however, but rather with what might be called "operational integration," the connectedness of the entities within an organism in the course of its normal moment-to-moment operation, or its activities. Operation includes behavior and physiology but is broader in that it also includes any nonfunctional processes (e.g., any undesirable but unavoidable biochemical side reactions).

Parts

A part is a pattern of high integration within a surround of lesser integration (see Campbell 1958; Bonner 1988; McShea and Venit 2001). In figure 2.3, the small circles are entities, and the arrows show interactions among them, as in figure 2.2; here, in addition, the thickness of the arrows indicates the strength of the interactions. The dashed circles highlight three parts: each is a cluster of entities in which integration is high internally, in that connections among entities are many or strong or both, and

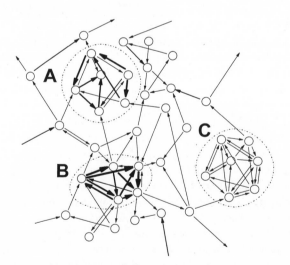

Fig. 2.3. Parts. *Small circles* are entities, and *arrows* show interactions among them; thickness of arrow corresponds to strength of interaction. *Dashed curves* enclose parts: part *B* has few internal interactions, but they are strong relative to those with external entities; part *C* has weaker internal interactions, but they are many relative to external interactions; part *A* is intermediate.

low externally, in that connections with other entities outside the cluster are few or weak or both. Thus, being a part involves a certain amount of integration and of its opposite, isolation. A similar notion was used by Wimsatt (1974) to show how systems may be "decomposable" (in the sense of Simon [1962]) into subsystems.

By these criteria, an iron bar is a part in that, first, it is internally integrated in the sense discussed. Second, it is relatively isolated from its surround, in that the bonds between the iron atoms and the atoms in the air, or the atoms in the surface on which the bar rests, are much weaker than the internal bonds.

Importantly, not every physical entity with contiguous components is a part. A row of books on a shelf might seem to be a part, one in which individual books are subparts. But it is only very weakly a part, because the books do not interact much. Ordinarily, they are not bonded to each other, and they exchange no signals. Their proximity to each other is the result of forces external to them. Likewise, a group of organisms is not a part simply by virtue of occurring adjacent to each other.

Thus, spatial proximity of entities is not a sufficient condition for their constituting a part. Neither is it a necessary condition. (Figure 2.3 is somewhat misleading in this respect.) The people interacting in a conference call are not ordinarily spatially contiguous, but a conference call is a part to the extent that the interactions among the participants are many or strong and external interactions are fewer or weaker.

Parts and Levels. All systems can be decomposed into a nested hierarchy of parts, subparts, sub-subparts, and so forth, but not all levels are expected to be equally relevant to the focal level. In particular, adjacent hierarchical levels should typically be more strongly connected to each other than to more distantly separated ones (Eldredge and Salthe 1984), in part on account of their more similar characteristic rates (O'Neill et al. 1986). More generally, higher levels tend to isolate or "screen off" lower levels, rendering the lower levels invisible to forces—such as selection—acting at higher levels (Brandon 1996; for further discussion, see McShea and Venit 2001). Therefore, in the case of cells combining to form multicellular organisms, the relevant parts—including those that the hierarchy hypothesis predicts will be lost—are most likely to be those just below the focal level but not lower, in other words, parts but not subparts. In other words, selection at the level of the multicellular organism may be felt at lower levels, but the effect is expected to attenuate, on average, as differences in level increase.

A perhaps surprising consequence is that number of genes or gene products may be poor proxies for number of cell parts. Genes are certainly functional parts at some level, but most molecules may be sufficiently distant (hierarchically) from the cellular level, that selection for loss of functional capacity might not be expressed there. For example, in principle, a cell organelle might well be lost with no change, or even with an increase, in number of expressed genes.

Parts and Function. Like integration, the degree to which a system is a part—its "partness," so to speak—is independent of functionality. A cell (which in a larger organism may be functional) and a cloud (which ordinarily is not) are equally candidates for being parts. This is a conceptual separation only: as mentioned, the hierarchy hypothesis assumes that, within organisms, a good correlation in fact exists between number of part types and number of functions (McShea 2000). However, one virtue of maintaining a clean *conceptual* separation between parts and functions is that doing so enables us to study their relationship empirically. In other words, the existence of a correlation becomes testable, at least in principle.

Parts and Development. The present conceptual scheme parallels, and is largely inspired by, another suggested by Wagner and Altenberg (1996; Wagner 1996; Raff 1996) concerning "modules" or what might be called "developmental parts." Consider a genotype-phenotype map in which genes (and gene products) interact to produce structures, or characters. In organisms, it is not the case that all genes affect all characters; rather, development is organized to some extent into "modules" in which a group of genes interact strongly among themselves in the development of single characters, and pleiotropic interactions among modules are fewer and weaker. Wagner and Altenberg argue that modular organization confers evolvability, in that it allows characters to vary independently in development and therefore to vary independently in evolution. Developmental modularity is therefore expected to be favored by selection for evolvability.

The connection with the Wagner and Altenberg view is this: parts are modules, but they are "operational modules," not developmental modules. In other words, parts are the units in the operation of an organism, in its activities (e.g., behavior, physiology) rather than in its generation. Mittenthal et al.'s "dynamic modules" (1992) include both developmental and operational entities, that is, both parts and developmental modules.

Parts and Genetic Relatedness. Just as parts are, in principle, independent of development, they are also independent of genetic relatedness. For example, a phoretic association consisting of a number of different species that interact ecologically and travel together (Wilson and Sober

1989) might be a part. But a group of genetically identical modules in a clonal organism—say, polyps in a coral—might not be. Again, the degree to which a system is a part depends only on the integration and isolation of its subparts in their present operation. A common genome among polyps in a coral colony creates a historical connection among the polyps, but does not necessarily produce any present physiological and behavioral connection among them. For parts, the relationships at issue are ecological, not genealogical.

Language Issues. Subparts are also parts, as discussed, but at a lower hierarchical level. Wholes are parts, too, but at a higher level (even if—like a free bar of iron—they are not obviously *part of* any larger entity). Thus, a whole is just a pattern of interaction and isolation. Notice that because parts, subparts, and wholes are all defined simply in terms of connectedness and isolation, the classical ontological issues—such as whether the properties of the whole are reducible to that of its components, whether they are emergent, and so on—do not arise.

I chose the word *part* because its colloquial meaning is fairly unspecific. An alternative was *individual,* but this word has usually been used to refer exclusively to organisms (e.g., Buss 1987), and the overlap in meaning with *part* is therefore imperfect. *Entity, component,* and *system* I reserve for nontechnical usage.

Counting Parts

Do Parts Exist?

We have at least three reasons to think that organisms are to some extent organized as systems of parts. The first is that organisms seem to be anatomically well compartmentalized. In multicellular organisms, for example, the fact that we can identify more or less discrete tissues, organs, and organ systems suggests considerable partness, assuming that morphological boundaries are good indicators of isolation (or at least of reduced integration). Second, in multicellulars at least, the consequences of injuries sometimes seem to be fairly localized, to be largely limited to directly impacted tissues, organs, or organ systems, suggesting some degree of isolation among them. Third, in simulated experiments with fitness landscapes, Kauffman (1993) discovered that in fully connected networks—in which the fitness of the whole was sensitive to variation in every component (or gene, in Kauffman's version) and parts were therefore absent—organisms tended to become trapped at low fitness values. In present terms, selection opposes full connectedness and therefore favors at least some degree of partness (see below). In any case, the assumption here is that organisms do have parts. Given that they do, can parts be counted in a meaningful way?

Difficulties in Counting

Some parts are obvious. In a vertebrate, for example, the circulatory system is sufficiently distinct from other structures that it would strike most people as an unambiguous part. Also, it is not hard to see how a structure like the circulatory system might meet the integration and isolation criteria for parts in the scheme above. But even the barest acquaintance with the internal structure of organisms is enough to suggest that counting parts in a consistent and objective way will usually be difficult. Consider these potential problems.

Invisible Parts. Many parts are manifest as objects and can be detected by visual inspection. However, some objects are too small to detect even with magnification (e.g., molecular and atomic parts). Also, some parts are not manifest as objects, for example, spatially distributed parts like some hormone-mediated systems. Or consider a small set of neurons that stimulate each other, but whose cell bodies are widely distributed throughout a brain. Imagine further that the pattern of activation is such that spatially adjacent neurons are not much affected. Although invisible at the gross anatomical level, such a system would be a part—a kind of neural conference call. (Neural parts of this sort may be manifest as behaviors, but they need not be.)

Part versus Subpart. An apparent part may actually be a subpart if there is an entity at an intervening level that is spatially distributed or otherwise invisible. Conversely, an apparent subpart may be a part if the intervening entity is, despite appearances, poorly integrated.

Another potential difficulty is that parts at different scales may overlap in their patterns of interaction, so that a single entity may be simultaneously both a part and a subpart. In a clam, for example, the shell might be both a part insofar as it operates, say, in defense, and a subpart in the organism's calcium storage and regulation system.

Partial Parts. As discussed, partness comes in degrees. In many cases, the answer to the question of whether two entities are full parts (as opposed to subparts within the same part) may be that they are somewhat parts, that they are "partly parts," so to speak.

Part Types. Differentiation among part types varies continuously, and even where partness is unambiguous, distinctiveness of types may not be. For example, if an arm is a part, is a left arm a different part type than a right arm?

None of these potential problems is a problem in principle. All parts would be equally visible and their levels apparent if we knew the complete pattern of interaction within an organism, that is, a complete "wiring diagram," perhaps like figure 2.3, showing every interaction: mechanical, chemical, electrical, and so forth. Hierarchical overlap also presents no problem in principle: entities may be both parts and subparts without contradiction. Finally, partial differentiation and imperfect distinctiveness among parts present formidable measurement problems, but no conceptual ones. (See McShea and Venit 2001 for further discussion.)

Differing Theoretical Perspectives. One apparent conceptual problem seems more serious: Wimsatt (1974) noted that the decomposition of organisms into parts seems to vary with theoretical perspective. For example, viewed as a device for locomotion, a tetrapod limb is decomposable into one set of parts, but viewed as a device for thermoregulation, it might have a different set, and the two sets might well have noncoincident boundaries (Wimsatt 1974). In present terms, however, his point is that an organism has multiple noncoincident *functional* decompositions; this does not deny that it has a single pattern of connectedness, a pattern that, if known, would reveal a single, unique decomposition into parts.

Counting Parts: Suggested Protocols

In the absence of complete wiring diagrams for detecting invisible parts and operational metrics for measuring partial parts, we may yet make a crude beginning. We can do this by using morphological boundaries, or physical demarcations, as indicators of reduction in degree of integration. Parts identified using such demarcations will be called "object parts." The proposal is that we can use counts of object-part types as proxies for counts of true part types.

An object part is understood here as an entity that is spatially separated from all others (i.e., free floating) or that is attached but demarcated by a local difference in composition or in configuration (shape) or both. Certain membrane-bound organelles would be considered free-floating object parts. For attached entities, first consider those demarcated by differences in composition. Desmosomes are junctions between cells in metazoans, appearing as regions of proliferation of certain transmembrane proteins adjacent to and within the cell membrane (Kowalczyk et al. 1999). Thus, insofar as a desmosome is compositionally distinct from the membrane, it would count as a distinct part. The compositional difference suggests some degree of isolation, that is, some degree of dynamical independence from the membrane in which it is invested. (One might argue that intercellular junctions should not count as cell parts because they lie at least partly outside the cell membrane. The point is debatable, but they are counted here because they are present mainly in the cells of multicellular organisms and therefore tend to bias the data *against* the hypothesis.)

Now consider attached parts demarcated by shape differences. For example, a pseudopod in *Amoeba* would count as an object part, one that is marked mainly by a local variation in membrane shape. The shape difference depends on compositional differences in the cytoplasm, but these are not visible ordinarily at low magnification. (In this case, the dynamical independence of the pseudopod, its ability to move at least somewhat independently of the rest of the cell, also suggests that it is a part.)

In principle, we might be able to order the types of object parts according to the probability that they correspond to local reductions in integration and therefore that they are true parts. Free-floating objects seem more likely to be true parts than those marked by compositional or shape differences. And boundaries marked only by a difference in composition seem more significant than those marked only by difference in shape. In this preliminary study, however, no ordering is assumed; all boundaries are weighted equally.

Here, the hierarchical level at which parts occur is assessed based on topology. Parts that appear to be physically nested within other parts—as the cristae are within mitochondria—are considered subparts. Likewise, nested changes in composition and shape correspond to subparts, although these are frequently more difficult to identify with confidence.

These protocols were developed especially to count part types within cells and may have to be modified to count at other levels. Also taxon-specific approaches may be necessary to count part types within multicellular organisms (McShea and Venit 2001).

Weaknesses. This approach has two major—but presently unavoidable—weaknesses. First, while the compositional and configurational criteria lend a degree of objectivity to parts identification, considerable subjectivity remains. In particular, decisions about how much and what sort of compositional and shape changes are sufficient for part status are somewhat subjective. Decisions about whether two parts are sufficiently different to count as distinct types also remain subjective. Second, in using object parts as a proxy for true parts, the assumption is that object parts are an unbiased sample of all true parts, or more precisely, that numbers of object-part types and true-part types are well correlated in organisms. I do not know how to test this assumption; the only justification that can be offered now is that we have no a priori reason to believe that object parts are *not* an unbiased sample.

In defense of this approach, it should be pointed out that these weaknesses are only sources of noise, not of spurious signal. In other words, they are not biases of the sort that should raise doubts about any consistent pattern or correlation that is discovered (although such biases might exist; see below) between number of part types and other variables.

Counting Parts in Cells

Methods

The protocols can now be applied in a test of the proposed cell-level pattern. The test consists of a comparison of numbers of object-part types in samples of free-living eukaryotic cells and of metazoan cells. The point is to illustrate the sort of empirical treatment that the conceptual scheme makes possible; results are preliminary.

Part-type counts were based on cell descriptions in the cytological literature, specifically, descriptions of cells (for free-living eukaryotes) or cell types (for metazoans) in a single species. Excluded were general descriptions of particular cell types in higher taxa (e.g., a general description of mammalian oocytes).

One problem is that coverage in the literature is uneven in that certain cells, such as *Amoeba* and *Euglena,* receive disproportionate attention. As a result, a strategy in which all available descriptive material is sought for each cell or cell type would tend to underestimate counts in the less well studied cells. Thus, in an attempt to introduce some consistency, only journal articles were used (thereby excluding books, long monographs, etc.), and only one article was used per cell type.

Object parts were inferred from descriptions in the text of an article and in figure captions, from the figures themselves, and from photographs, usually electron micrographs.

Cell contents typically vary in time, on account of either a cell's complex life cycle or the varying demands of development and physiology. No attempt was made to estimate a cell-lifetime average number of part types; instead, only descriptions of a single cell stage were used. Such single-stage descriptions can be understood as snapshots, so to speak, which amount to random samples in time, of a cell's contents.

One difficulty with using journal articles is that descriptions of specialized cells tend to focus on their specialized parts, often to the exclusion of others. Thus, a description of retinal cells might focus on the light-sensing apparatus and fail to mention that the cell has a Golgi apparatus (dictyosomes). In an attempt to solve this problem, I developed a list of cell part types and part-type categories—what I call the "standard set"—

Table 2.1. The "standard set" of cell parts

Plasma membrane
Nucleus
Mitochondria (or "hydrogenosomes")
Endoplasmic reticulum
Ribosomes
Golgi apparatus
Any microtubular structures (e.g., centrioles)
Any internally undifferentiated spherical (or subspherical) structures (e.g., lysosomes, peroxisomes, phagosomes, vacuoles, vesicles, concretions, granules, glycogen bodies, etc.)

that occur in almost all cells, both free-living and metazoan (table 2.1). Every cell is assumed to have every member of the standard set unless an absence is explicitly noted in the description. Thus, part-type counts represent not total cell contents but numbers of part types *in addition to* the standard set. (Part-type absences were noted only rarely, but when they were, one part type was subtracted from the total for each; notice that a consequence of this procedure is that, in principle, negative part-type counts are possible.)

Notice that the standard set is actually much larger than a glance at table 2.1 might suggest, because two entries, microtubular structures and undifferentiated spherical structures, cover a great number of different part types.

Appropriate descriptive articles were found mainly by scanning tables of contents of major journals, such as *Protozoology* and *Journal of Morphology*. Certain cells were excluded, such as—for free-living eukaryotes—cells that obligately or facultatively spend any portion of their life cycle within a larger organism, whether the relationship is parasitic, commensal, or mutualistic. Also ruled out were studies of natural or artificially induced cell pathologies, and studies of cells subjected to special procedures or conditions, such as those maintained in artificial conditions.

Overall, the strategy was not to try to count all of the object-part types in any cell, but rather to sample parts in a consistent way. The assumption is that the sampling procedure is unbiased.

Choice of Taxa

For the free-living cells, some attempt was made to cover a wide taxonomic range; the analysis was conducted roughly at the level of the phylum, in the sense of Corliss (1994; see also Margulis et al. 1990). That is, most phyla were represented by only one species, and part-type counts were averaged within phyla for those represented by more than one.

Phylogenetic relationships among taxa were not taken into account in this preliminary study.

For the metazoan cells, taxonomic diversity was a lower priority; the higher priority was cell-type diversity. No cell type was represented more than once. The reason this was done is that, apparently, most of the differentiation among cells in metazoans occurs among cell types, not among taxa. Some evidence for this is presented in table 2.2, which shows part-type counts for cells within two higher taxa, sponges and insects. The data were gathered using a somewhat different method than outlined above—using a single monographic treatment of cell types for each group (Simpson 1984; Smith 1968)—and thus the absolute values of the part-type counts may not be comparable with those used in the comparison

Table 2.2. Within-group variance

Insects[a]	Part Count
Salivary gland	4
Midgut	4
Hindgut rectal epithelial	5
Medullary of hindgut	1
Malphigian tubule	3
Mycetocytes	0
Fat body of larva	0
Oenocytes	1
Pericardial	4
Heart muscle	2
Retinular	3
Corpus allatum	0
Corpus cardiacum	1
Muscle	2
Mean	2.1
Standard deviation	1.7

Sponges[b]	Part Count
Exopinacocytes	4
Choanocytes	4
Archeocytes	1
Myocytes	2
Polyblasts	0
Thesocytes	1
Spongocytes	1
Mean	1.9
Standard deviation	1.6

Note: Part counts are numbers of part types in addition to the standard set.
[a] Data from Simpson 1984.
[b] Data from Smith 1968.

Table 2.3. Part counts

	Metazoa			
Higher Taxon	Species	Cell Type	Part Count	Max
Chordata	*Gallus domesticus*	Lipid-containing of anterior pituitary gland	1	1
Chordata	*Homo sapiens*	Melanocytes	3	3
Chordata	*Mus*	Epithelial of fundus gland of gastric mucosa	3	3
Chordata	*Oryctolagus cuniculus*	Obplacental giant	3	3
Chordata	*Lepidosiren paradoxa*	Agranular "cavity-boundary"	5	6
Arthropoda	*Periplaneta americana*	Hemocytes	3	3
Arthropoda	*Gammarus setosus*	Attachment of organ of Bellonci	6	6
Arthropoda	*Leucania loreyi*	"Isolated type" of corpus allata	1	1
Arthropoda	*Locusta migratoria*	"Type B" of rectal pad epithelium	3	4
Porifera	*Oscarella lobularis*	Vacuolar	1	1
Echinodermata	*Eupentacta quinquesemita*	Morula	1	1
Sipuncula	*Phascolosoma granulatum*	Nephridial granular (dark)	1	1
Nemertea	*Riseriellus occultus*	Sensory	4	8
Mollusca	*Tridacna crocea*	Smooth muscle in adductors	1	1
Annelida	*Haementeria ghilianii*	Anterior salivary	2	2
Annelida	*Eisenia foetida*	Cocoon-producing	1	2
Mean			2.4	2.9
Standard deviation			1.6	2.2

Continued on next page

between free-living and metazoan cells (table 2.3). However, the variances should be comparable. The table shows that the standard deviation of part-type counts both within insects (1.7) and within sponges (1.6) is very close to that within metazoans as a whole (1.6; see table 2.3), suggesting that most of the variance occurs among cell types.

Notice that, as a result of the way that metazoan cell types were sampled, with each cell type represented once, the focal comparison is properly understood as that between a group, the free-living eukaryotes, and what might be thought of as a single *composite* metazoan. In other words, the Metazoa are understood as a single data point, a single instance of the evolution of multicellularity.

Results and Discussion

Table 2.3 shows part-type counts. The first column of numbers shows counts where parts can be identified unambiguously (or nearly so) using the criteria above. The second column of numbers shows maximum counts, which include entities whose status is questionable because (1) the existence of a composition or shape boundary is uncertain; (2) the

Table 2.3. *(continued)*

				Higher Taxon Means[a]	
Higher Taxon	Species	Part Count	Max	Part Count	Max
Chlorarachnida	*Chlorarachnion reptans*	2	3	2	3
Chrysophyceae	*Chromulina placentula*	5	6	6	7
Chrysophyceae	*Ochromonas tuberculatus*	6	7		
Chrysophyceae	*Chrysococcus rufescens*	7	8		
Pedinellophyceae	*Pedinella hexacostata*	6	8	6	8
Ciliophora	*Perispira ovum*	8	9	8	9
Cryptophyta	*Rhodomonas lacustris*	6	7	6	7
Dinoflagellata	*Aureodinium pigmentosum*	3	4	4	6
Dinoflagellata	*Woloszynskia micra*	5	8		
Diplomonadida	*Trepomonas agilis*	2	4	2	4
Euglenida	*Khawkinea quartana*	5	6	4.5	6
Euglenida	*Paranema trichophorum*	4	6		
Eustigmatophyta	*Chlorobotrys regularis*	1	2	2.5	3.5
Eustigmatophyta	*Pseudocharaciopsis texensis*	4	5		
Foraminifera	*Globigerinoides ruber*	6	8	6	8
Glaucocystophyta	*Glaucocystis nostochinearum*	2	3	2	3
Prymnesiophyta	*Diacronema vlkianum*	3	5	3.5	4.5
Prymnesiophyta	*Chrysochromulina megacylindra*	4	4		
Kinetoplastida	*Bodo curvifilus*	3	4	3	4
Raphidophyta	*Olisthodiscus luteus*	2	4	2	4
Rhizopoda	*Trichosphaerium micrum*	2	3	2	3
Rhodophyta	*Rhodella maculata*	3	3	2	2
Rhodophyta	*Dixoniella grisea*	1	1		
Mean				3.8	5.1
Standard deviation				2.0	2.2
		Difference between means		1.4	2.3
		$p \leq$		0.012	0.001

Note: Part counts are numbers of part types in addition to the standard set.

[a] The mean for each higher taxon appears in the row containing the first species in that taxon; e.g., the mean for all three Chrysophyceae species (6 parts; 7 max) appears in the row for *Chromulina*.

part's hierarchical level is unclear, that is, it may be subpart instead; or (3) the part is quite similar to some other part and thus not clearly a distinct type. (These are listed under "possible parts" in the appendix.) On average, free-living eukaryotic cells had 1.4 more part types per cell ($p \leq .012$, using a bootstrap test of the difference between means) than a composite metazoan, or 2.3 more part types ($p \leq .001$) using maximum counts. Sample sizes are low, but p values will likely only decrease—that is, the correlation will improve—as sample sizes increase.

The main point of this preliminary empirical treatment is to show how the counting of part types might be operationalized. It is interesting that the preliminary data support the impressionistic pattern—free-living cells having more part types, on average—but it would be premature to draw any conclusions about evolutionary pattern and mechanism. Similar comparisons need to be done for plant cells, fungal cells, and perhaps other taxa in which a higher level above the cell has emerged to some extent (e.g., *Volvox*). Also, comparisons need to be done in a phylogenetic context, ideally as independent sister-group comparisons.

Finally, various biases can be imagined. One possibility is that, because counts of part types are somewhat subjective, my counts may have been biased by my foreknowledge of the expected pattern. To test this, I asked a student to repeat the counts for a number of taxa. The student was given the same papers I used but was not told about the expected pattern or the hierarchy hypothesis. Her mean counts were systematically lower—1.1 part types per cell lower for metazoan cells and 0.7 lower for free-living eukaryotes—but the difference was in the same direction and highly significant. But other biases are possible and need to be investigated.

Theory

The Hierarchy Hypothesis

The argument of the hypothesis is this: in species in which organisms have combined to form a higher-level entity, to the extent that this entity is functional (presumably as a result of higher-level selection), functional demands on the component organisms are reduced. And to the extent that functional demands are reduced, selection will favor a reduction in part types in the interest of economy. In principle, this selective force could have acted at all hierarchical levels: on prokaryotic cells, when they combined to form the first eukaryotic cell, on eukaryotic cells, as they combined to form multicellular organisms, and so on.

The hypothesis is a generalization of a more familiar argument in biology that is often invoked to explain the apparent reduction in internal structure of many parasites compared to their free-living relatives. Parasites are protected by the host and often sustained by the host's metabolic and homeostatic functions. Species in which the structures performing these functions are lost save the metabolic cost of producing them, and therefore such losses are favored by selection.

The rationale for the hierarchy hypothesis is so familiar and straightforwardly selectionist that its speculative nature is worth emphasizing. First, existing evidence is equivocal at all scales. Even the notion that parasites are structurally simpler than their free-living relatives has been doubted (Brooks and McLennan 1993). And of course, the patterns with which this chapter was introduced are still undemonstrated. Also, the selective force it invokes may be present but weak, and perhaps nowhere detectable. Developmental constraints may resist or prevent part-type losses. And other forces (perhaps selective also) may exist that tend to increase numbers of part types, overwhelming selection for economy.

A clarification is in order here. "Reduction" is understood here as a relative term. That is, a reduction in number of part types could be manifest as a decrease in absolute number, or as a reduction relative to a background tendency to increase, if such exists. Another way to say this is that the rationale for the hypothesis is indifferent to whether part-type counts were primitively high or low. For example, the ancestor of modern free-living eukaryotic cells might have had few part types, with increases occurring among all of its free-living descendants but not—or at a lower rate—in multicellular organisms. Or the ancestor might have had many part types, with little or no increase in its free-living descendants, but with a reduction occurring in the multicellulars. The data above are consistent with either alternative. Other possibilities can also be imagined.

An Alternative Version

Another possible explanation of the apparent pattern is that, as organisms combine and higher-level functionality emerges, a shift in part *size* occurs. That is, the macroscopic parts involved in the functions assumed by the higher-level entity might be lost, while smaller—and mostly invisible—parts are gained. The argument would be that the emergence of function at a higher level requires an increase in integration among the component organisms, integration can take the form of signaling, and signaling is often mediated by smaller parts. In concrete terms, metazoan cells might be expected to have fewer macroscopic part types, but more molecule-sized part types than free-living cells, because metazoan development and physiology require considerable coordination of cellular activity, most of which is likely to be mediated by molecular signals.

One problem with this version of the hypothesis is that—in the absence of data on numbers of small part types—a fairly compelling counterargument can be made. A free-living cell may not require any local cell-to-cell signaling and coordination, but molecular cues are doubtless extremely important in detecting and orienting to a variety of food sources, avoiding extreme or otherwise dangerous external conditions (such as toxic substances), detecting and avoiding predators, and so on. Given all these requirements, it is not obvious that a free-living cell ought to require fewer molecule-sized parts than a metazoan cell.

In any case, the cell data here include only large parts, and therefore are equally compatible with both versions of the hierarchy hypothesis.

A Required Assumption

The hierarchy hypothesis requires an assumption, namely that number of part types is well correlated with number of functions in organisms, that is, that functions tend to be isolated to some extent in parts. This assumption is required so that reductions in functional demands also produce reductions in numbers of part types. One reason that such a correlation might be expected is that, in order to carry out a function (e.g., walking, chewing), the components involved in that function need to coordinate their activity, and therefore need to be fairly integrated. But as a group, they also need to be isolated to some extent, so that the other components do not interfere with their operation. In other words, functions should be isolated to some extent in parts. Thus, in modern birds, selection is expected to have favored a separation of the structures involved in flying from those involved in catching insects, so that, for example, beak movements do not cause wing spasms. That is, flying and insect-catching should be accomplished by distinct parts. The argument parallels the rationale that was discussed earlier for the evolution of developmental modules (Wagner and Altenberg 1996).

Importantly, complete isolation is not expected; the relationship between part types and functions is not expected to be one to one. Overlap can be engineered, especially where the sequences of states or activities that are demanded by the overlapping functions (their "state cycles") are simple or short and therefore unlikely to conflict. Thus, a clam shell can function both for defense and for burrowing simultaneously, because the demands that both functions make on the shell are simple and compatible: defense might demand rigidity, while burrowing might demand that the shell maintain a certain shape. Overlap of more complex state cycles is also possible if functions can be coordinated, so that both make similar demands on the same subparts at the same time. For further discussion, see McShea (2000).

Testing

The preliminary test here took advantage of the fact that we can identify certain extremes of upper-level functionality. Specifically, I compared cells in a multicellular organism, which is obviously highly functional, with free-living cells, which have no discernible group-level functionality (because there is no group). However, more sensitive tests may be possible using comparisons over a range of degrees of upper-level functionality. In bryozoans, for example, the degree to which the colony as a whole is functional—the degree of "colony dominance"—varies throughout the group, and criteria for assessing colony dominance have been devised (Boardman and Cheetham 1973). Thus, the prediction would be that the number of lower-level part types (i.e., part types within zooids) should be inversely correlated with degree of colony dominance.

Importantly, tests at all levels must meet certain requirements: (1) Entities at the focal level must be organisms, or homologous with organisms; cells in multicellular organisms are presumably homologous with free-living cells, sharing their organization by common descent. If the focal entity is not an organism (or an organism homologue), it is not expected to have functional parts in the first place, and therefore is not expected to lose parts when higher-level functionality emerges. Also, the organisms and their homologues in the groups compared must occupy the same hierarchical level; eukaryotic cells must be compared with eukaryotic cells.

(2) The level at which part types are identified and counted should be chosen so that parts lie below the focal level but as close to it as possible. Arguably, there is some room for error here, because selection for economy will tend to remove part types at all lower levels. However, as discussed, the effect is expected to be most direct, and strongest, at the level just below.

(3) Higher levels should also be close to the focal level, but not any higher level will do. In order to have the effect predicted by the hierarchy hypothesis, the higher level should be one at which selection has operated to produce functionality (Buss 1987). Thus, the organ and tissue levels lie above and adjacent to the cellular level, but they are not the relevant higher level above cells. (Presumably, organs and tissues are produced by selection acting at a higher level yet, the level of the whole organism.)

In some cases, identifying appropriate upper levels will not be a trivial exercise. For example, it might seem obvious that, if the focal level is the multicellular organism, then the next higher level at which selection could have operated is the colony. But why could not the next highest selected level be the sexual pair, for example? (See the list of levels and transitions in Maynard Smith and Szathmáry 1995, 6.) Actually, tests may be able to tolerate some error here. Consistent with the hypothesis, in order for multicellular organisms to experience selection for loss of part types, it is only necessary that functionality has emerged at *some* higher level. Of course, the closer the higher level is to the focal level, the stronger the expected effect.

Summary

Impressionistically, cells in multicellular organisms have fewer types of parts than their free-living relatives, and the modules in certain, highly integrated, colonial, marine invertebrate species have fewer part types than their more solitary relatives. These patterns are interesting in that they raise the possibility of a consistent relationship between the emergence of functionality in higher-level entities and numbers of lower-level part types. One possible explanation, here called the hierarchy hypothesis, is that, as organisms combine to form higher-level wholes and as functionality emerges in those wholes, functional demands on the organisms decrease. Selection for economy then favors the reduction in functionality in those organisms, and therefore the loss of part types (assuming that numbers of part types and numbers of functions are well correlated). In principle, this mechanism could work at any level, from prokaryotic cell to colony, and perhaps higher. The argument is similar to, indeed a generalization of, the standard Darwinian argument for the putative simplification of structure in parasites.

An alternative version of the hypothesis is consistent with the same patterns: instead of eliminating part types, the emergence of higher-level functionality might instead produce a shift in part size. That is, it might reduce the number of macroscopic part types involved in large-scale functions, but increase the number of smaller and mostly invisible part types involved in integration (i.e., communication) among the organisms at the focal level.

The objectives here are as follows: (1) to raise the patterns and the hypothesis as matters of possible empirical and theoretical interest; (2) to provide a conceptual scheme within which investigation of both pattern and hypothesis is possible; and (3) to apply the scheme in a preliminary way, comparing part-type counts in small samples of metazoan cells and free-living eukaryotic cells; the preliminary data seem to support the conclusion that metazoan cells have fewer part types.

Acknowledgments

I thank F. K. McKinney, V. L. Roth, and two anonymous reviewers for careful readings and critiques of the manuscript, and D. Raup and L. Van Valen for comments on an earlier version. For discussions, I thank R. Brandon, A. Dajer, M. Mitchell, C. Cunningham, R. LeGrand, F. Nijhout, M. Nijhout, D. Pope, L. Roth, V. Simon, E. Venit, G. Wagner, W. Wimsatt, and the participants in Biology and Philosophy Discussion Group at Duke University. V. Simon kindly carried out the parts recount.

Appendix

Table 2A.1. Cell parts

Species	Cell Type	Metazoan Cells			Source
		Parts		Possible Parts	
Gallus domesticus	Anterior pituitary	1 Multilobed structures			Raymond et al. 1993
Homo sapiens	Melanocytes	1 Isolated cilia			Masuda et al. 1994
		2 Melanocyte dendrites			
		3 Gap junctions			
Mus	Epithelial of fundus gland of gastric mucosa	1 Microvilli			Helander and Ekholm 1959
		2 Convolutions of membrane			
		3 Intracellular secretory canaliculi			
Oryctolagus cuniculus	Obplacental giant	1 Desmosomes			Blackburn et al. 1989
		2 Elongate cell processes			
		3 Intracellular canaliculi			
Lepidosiren paradoxa	Adenohypophysis agranular	1 Cytoplasmic processes			Chiba et al. 1991
		2 Junctional complexes			
		3		Desmosomes	
		4 Microvilli			
		5 Cilia			
		6 Tiny processes			
Periplaneta americana	Hemocytes	1 Pseudopod-like cytoplasmic extensions			Baerwald and Boush 1970
		2 "Cylinder" inclusions			
		3 Large "inclusion" with "packed subunits"			
Gammarus setosus	Attachment of organ of Bellonci	1 Dorsal processes/branches with microtubules			Steele and Oshel 1989
		2 Dorsal processes/branches with vesicles			

Continued on next page

Table 2A.1. (continued)

Species	Cell Type	Metazoan Cells		
		Parts	Possible Parts	Source
Leucania loreyi	"Isolated type" of corpora allata	3 Convoluted interdigitations 4 Septate junctions 5 Desmosomes		Kou et al. 1995
Locusta migratoria	"Type B" of rectal pad epithelium	6 Pits with dense areas 1 Interdigitations at cell periphery 1 Septate desmosomes 2	Convoluted membrane near desmosomes	Peacock 1979
Oscarella lobularis	Vacuolar	3 Hemidesmosomes 4 Apical microvilli		Gaino et al. 1986
Eupentacta quinquesemita	Morula	1 Fibril-containing inclusions 1 Secretory vesicles (differentiated)		Byrne 1986
Phascolosoma granulatum	Nephridial granular (dark)	1 Pseudopod-like projections		Serrano et al. 1990–91
Riseriellus occultus	Sensory	1 Cilium 2 3 4 Dendrite 5 6 7 Zonula adherens 8 Septate junctions	Microvilli around cilium Bulblike tip of cilium Bulbous tip of dendrite Axon	Montalvo et al. 1996
Tridacna crocea	Smooth muscle in adductors	1 Thick myofilaments		Matsuno and Kuga 1989
Haementeria ghilianii	Anterior salivary gland	1 Surface infoldings 2 Long cytoplasmic processes		Walz et al. 1988
Eisenia foetida	Cocoon-producing	1 Microvilli with electron-dense rods 2	Cuticular pore	Morris 1983

Free-Living Eukaryotic Cells

Higher Taxon	Species	Parts	Possible Parts	Source
Chlorarachnida	*Chlorarachnion reptans*	1 Chloroplasts 2 Pseudopodia 3	Extrusomes	Hibberd and Norris 1984
Chrysophyceae	*Chromulina placentula*	1 Chromatophore 2 Flagellum 3	Short flagellum	Belcher and Swale 1967
		4 Stigma 5 Contractile vacuoles 6 Elliptical scale layer (on membrane)		
Chrysophyceae	*Ochromonas tuberculatus*	1 Plastid 2 Contractile vacuoles 3 Eyespot 4 Long flagellum 5 6 Discobolocysts 7 Periplast	Short flagellum	Hibberd 1970
Chrysophyceae	*Chrysococcus rufescens*	1 Lorica 2 Long flagellum 3 4 Chromatophore 5 Stigma 6 Contractile vacuole 7 Cytoplasmic projections 8 "Peripheral system"	Short flagellum	Belcher 1969
Pedinellophyceae	*Pedinella hexacostata*	1 Indentation at anterior pole 2 3 Chromatophores	Indentation at posterior pole	Swale 1969

Continued on next page

Table 2A.1. (continued)

Higher Taxon	Species	Free-Living Eukaryotic Cells		
		Parts	Possible Parts	Source
Ciliophora	Perispira ovum	4 Contractile vacuoles 5 Flagellum 6 7 Tentacles 8 Peduncle	Second flagellum	Johnson et al. 1995
		1 Micronucleus 2 Contractile vacuole 3 Rhomboid mucocysts 4 5 Pellicle 6 Spiral ridge 7 Cytopharynx 8 Cilia 9 Toxicysts	Spherical mucocysts	
Cryptophyta	Rhodomonas lacustris	1 Ventral gullet 2 Flagellum 3 4 Chloroplast 5 Contractile vacuole complex 6 Periplast 7 Ejectosomes	Shorter flagellum	Klaveness 1981
Dinoflagellata	Aureodinium pigmentosum	1 Longitudinal flagellum 2 3 Chloroplast 4 Theca	Transverse flagellum	Dodge 1967

Group	Species	Characters	Reference
Dinoflagellata	*Woloszynskia micra*	1 Chloroplast 2 Pusule 3 Trichocysts 4 Transverse flagellum 5 Longitudinal flagellum 6 Theca 7 Girdle 8 Sulcus	Leadbeater and Dodge 1966
Diplomonadida	*Trepomonas agilis*	1 Locomotor flagellum 2 Oral flagellae 3 Oral grooves 4 Contractile vacuole 5 Symmetric membrane −1 Absent: Golgi −1 Absent: microbodies −1 Absent: mitochondria	Eyden and Vickerman 1975
Euglenida	*Khawkinea quartana*	1 Reservoir 2 Long flagellum 3 Short flagellum 4 Eyespot 5 Contractile vacuole 6 Pellicle	Schuster and Hershenov 1974
Euglenida	*Paranema trichophorum*	1 Leading flagellum 2 Trailing flagellum 3 Pellicle 4 Pellicle groove 5 Contractile vacuole 6 Reservoir	Roth 1959
Eustigmatophyta	*Chlorobotrys regularis*	1 Mucilage layers 2 Chloroplast	Hibbard 1974

Continued on next page

Table 2A.1. (continued)

Higher Taxon	Species	Free-Living Eukaryotic Cells		Source
		Parts	Possible Parts	
Eustigmatophyta	*Pseudocharaciopsis texensis*	1 Chloroplast (incl. pyrenoid) 2 Basal disk (with stalk) 3 Eyespot 4 Long flagellum 5	 Short flagellum	Lee and Bold 1973
Foraminifera	*Globigerinoides ruber*	1 Test 2 Primary aperture 3 4 Vesicular system 5 Cryptosome 6 Somatic nuclei 7 "Strange multinuclear inclusion," type 1 8	 Secondary aperture "Strange multinuclear inclusion," type 2	Lee et al. 1965
Glaucocystophyta	*Glaucocystis nostochinearum*	1 Equatorial lamellar ring 2 Flagellar bases 3	 Indentate cell membrane	Echlin 1967

Group	Organism	Features	Reference
Prymnesiophyta	*Diacronema vlkianum*	1 Long flagellum; 2 Short flagellum; 3 Plastid; 4 Muciferous body	Fournier 1969
Prymnesiophyta	*Chrysochromulina mega-cylindra*	1 Plastids; 2 Flagellae; 3 Haptonema; 4 Surface scales; 5 "Nuclear foreign body"	Manton 1972
Kinetoplastida	*Bodo curviflus*	1 Anterior flagellum; 2 Posterior flagellum; 3 Contractile vacuole; 4 Cytopharynx	Burzell 1975
Raphidophyta	*Olisthodiscus luteus*	1 Anterior flagellum; 2 Posterior flagellum; 3 Chloroplasts; 4 Swollen perinuclear space	Leadbeater 1969
Rhizopoda	*Trichosphaerium micrum*	1 Pseudopodium; 2 Lobopodium; 3 Test	Angell 1975
Rhodophyta	*Rhodella maculata*	1 Chloroplast; 2 Mucilage layer (outside cell membrane); 3 Nuclear membrane projection into pyrenoid	Evans 1970
Rhodophyta	*Dixoniella grisea*	1 Chloroplasts	Scott et al. 1992

References

Allen, T. F. H., and T. B. Starr. 1982. *Hierarchy: Perspectives for ecological complexity.* Chicago: University of Chicago Press.

Angell, R. W. 1975. Structure of *Trichosphaerium micrum* sp. n. *Journal of Protozoology* 22:18–22.

Baerwald, R. J., and G. M. Boush. 1970. Fine structures of the hemocytes of *Periplaneta americana* (Orthoptera: Blattidae) with particular reference to marginal bundles. *Journal of Ultrastructure Research* 31:151–61.

Beklemishev, W. N. 1969. *Principles of comparative anatomy of invertebrates.* Volume 1, *Promorphology,* ed. Z. Kabata. Translated by J. M. MacLennan. Chicago: University of Chicago Press.

Belcher, J. H. 1969. A morphological study of the phytoflagellate *Chrysococcus rufescens* Klebs in culture. *British Phycological Journal* 4:105–17.

Belcher, J. H., and E. M. F. Swale. 1967. *Chromulina placentula* sp. nov. (Chrysophyceae), a freshwater nannoplankton flagellate. *British Phycological Bulletin* 3:257–67.

Bell, G., and A. O. Mooers. 1997. Size and complexity among multicellular organisms. *Biological Journal of the Linnean Society* 60:345–63.

Blackburn, D. G., K. G. Osteen, V. P. Winfrey, and L. H. Hoffman. 1989. Obplacental giant cells of the domestic rabbit: Development, morphology, and intermediate filament composition. *Journal of Morphology* 202:185–203.

Boardman, R. S., and A. H. Cheetham. 1973. Degrees of colony dominance in stenolaemate and gymnolaemate Bryozoa. In *Animal colonies: Development and function through time,* ed. R. S. Boardman, A. H. Cheetham, and W. A. Oliver, Jr., 121–220. Stroudsburg, PA: Dowden, Hutchinson, and Ross.

Bonner, J. T. 1988. *The evolution of complexity.* Princeton: Princeton University Press.

Brandon, R. N. 1996. *Concepts and methods in evolutionary biology.* Cambridge: Cambridge University Press.

Brooks, D. R., and D. A. McLennan. 1993. Macroevolutionary patterns of morphological diversification among parasitic flatworms (Platyhelminthes: Cercomeria). *Evolution* 47:495–509.

Burzell, L. A. 1975. Fine structure of *Bodo curvifilus* Griessmann (Kinetoplastida: Bodonidea). *Journal of Protozoology* 22:35–39.

Buss, L. W. 1987. *The evolution of individuality.* Princeton: Princeton University Press.

Byrne, M. 1986. The ultrastructure of the morula cells of *Eupentacta quinquesemita* (Echinodermata: Holothuroidea) and their role in the maintenance of the extracellular matrix. *Journal of Morphology* 188:179–89.

Campbell, D. T. 1958. Common fate, similarity, and other indices of the status of aggregates of persons as social entities. *Behavioral Science* 3:14–25.

Cheverud, J. M. 1995. Morphological integration in the saddle-back tamarin *(Saguinus fuscicollis)* cranium. *American Naturalist* 145:63–89.

———. 1996. Developmental integration and the evolution of pleiotropy. *American Zoologist* 36:44–50.

Chiba, A., S. Oka, Y. Honma, and M. Ishiyama. 1991. Ultrastructure of the agranular cells in the adenohypophysis of the South American lungfish, *Lepidosiren paradoxa. Journal of Morphology* 207:73–79.

Coates, A. G., and J. B. C. Jackson. 1985. Morphological themes in the evolution of clonal and aclonal marine invertebrates. In *Population biology and evolution of clonal organisms,* ed. J. B. C. Jackson, L. W. Buss, and R. E. Cook, 67–106. New Haven: Yale University Press.

Coates, A. G., and W. A. Oliver, Jr. 1973. Coloniality in zoantharian corals. In *Animal colonies: Development and function through time,* ed. R. S. Boardman, A. H. Cheetham, and W. A. Oliver, Jr., 3–27. Stroudsburg, PA: Dowden, Hutchinson, and Ross.

Corliss, J. O. 1994. An interim utilitarian ("user-friendly") hierarchical classification and characterization of the protists. *Acta Protozoologica* 33:1–51.

Dodge, J. D. 1967. Fine structure of the dinoflagellate *Aureodinium pigmentosum* gen. et sp. nov. *British Phycological Bulletin* 3:327–36.

Echlin, P. 1967. The biology of *Glaucocystis nostochinearum. British Phycological Bulletin* 3:225–39.

Eldredge, N., and S. N. Salthe. 1984. Hierarchy and evolution. *Oxford Surveys in Evolutionary Biology* 1:184–208.

Evans, L. V. 1970. Electron microscopical observations on a new red algal unicell, *Rhodella maculata* gen. nov., sp. nov. *British Phycological Journal* 5:1–13.

Eyden, B. P., and K. Vickerman. 1975. Ultrastructure and vacuolar movements in the free-living diplomonad *Trepomonas agilis* Klebs. *Journal of Protozoology* 22;54–66.

Fournier, R. O. 1969. Observations on the flagellate *Diacronema vlkianum* Prauser (Haptophyceae). *British Phycological Journal* 4:185–90.

Gaino, E., B. Burlando, and P. Buffa. 1986. The vacuolar cells of *Oscarella lobularis* (Porifera: Demospongiae): Ultrastructural organization, origin, and function. *Journal of Morphology* 188:29–37.

Gerhart, J., and M. Kirschner. 1997. *Cell, embryos, and evolution.* Malden, MA: Blackwell.

Harvell, C. D. 1994. The evolution of polymorphism in colonial invertebrates and social insects. *Quarterly Review of Biology* 69:155–85.

Helander, H., and R. Ekholm. 1959. Ultrastructure of epithelial cells in the fundus glands of the mouse gastric mucosa. *Journal of Ultrastructural Research* 3:74–83.

Hibberd, D. J. 1970. Observations on the cytology and ultrastructure of *Ochromonas tuberculatus* sp. nov. (Chrysophyceae), with special reference to the discobolocysts. *British Phycological Journal* 5:119–43.

———. 1974. Observations on the cytology and ultrastructure of *Chlorobotrys regularis* (West) Bohlin with special reference to its position in the Eustigmatophyceae. *British Phycological Journal* 9:37–46.

Hibberd, D. J., and R. E. Norris. 1984. Cytology and ultrastructure of *Chlorarachnion reptans* (Chlorarachniophyta divisio nova, Chlorarachniophyceae classis nova). *Journal of Phycology* 20:310–30.

Johnson, P. W., P. L. Donaghay, E. B. Small, and J. M. Sieburth. 1995. Ultrastructure and ecology of *Perispira ovum* (Ciliophora: Litostomatea): An aerobic, planktonic ciliate that sequesters the chloroplasts, mitochondria, and paramylon of *Euglena proxima* in the micro-oxic habitat. *Eukaryotic Microbiology* 42:323–35.

Karsai, I., and J. W. Wenzel. 1998. Productivity, individual-level and colony-level flexibility, and organization of work as consequences of colony size. *Proceedings of the National Academy of Sciences, USA* 95:8665–69.

Kauffman, S. A. 1993. *The origins of order*. New York: Oxford University Press.

Klaveness, D. 1981. *Rhodomonas lacustris* (Pascher & Ruttner) Javornicky (Cryptomonadida): Ultrastructure of the vegetative cell. *Journal of Protozoology* 28:83–90.

Kou, R., M. P. Tu, C. Y. Chang, and C.-M. Yin. 1995. Isolated cell type corpora allata in adults of the loreyi leafworm, *Leucania loreyi* Duponchel (Lepidoptera: Noctuidea). *Journal of Morphology* 225:369–76.

Kowalczyk, A. P., E. A. Bornslaeger, S. M. Norvell, H. L. Palka, and K. J. Green. 1999. Desmosomes: Intercellular adhesive junctions specialized for attachment of intermediate filaments. *International Review of Cytology* 185:237–301.

Leadbeater, B., and J. D. Dodge. 1966. The fine structure of *Woloszynskia micra* sp. nov., a new marine dinoflagellate. *British Phycological Bulletin* 3:1–17.

Leadbeater, B. S. C. 1969. A fine structural study of *Olisthodiscus luteus* Carter. *British Phycological Journal* 4:3–17.

Lee, J. J., H. D. Freudenthal, V. Kossoy, and A. Bé. 1965. Cytological observations on two planktonic Foraminifera, *Globigerina bulloides* d'Orbigny, and *Globigerinoides ruber* (d'Orbigny, 1839) Cushman, 1927. *Journal of Protozoology* 12:531–42.

Lee, K. W., and H. C. Bold. 1973. *Pseudocharaciopsis texensis* gen. et sp. nov., a new member of the Eustigmatophyceae. *British Phycological Journal* 8:31–37.

Leigh, E. G., Jr. 1983. When does the good of the group override the advantage of the individual? *Proceedings of the National Academy of Sciences, USA* 80:2985–89.

Lidgard, S. 1986. Ontogeny in animal colonies: A persistent trend in the bryozoan fossil record. *Science* 232:230–32.

Lidgard, S., and J. B. C. Jackson. 1989. Growth in encrusting cheilostome bryozoans: 1, Evolutionary trends. *Paleobiology* 15:255–82.

Manton, I. 1972. Observations on the biology and micro-anatomy of *Chrysochromulina megacylindra* Leadbeater. *British Phycological Journal* 7:235–48.

Margulis, L., J. O. Corliss, M. Melkonian, and D. J. Chapman. 1990. *Handbook of Protoctista*. Boston: Jones and Bartlett.

Masuda, M., K. Yamazaki, J. Kanzaki, and Y. Hosoda. 1994. Ultrastructure of melanocytes in the dark cell area of human vestibular organs: Functional implications of gap junctions, isolated cilia, and annulate lamellae. *Anatomical Record* 240:481–91.

Matsuno, A., and H. Kuga. 1989. Ultrastructure of muscle cells in the adductor of the boring clam *Tridacna crocea*. *Journal of Morphology* 200:247–53.

Maynard Smith, J. 1988. Evolutionary progress and levels of selection. In *Evolutionary progress*, ed. M. Nitecki, 219–30. Chicago: University of Chicago Press.

Maynard Smith, J., and E. Szathmáry. 1995. *The major transitions in evolution.* Oxford: Freeman.

McKinney, F. K., and J. B. C. Jackson. 1989. *Bryozoan evolution.* Chicago: University of Chicago Press.

McShea, D. W. 1996a. Complexity and homoplasy. In *Homoplasy,* ed. M. J. Sanderson and L. Hufford, 207–25. San Diego: Academic Press.

———. 1996b. Metazoan complexity and evolution: Is there a trend? *Evolution* 50:477–92.

———. 2000. Functional complexity in organisms: Parts as proxies. *Biology and Philosophy* 15:641–68.

McShea, D. W., and E. P. Venit. 2001. What is a part? In *The character concept in evolutionary biology,* ed. G. P. Wagner, 259–84. San Diego: Academic Press.

Michod, R. E. 1999. *Darwinian dynamics: Evolutionary transitions in fitness and individuality.* Princeton: Princeton University Press.

Mittenthal, J. E., A. B. Baskin, and R. E. Reinke. 1992. Patterns of structure and their evolution in the organization of organisms: Modules, matching, and compaction. In *Principles of organization in organisms,* ed. J. Mittenthal and A. Baskin, Santa Fe Institute Studies in the Sciences of Complexity, Proceedings, 13:321–32. Reading, MA: Addison-Wesley.

Montalvo, S., J. Junoy, C. Roldán, and P. Garcia-Corrales. 1996. Ultrastructural study of sensory cells of the proboscidial glandular epithelium of *Riseriellus occultus* (Nemertea: Heteronemertea). *Journal of Morphology* 229: 83–96.

Morris, G. M. 1983. The cocoon-producing cells of *Eisenia foetida* (Annelida: Oligochaeta): A histochemical and ultrastructural study. *Journal of Morphology* 177:41–50.

Olson, E. C., and R. L. Miller. 1958. *Morphological integration.* Chicago: University of Chicago Press.

O'Neill, R. V., D. L. DeAngelis, J. B. Waide, and T. F. H. Allen. 1986. *A hierarchical concept of ecosystems.* Princeton: Princeton University Press.

Pattee, H. H. 1970. The problem of biological hierarchy. In *Towards a theoretical biology,* ed. C. H. Waddington, 3:117–36. Edinburgh: Edinburgh University Press.

Peacock, A. J. 1979. Ultrastructure of the type "B" cells in the rectal pad epithelium of *Locusta migratoria. Journal of Morphology* 159:221–32.

Raff, R. A. 1996. *The shape of life.* Chicago: University of Chicago Press.

Raymond, C. M. L., R. W. Lea, P. J. Sharp, and M. H. Maxwell. 1993. Ultrastructure of the lipid-containing cells of the anterior pituitary gland of the domestic chicken, *Gallus domesticus. Anatomical Record* 237:506–11.

Roth, L. E. 1959. An electron-microscope study of the cytology of the protozoan *Peranema trichophorum. Journal of Protozoology* 6:107–16.

Salthe, S. N. 1985. *Evolving hierarchical systems.* New York: Columbia University Press.

———. 1993. *Development and evolution.* Cambridge: MIT Press.

Schopf, T. J. M. 1973. Ergonomics of polymorphism: Its relation to the colony as the unit of natural selection in species of the phylum Ectoprocta. In *Animal colonies: Development and function through time,* ed. R. S. Boardman, A. H.

Cheetham, and W. A. Oliver, Jr., 247–94. Stroudsburg, PA: Dowden, Hutchinson, and Ross.

Schuster, F. L., and B. Hershenov. 1974. *Khawkinea quartana,* a colorless euglenoid flagellate: 1, Ultrastructure. *Journal of Protozoology* 21:33–39.

Scott, J. L., S. T. Broadwater, B. D. Sanders, and J. P. Thomas. 1992. Ultrastructure of vegetative organization and cell division in the unicellular red alga *Dixoniella grisea* gen. nov. (Rhodophyta) and a consideration of the genus Rhodella. *Journal of Phycology* 28:649–60.

Serrano, M. T., E. Angulo, and J. Moya. 1990–91. Histochemistry and ultrastructure of the nephridial granular cells of *Phascolosoma granulatum* (Leuckart, 1928) (Sipuncula). *Biological Structures and Morphogenesis* 3:107–14.

Simon, H. A. 1962. The architecture of complexity. *Proceedings of the American Philosophical Society* 106:467–82.

Simpson, T. L. 1984. *The cell biology of sponges.* New York: Springer.

Smith, D. S. 1968. *Insect cells: Their structure and function.* Edinburgh: Oliver and Boyd.

Steele, V. J., and P. E. Oshel. 1989. Ultrastructure of the attachment cells of the organ of Bellonci in *Gammarus setosus* (Crustacea: Amphipoda). *Journal of Morphology* 200:93–119.

Swale, E. M. F. 1969. A study of the nannoplankton flagellate *Pedinella hexacostata* Vysotskii by light and electron microscope. *British Phycological Journal* 4:65–86.

Valentine, J. W., A. G. Collins, and C. P. Meyer. 1993. Morphological complexity increase in metazoans. *Paleobiology* 20:131–42.

Wagner, G. P. 1996. Homologues, natural kinds, and the evolution of modularity. *American Zoologist* 36:36–43.

Wagner, G. P., and L. Altenberg. 1996. Complex adaptations and the evolution of evolvability. *Evolution* 50:967–76.

Walz, B., K.-H. Schäffner, and R. T. Sawyer. 1988. Ultrastructure of the anterior salivary gland cells of the giant leech, *Haementeria ghilianii* (Annelida: Hirudinea). *Journal of Morphology* 196:321–32.

Wilson, D. S., and E. Sober. 1989. Reviving the superorganism. *Journal of Theoretical Biology* 136:337–56.

Wilson, E. O. 1968. The ergonomics of caste in social insects. *American Naturalist* 102:41–66.

Wimsatt, W. C. 1974. Complexity and organization. In *Philosophy of Science Association 1972,* ed. K. F. Schaffner and R. S. Cohen, 67–86. Dordrecht, Holland: D. Reidel.

Wood, R., A. Y. Zhuravlev, and F. Debrenne. 1992. Functional biology and ecology of Archaeocyatha. *Palaios* 7:131–56.

Zelditch, M. L. 1996. Historical patterns of developmental integration. *American Zoologist* 36:1–3.

Refuges Revisited **3**

Enemies versus Flow and Feeding as Determinants of Sessile Animal Distribution and Form

BETH OKAMURA, JEAN-GEORGES HARMELIN,
AND JEREMY B. C. JACKSON

Assemblages of encrusting marine organisms provide clear and important examples of the significance of biological interactions in past and present-day communities (e.g., Jackson 1983; McKinney and Jackson 1989; Jackson and McKinney 1990). The majority of organisms resident in these assemblages undergo asexual growth that results in high levels of intra- and interspecific competition for space among colonies of different sizes and shapes. The notable influence of colony morphology on the outcome of spatial competition and other biological interactions suggests that particular adaptive morphological strategies are associated with different growth forms of colonial invertebrates such as sponges, bryozoans, cnidarians, and ascidians (Jackson 1977, 1979; Buss 1979). These adaptive strategies describe ways in which colonies of different morphologies are adapted to persist in a spatially limited environment by either confronting or avoiding interactions with biotic enemies.

However, food is another potentially limiting resource requirement of encrusting organisms. All of these animals face the task of extracting tiny food particles present in dilute concentrations in the water column and depend on the movement of water to deliver food resources to their feeding surfaces. Empirical studies have demonstrated that colony form and feeding are interrelated, with intracolony patterns of feeding currents being highly influenced by colony morphology (e.g., Cook 1977; Winston 1978, 1979; Lidgard 1981; Okamura 1984, 1985; McKinney 1990) and ambient water-flow regime (e.g., Okamura 1984, 1985, 1988, 1992). The influence of morphology on feeding has recently been formalized by considering how colony integration in encrusting bryozoans influences feeding from boundary-layer flow regimes as predicted by a two-dimensional, advection-diffusion model (Okamura and Eckman 1997; Eckman and Okamura 1998). This modeling has confirmed the results of empirical studies, demonstrating that colony morphology has profound consequences for feeding.

Thus there are now two models that contrast in their interpretation of the significance of colony shape by focusing on the role of different resources. One model focuses on spatial limitation due to biotic interactions and the other on food acquisition. In this paper we aim to compare the applicability of these two views by considering the alternative predictions made by the models with regard to the distribution and abundance of encrusting bryozoans along environmental gradients of food availability, especially the steep gradients associated with caves and other cryptic habitats. We first discuss how the two models vary in their interpretation of the significance of colony morphology. We then contrast predictions of the models for bryozoan assemblages along an environmental gradient of food availability and assess empirical support for these predictions by considering Mediterranean cave assemblages which offer gradients in food concentration and water flow. We further extend the approach to consider patterns in other habitats characterized by similar environmental gradients. Overall, we show that patterns of distribution and abundance conform closely to predictions made by the food acquisition model, suggesting that food acquisition plays an important role in determining colony morphology, distribution, and abundance. In addition, we discuss how implications of this study lead us to reject the notion that primitive, anachronistic species have been relegated to marginal refugia such as submarine caves. Rather, we propose that life in such extreme environments requires sophisticated and specific adaptation.

The Spatial Limitation Model

The spatial limitation model has served as the major paradigm for interpreting the significance of colony morphology since its initial articulation (Buss 1979; Jackson 1977, 1979; McKinney and Jackson 1989). The model proposes that highly integrated, sheetlike forms employ a confrontational tactic and thus are committed to colony defense and maintenance (see figs. 3.1A and 3.1B). Among bryozoans, such colonies typically possess polymorphic nonfeeding zooids that function in colony defense and brooding of larvae. Colonies are often highly calcified and show indeterminate growth. They generally possess complex zooidal or multizooidal budding patterns at colony margins, both of which entail significant outward growth extensions and nutritional support from parental zooids before interior walls and fully functional zooids are formed (Lidgard 1985). These various traits all represent energetically costly activities that can be related to colony defense and maintenance in highly integrated sheetlike colonies.

In contrast, poorly integrated, runnerlike forms escape expending energy on colony defense because they avoid enemies through the rapid location and exploitation of unpredictable spatial refuges (see fig. 3.1C). Consequently runners show little development of the energetically costly structures and activities typical of sheets. Treelike colonies also avoid biotic enemies by exploiting the vertical dimension (see fig. 3.1D), although they may also achieve competitive dominance by growing in dense

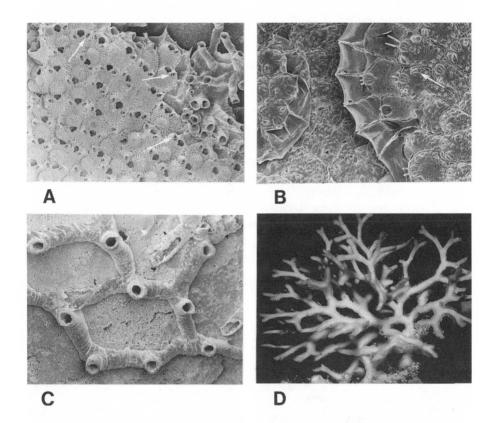

Fig. 3.1. Different morphological types of bryozoan colonies identified by the adaptive strategy model. *A*, A sheetlike colony of *Schizoporella errata* (Waters 1878) beginning to overgrow the cyclostome bryozoan *Tubulipora* sp. Note that multiserial growth results in rows of contiguous zooids that produce a sheetlike growth form. *Arrows* point to defensive, nonfeeding avicularia (polymorphic zooids), which are regularly deployed by autozooids (normal feeding zooids) in this colony. Ovicells (brood chambers for larvae) can be seen as swollen regions distal to orifices of maternal zooids. Magnification = 25×. *B*, Close-up of budding zone of sheetlike *Schizomavella linearis* (Hassall 1841). Note budding zone entails significant outward extensions (seen here in different stages of development) requiring nutritional support from functional zooids near edges of colony. *Arrow* points to avicularium. Magnification = 41×. *C*, Runnerlike colony of *Stomatopora* sp. Note lack of zooid polymorphism and the isolation of zooids resulting from uniserial growth. Magnification = 35×. *D*, Treelike colony of *Smittina cervicornis* (Pallas 1766). Note that upright growth often requires extensive calcification as shown here to support colony above the substratum. Magnification = 0.8×.

stands whereby they resist overgrowth on their flanks or smother or starve animals growing beneath their branches (e.g., Buss 1981). This often requires expending energy on colony development and support, since colonies spread from a relatively small area of attachment and vertical growth may expose colonies to breakage from drag forces (Cheetham and Thomsen 1981; Cheetham 1986). Thus many treelike forms are highly calcified and possess numerous nonfeeding polymorphs involved with structural support and attachment. However, another less costly strategy employed by some trees is to avoid drag by accommodating bending (Harvell and LaBarbera 1985). Such species are at most only lightly calcified.

Some time later, Bishop (1989) identified a fourth morphological category of small "spot" colonies (see also Winston 1985; McKinney and Jackson 1989). Bishop argued that spots are adapted primarily to exploit particular physical configurations of substrata such as those offered by disarticulated bivalve shells, flexible algal substrata, and sand- or gravel-sized particulate seabeds. Diagnostic features of spots that allow them to exploit such substrata are determinate growth to a small size, sexual maturity at small size, larval settlement behaviors that achieve colonization of appropriate substrata, and growth generally by intrazooidal budding (see fig. 3.2). However, recent investigation of spots growing under

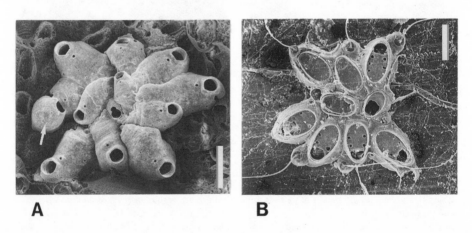

A   **B**

Fig. 3.2. Determinately growing spot colonies. *A*, Colony of *Haplopoma sciaphilum* (Silén and Harmelin 1976) collected 40 m from entrance of 3PP Cave. Note tiny size of colony and the presence of the hoodlike ovicell distal to one of the zooids noted by *arrow*. *Scale bar* = 0.25 mm. *B*, Colony of *Setosella cavernicola* (Harmelin 1977) collected 40 m from entrance of Trémies Cave. The long, mobile whiplike structures are modified opercula (lids that cover orifices of autozooids) on nonfeeding polymorphic zooids known as vibracula. Vibracula play a locomotory role in free-living colonies in high-energy sedimentary habitats. Their function in sessile colonies is speculated to be one of cleaning colony surfaces. *Scale bar* = 0.25 mm.

unusual environmental conditions reveals that some spots will grow to larger sizes than the fixed sizes achieved in other habitats (McKinney in press). In addition, in spotlike cyclostomes, the onset of reproduction is variable and can be delayed until some environmental signal is detected, such as approach to or contact with another encrusting organism (McKinney and Taylor 1997). The interpretation of spots according to the spatial limitation model suggests an alternate view to that proposed by Bishop. Thus spots employ a similar strategy to that of runners in exploiting substrata that represent refuges from biotic enemies, although in the case of spots, these refuges are predictable. Spots require these refuges since they are overgrown by superior spatial competitors in other habitats (Kay and Keough 1981; Jackson and Winston 1982; Keough 1984). However, neither Bishop's explanation (as he acknowledges) nor the spatial limitation view easily accommodates the presence of typical spots in cryptic habitats such as caves, crevices, and the undersurfaces of stones. Evidently, spots are not just confined to the exploitation of small or flexible substrata. Indeed, Harmelin (1986) has suggested that spots in caves are favored by their low energy requirements in such oligotrophic conditions.

The Food Acquisition Model

The food acquisition model proposes that feeding opportunities have had an important influence on the ecology and evolution of colony form. It is based on the development and predictions of a two-dimensional numerical model of suspension feeding by encrusting bryozoans that was designed to predict spatial and temporal variability of suspended food in a fluid in motion (Okamura and Eckman 1997; Eckman and Okamura 1998). The principal independent parameters in the model are the boundary-layer flow regime in which feeding occurs, feeding currents produced by individual tentacular crowns (lophophores) in colonies, and the spacing of modules (zooids) within colonies. Because it is conceptually more difficult, and has only recently been formulated, the food acquisition model and its interpretations require somewhat greater development and explanation than the spatial limitation model.

Although the ocean is a dynamic environment, encrusting organisms are largely buffered from the direct effects of tides, ambient currents, and waves because they are attached to the substratum. This means they live within benthic boundary-layer flow regimes that develop near surfaces due to the frictional effect of the surface on flow (Nowell and Jumars 1984; Denny 1988). In boundary layers, the speed of flow increases exponentially until mainstream flow is achieved at some height above the surface. In turbulent boundary-layer flow regimes, turbulent mixing increases with height above the substratum as does velocity. In laminar boundary-layer flow regimes, turbulent mixing is negligible, and flow follows smooth streamlines. Turbulent boundary-layer flow regimes characterize the majority of inshore and continental shelf microhabitats (Nowell and Jumars 1984; Denny 1988), including many inhabited by bryozoans (Eckman and Okamura 1998); however, laminar boundary-layer flow conditions may prevail in very calm habitats such as may occur deep within caves or in crevices, subtidally below rocks, or perhaps during transient tidal conditions.

The two-dimensional, advection-diffusion model was developed to describe the distribution of particles within boundary-layer flows over encrusting bryozoan colonies and to predict patterns of particle capture at both the zooid and colony level for colonies with different degrees of integration as measured by zooid spacing. Zooid spacing within colonies is predicted to have an important influence on feeding due to different patterns of processing water that result when lophophores are tightly packed (when there are no gaps between the everted lophophores) or more widely spaced (when gaps varying from one to five times the lophophore diameter are present between everted lophophores). The presence of these gaps in colonies with widely spaced lophophores means that feeding currents created by cilia on the tentacles of the lophophore will be isolated and will result in a feeding zone that develops above each lophophore as water is drawn through the lophophore and exits at the base between the tentacles (see fig. 3.3A). The volume of this feeding zone expands in cross-sectional area with increasing height above the lophophore until the strength of external flow overwhelms the strength of feeding currents. Thus a conical feeding zone results, in which local feeding current speed decreases with height above the lophophore while horizontal flow increases above the lophophore. Excurrent water that exits at the base of widely spaced lophophores does not escape from colony surfaces and therefore, as discussed later, presents the potential for refiltration.

In contrast, when zooids are closely spaced, everted lophophores will be tightly packed and no gaps will be present between them. As a result, ciliary feeding currents will be directed vertically at all points due to tight packing, and the volume of water drawn toward the lophophore will not expand in cross-sectional area with height (see fig. 3.3B). Hence the speed of the vertical feeding current is constant with height, and a cylindrical feeding zone develops, which will extend to greater heights above lophophores until it is overwhelmed by external flow. Note that peripheral lophophores on the edges of colonies will produce partially conical feeding zones whose effects are not included in the model. In colonies

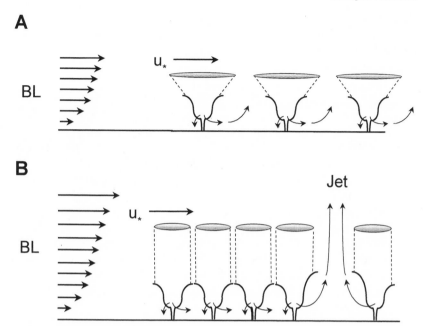

Fig. 3.3. Processing of water by widely spaced *(A)* and tightly packed *(B)* lophophores. Feeding occurs within the benthic boundary layer *(BL)* in which flow speeds increase with height above the surface as indicated by *arrows*. In turbulent boundary layers turbulent mixing also increases with height. Ciliary feeding currents produced by widely spaced lophophores draw water in from a volume that increases with height from the lophophore, while those produced by tightly packed lophophores are constrained. Thus colonies with tightly packed lophophores develop cylindrical feeding zones which divert water into the lophophore from greater heights. Excurrent flow is directed between tentacles at the base of the lophophore (shown by *smaller arrows*). Note that excurrent flow will be available for refiltration in colonies with widely spaced lophophores, but in colonies with tightly packed lophophores, flow travels below lophophore arrays until it is ejected as jets in regions where lophophores are not deployed. In both cases, feeding currents become ineffective when the strength of horizontal flow past the lophophores exceeds the strength of feeding currents (see text for further discussion). Note that in reality external flow will deform these idealized shapes of feeding zones (see text for further discussion).

with tightly packed lophophores, excurrent flow is observed to be directed below the array of everted lophophores and subsequently directed as a jet away from colony surfaces at "chimneys" (Banta et al. 1974; Cook 1977; Winston 1978, 1979) (see fig. 3.3B). Chimneys represent small areas of colonies where lophophores are not everted and develop when the pressure beneath the array of lophophores exceeds a certain limit (Dick 1987). Excurrent flow will also find exit at the periphery of colonies. Refiltration of excurrent flow within colonies with tightly packed lophophores will depend on the extent to which excurrent flow leaks back into the system and is not vented through chimneys. The effects of such refiltration were tested by setting parameters to model 100% refiltration and 100% venting through chimneys (Eckman and Okamura 1998).

In reality, conical and cylindral feeding zones will be deformed due to external flow. This deformation will result in a shift in the horizontal location in the water column from which water is pulled through the lophophore and remixed, relative to the position of the lophophore. However, such a deformation of these idealized feeding zones should not affect the model predictions of capture rates (Eckman and Okamura 1998). Naturally, a continuum of colonies exists, with colonies having different degrees of spacing between lophophores. The continuous effect of lophophore spacing was not assessed, as the model used discrete values of lophophore diameters to predict feeding patterns.

Another assumption of the advection-diffusion model is that lophophores operate at a similar height above the substratum. If they do not, then feeding zones will not be the simple conical and cylindrical shapes described above. Thus surface rugosity may complicate feeding patterns. While this no doubt occurs, the notable posturing employed by tightly packed lophophores indicates that many colonies may deploy lophophores at similar levels to effect cylindrical feeding zones and coordinated feeding even when growing on fairly uneven surfaces (Winston 1978, 1979). Another means by which colonies may deploy lophophores at similar levels is through altering the sizes and shapes of individual lophophores. Such phenotypic plasticity is notable when new chimneys develop within colonies (e.g., through damage or overgrowth by other organisms), and lophophores surrounding these new chimneys become larger and more elliptical in shape (Dick 1987). In addition, surface rugosities probably act as foci for the development of excurrent chimneys, and hence do not disrupt the integrated feeding currents produced by neighboring lophophores that can be deployed at similar levels.

In summary, the advection-diffusion model predicts the patterns of feeding by colonies with different degrees of zooid spacing that result in processing water as described above. In addition to considering the effects of zooid spacing, the model makes a series of predictions as to the effects of the strengths of both feeding currents and of external flow in the boundary layer as measured by the shear velocity u_* (fig. 3.4; Okamura and Eckman 1997; Eckman and Okamura 1998). Here we summarize key results obtained from model simulations with regard to particle capture at the colony and zooid levels and the importance of turbulence in suspension feeding.

Food availability is a function of

1) rate of food transport (u_*)

2) concentration of food (c) $\Big\}$ **= flux of food**

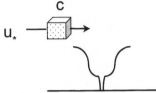

Fig. 3.4. Food availability to lophophores is determined by the rate of transport of food to lophophores (which is a function of shear velocity in the boundary layer, u_*) and the concentration of food in the water column as shown schematically. Note that this is equivalent to the flux of food to lophophores.

Colony-Level Feeding

The model predicts that colonies with tightly packed lophophores that are able to vent excurrents entirely through chimneys will have much higher feeding rates than colonies with lophophores more widely spaced. This effect is diminished as the degree to which excurrents are vented away from colony surfaces decreases because of local leakage of excurrents and hence refiltration. In nearly all conditions, however, leaky but tightly packed colonies still show higher feeding rates than colonies with lophophores more widely spaced. This is because the cylindrical feeding zones of tightly packed colonies reach higher into the boundary layer where turbulent remixing replenishes food at greater rates than is the case for colonies with more widely spaced lophophores (fig. 3.3). The enhanced feeding of colonies with tightly packed lophophores increases with advection ratio (i.e., the ratio of feeding current speed to the strength of external flow past the lophophore as measured by the shear velocity, u_*). Thus, all else being equal, colonies with tightly packed lophophores that produce strong feeding currents or live in weak flow environments will tend to experience maximal feeding rates. If ambient flow is extremely low, however, colonies will continuously refilter the same water regardless of the arrangement of their lophophores.

Indirect empirical evidence to confirm model predictions comes from comparisons of feeding rates of two encrusting species (Okamura and Eckman 1997). When feeding from relatively slow flow (and hence high advection ratio), *Conopeum reticulum* experienced significantly greater feeding rates than *Electra pilosa*. The former produces sheetlike colonies with tightly packed lophophores, while in the latter, lophophores are more widely spaced. As the advection ratio decreased, however, *C. reticulum* showed greater declines in feeding rates than *E. pilosa*. Both the greater feeding rates at high advection ratios and the greater constraint in feeding with decreased advection ratios are in keeping with the feeding responses of colonies according to model predictions.

Zooid-Level Feeding

Model simulations reveal diminished feeding by downstream zooids due to refiltration of excurrent-processed water. Thus refiltration of excurrents results in feeding interference within colonies. However, this interference decreases with increased zooid spacing. Feeding interference is therefore highest in tightly packed, sheetlike colonies and lowest in widely spaced runners. The great exception is for tightly packed colonies that successfully vent all excurrents through chimneys, in which case all zooids obtain equal food rations. In addition, interference increases with advection ratio and thus is greatest when feeding currents are strong and/or when colonies live in weak flow environments.

The observation of lower feeding rates for zooids in central regions of arborescent colonies compared with zooids in upstream regions (Okamura 1984) provides indirect empirical support of feeding interference as predicted by the advection-diffusion model. Further theoretical evidence for feeding interference is provided by Grünbaum (1995), who showed that negative hydrodynamic interactions among tightly packed lophophores in encrusting colonies serve to reduce rates of flow through lophophores in the interior of colonies and that feeding interference should therefore result. Despite these negative hydrodynamic interactions within tightly packed colonies, however, feeding rates predicted by the two-dimensional advection-diffusion model are still considerably higher for zooids in tightly packed colonies than for zooids in colonies with more widely spaced lophophores under nearly all conditions (Eckman and Okamura 1998). This result of the model reflects feeding from regions of greater turbulent replenishment that is achieved by cylindrical feeding zones and venting of excurrents through chimneys.

The Importance of Turbulence

For widely spaced and tightly packed but leaky colonies, feeding rates are increased by orders of magnitude in turbulent boundary-layer flow regimes relative to feeding rates in laminar boundary-layer flow regimes with equivalent flow speeds (Eckman and Okamura 1998). This increase reflects the turbulent mixing of suspended food from higher in the water column into regions near colonies where food is depleted by feeding.

However, the great exception to this result is feeding by tightly packed colonies that vent excurrents entirely through jets. Feeding in such colonies is equivalent in turbulent and laminar flows of equivalent speeds because diversion of excurrents enables them to avoid refiltering food-depleted water.

Contrasting Predictions of the Models

The spatial limitation and food acquisition models lead to partially different predictions about the distributions and abundances of bryozoans along an environmental gradient of food availability. These differences should allow us to test predictions of the models by considering empirical evidence from appropriate habitats. To better understand these differences, we first define food availability for suspension feeders.

Food Availability

As indicated earlier, suspension feeders rely on ambient water flow to introduce suspended particles to their microhabitats. Thus food availability will be a function of the rate of transport of food to feeding surfaces (which is determined by the shear velocity) and the concentration of food in the feeding region (see fig. 3.4). Together, these factors will determine the flux of food to lophophores. The extent to which lophophores will capture suspended particles depends on the velocity of the feeding currents. Thus food availability will be a function of the advection ratio, which is the ratio of feeding current velocity to shear velocity in the boundary layer (fig. 3.5).

Predictions of the Models

Predictions based on the two models are summarized in table 3.1. As food availability decreases, both models predict decreases in percentage cover and biomass as well as decreases in the representation of energetically costly colony forms. Thus trees, with their highly developed structures for support, and sheetlike colonies with high levels of morphological defenses, should diminish in absolute abundance and relative to other colony forms. Similarly, levels of costly, nonfeeding polymorphic zooids should decrease.

The two models differ, however, regarding the comparative success of other colony forms with decreasing food availability. The spatial limitation model predicts that competitively inferior runners and spots should increase because of increased availability of space. In contrast, the food acquisition model predicts that runners should not increase and may even decrease because of feeding interference that should affect even those colonies with widely spaced zooids in environments with low food flux.

Trees that do not have high support costs, such as *Crisia,* should experience similar interference. Such forms are essentially erect runners, and refiltration may occur as some lophophores will process water previously depleted by those on neighboring branches. However, staggering of lophophores at different levels above the substratum may allow inexpensive trees to persist in environments with low availability of food to a greater extent than runners.

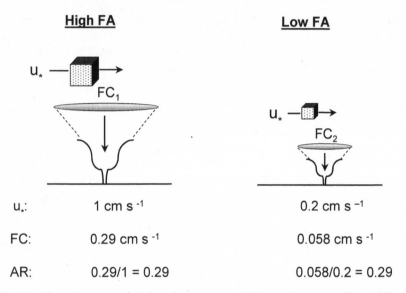

Fig. 3.5. The maintenance of similar advection ratios *(AR)* in environments offering different levels of food availability *(FA)* can ensure that external flow is sufficient to replenish water depleted of suspended food by feeding. This can be achieved by altering feeding currents *(FC)* in response to alterations in flows (u_*) in the boundary layer. Values for feeding currents and shear velocities based on known ranges of feeding current velocities (reviewed in McKinney 1990) and boundary-layer flow regimes (Eckman and Okamura 1998).

Table 3.1. Predictions of the two models as food availability decreases

	Spatial Limitation Model	Food Acquisition Model
Percentage cover	Decreases	Decreases
Biomass	Decreases	Decreases
Energetically costly forms	Decrease	Decrease
Competitively inferior forms	Runners and spots increase	Runners remain same or decrease; spots greatly increase
Colony size	Decreases (due to relative increase in spots)	Decreases greatly (due to constraint of refiltration)
Zooid size	No change	Decreases

The food acquisition model predicts that spots may be particularly favored in conditions of low food availability because they should benefit from the feeding advantage of tightly packed lophophores. However, their small size may entail two other benefits. First, small size should minimize hydrodynamic interference that decreases feeding rates achieved by lophophores within larger colonies (Grünbaum 1995). Second, minimization of the excessive backpressure that develops in larger sheets may preclude the development of excurrent jets. Jetting through chimneys may be unfavorable in environments with low availability of food for two reasons. First, jets require space and are thus expensive to produce. Jets make up 14% of the surface area of *Membranipora membranacea* colonies (Lidgard 1981), and many diminutive spotlike colonies are of a size similar to single excurrent jets of large spreading sheets. Second, refiltration of jets voided directly above colony surfaces may be problematic in environments with low flow since jets will only be very slowly displaced and diluted downstream. Since food availability is determined by food flux, which is a function of ambient flow, environments with low flow will generally entail low levels of food availability. Thus, in such environments, it may be better to dribble excurrents away from colony edges into the viscous sublayer than to send them back up above the colony. However, the production of single excurrent jets in spotlike colonies of the cyclostome *Disporella ovoidea*, which lives on cave or tunnellike surfaces (Winston 1985), indicates that single excurrent jets can be tolerated by spot colonies in some presumably low-food environments.

The spatial limitation model predicts that energetically inexpensive sheetlike colonies should be the most common colony form in conditions of decreased food availability. Thus, as long as sheetlike colonies can be produced relatively cheaply, there will be no constraint on colony size with decreased food availability. Such sheets may persist, for instance, by reducing allocation to energetically costly structures or behaviors. However, colony size overall will decrease due to increased representation of spots that are not limited by spatial competition. Greater availability of space with decreased food availability will result from the reduction in cover of both trees and sheetlike colonies that are unable to produce energetically inexpensive colonies, as well as reduction in cover of a variety of other sessile organisms (e.g., cnidarians) that are abundant in conditions of greater food availability. In contrast, the food acquisition model predicts great decreases in the sizes of sheetlike colonies, even if they are relatively energetically inexpensive, because of negative hydrodynamic interactions and refiltration. Sheets may persist along with the greatly increased number of true spots, but such sheets will show semi-determinate growth. By this scenario, energetically inexpensive sheetlike colonies will occur at increasingly smaller sizes along a gradient of decreased food availability even though they may grow to indeterminate and very large sizes when food is not limiting.

The spatial limitation model makes no particular predictions about zooid size changes along gradients in food availability. However, the food acquisition model predicts decreases in zooid size as food availability decreases due to changing relations of supply and demand embodied by the advection ratio. As the advection ratio increases, because feeding currents become stronger or ambient flows become weaker, the external flow will have a diminished capacity to replenish water depleted of suspended food by feeding. The strength of the feeding currents is a function of lophophore size (Best and Thorpe 1986; Riisgård and Manríquez 1997) which, in turn, is a function of zooid size (McKinney and Jackson 1989, and references therein). Therefore, we would expect a decrease in zooid size as food availability diminishes. These considerations suggest that advection ratios will tend to be conserved across gradients in food availability, so that large lophophores in regions of high food availability and small lophophores in habitats of low food flux will experience similar advection ratios. Such variation in size would result in the maintenance of something approximating an optimal advection ratio in which feeding rates are closely matched to rates of environmental supply. The known ranges of feeding current velocities and boundary-layer flow regimes near bryozoans suggest this is feasible (see fig. 3.5), but the issue merits further investigation. Note that size changes will ultimately be constrained by other factors. Thus the mechanics of generating effective feeding currents will eventually limit decreases in size. Increases in lophophore size may be constrained by drag forces that act to deform lophophores at high flow speeds that may be associated with low advection ratios, or by the associated high fluxes of food that may result in clogging of the lophophore. Size increase also may be constrained by interactions between functional surface areas and total mass that determine whether metabolic energy demands of the colony are met.

Environmental Gradients in Caves

Submarine caves are an ideal system to test the predictions of the two models because they are characterized by strong gradients in flow of water, concentration of food, and occupation of space by encrusting organisms (Harmelin 2000). Such caves occur throughout the world and are created by a variety of processes. In the Mediterranean, numerous karstic caves have arisen through subaerial dissolution of limestone and subsequent submergence by rising sea level at the beginning of the Holocene about 10,000 years ago. These karstic caves have been the focus of long-term investigations by Harmelin and coworkers in order to characterize both the biotic assemblages and the steep environmental gradients in light, food, and circulation of water from the outermost zone to the innermost cave recesses.

Data collected for Trémies Cave (fig. 3.6) show that the residence time of water increases from 1 to 8 days from the entrance to the rear of the cave (Fichez 1989, 1991b), indicating increasingly sluggish flows and

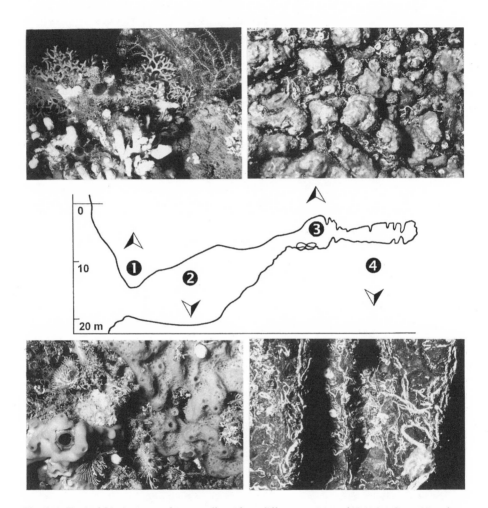

Fig. 3.6. Typical biotic cover of cave walls at four different regions of Trémies Cave. Note how cover diminishes with distance from the entrance, with the loss of arborescent forms being particularly notable. Bare rock is visible between encrusting colonies and serpulid worm tubes in zone 3, and is even more apparent in zone 4. Data on mean percentage cover (and standard deviation) in each region are also provided in table 3.2. Actual width of photographed wall portions: 1 = 50 cm; 2–4 = 20 cm.

larger boundary layers with laminar flow regimes. Light levels diminish rapidly within the cave (Passelaigue 1989), and food levels measured in four ways (chlorophyll a content of the water, concentration of suspended particulate material, and organic carbon and nitrogen fluxes provided through sedimentation) fall dramatically (table 3.2; Fichez 1991b). In keeping with these characteristics, rates of colonization (Harmelin 1997; Hugues-Dit-Ciles 1997) decrease significantly with increasing distance from the open sea, both in terms of species richness and colony number (see also Jackson 1977 for colonization rates in small experimental caves). These low rates imply that growth may occur very slowly. For example, the ages of 2.5-mm-high nodules of layers of encrusting colonies that develop within caves are likely to exceed one century, since each nodule layer requires several years for completion (Harmelin 2000). Slow growth is further supported by the very low metabolic activity of the benthos in the dark recesses of caves where respiration rates are similar

Table 3.2. Gradation of abiotic and biotic features from four zones ranging from just outside Trémies Cave to deep within cave recesses

	Zone 1	Zone 2	Zone 3	Zone 4
Distance to entrance (m)	0–10	10–30	30–45	45–60
Light level (W m^{-2})[a]		1×10^{-2}–10^{-3}	3×10^{-6}	$<10^{-8}$
Temperature (°C)[b]	15.6	15.6	16.9	17.2
	(3.5)	(3.5)	(3.8)	(3.9)
Chlorophyll a (ng l^{-1})[c]	482	461	119	78
	(645)	(650)	(155)	(103)
Organic carbon flux	—	107.6	26.8	22.2
(Mg m^{-2} d^{-1})[d]		(108.1)	(25.9)	(21.4)
Organic nitrogen flux	—	11.3	3.0	2.2
(Mg m^{-2} d^{-1})[d]		(11.2)	(2.9)	(2.1)
Water residence time (d)[e]	1	4	—	8
Percentage cover of walls[f]	100	100	60.2	12.1
			(9.5)	(10.3)
Biomass (g m^{-2})[g]	360	250	38	26
	(260)	(21)	(10)	(6)

Note: Standard deviations associated with mean values in brackets.
[a] Passelaigue 1989.
[b] Fichez 1991b.
[c] Fichez 1990b.
[d] Fichez 1990a.
[e] Fichez 1989.
[f] Percentage cover between zones 3 and 4 decreases significantly ($F_{1,10} = 48.066, p \ll 0.001$ on arcsine-transformed unpublished data from Harmelin). Percentage cover of walls in zones 1 and 2 is always 100% (Harmelin 1986 and unpublished data).
[g] Dry weight of organic matter decreases significantly ($F_{14,19} = 451.03, p \ll 0.0011$; Fichez 1989, 1991b).

to levels recorded for oligotrophic environments at 1,000–2,000-m depth (Fichez 1991a).

Testing Predictions of the Models in Caves

Predictions about Coverage, Biomass, and Energetically Costly Morphologies

Both models correctly predict striking and highly significant reductions in percentage cover and biomass on cave walls with the increasingly oligotrophic conditions and low rates of mixing toward the backs of caves (table 3.2). The incidence of energetically costly colony forms (trees) and zooidal polymorphisms also decreases (fig. 3.7). Both the proportion and total number of treelike bryozoan species is highest just outside the cave, decreases dramatically within it, and only a very small number remain in the rear (goodness-of-fit test based on null hypothesis that the number of tree species is equally distributed among the four cave regions: $G = 72.1$, df $= 3$, $p < 0.001$). Moreover, treelike colonies of *Crisia* that occur in the most remote regions of caves are poorly branched and never fertile

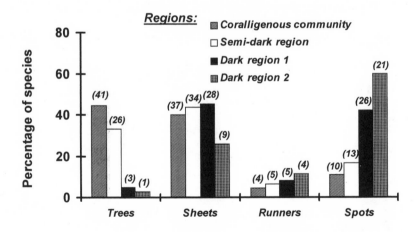

Distribution of forms in different regions

Fig. 3.7. The percentage of species (number of species in parentheses) of colonies of different growth forms from Mediterranean coralligenous environments just outside of caves (= coralligenous community), in semidark regions (approximately 8–10 m from cave entrances), and from two dark regions toward the rear of caves. Colonies in dark region 1 are from approximately 25–40 m from cave entrances, while those from dark region 2 are >40 m from cave entrances. Data for colonies from ten submarine caves from the Provence region (most from Trémies and 3PP Caves) (Harmelin 1976, 2000).

(Harmelin 2000). Thus trees deep in cave recesses may contribute little or nothing to the regional reproductive pool.

The total number of species that form large spreading sheets also decreases within caves, although the proportion of sheets does not decrease until the rear of caves, due to the great decrease in trees (fig. 3.7; G = 21.3, df = 3, $p < 0.001$). Among these sheet species, the frequency of species with complex zooidal budding also diminishes within caves (fig. 3.8; $F_{2,27} = 23.758$, $p \ll 0.001$), while that of species with intrazooidal budding increases ($F_{2,27} = 35.940$, $p \ll 0.001$; ANOVA on ln-transformed data). Finally, there is a decrease in the number and proportion of species producing adventitious avicularia (defensive polymorphs) from coralligenous regions outside caves to mid- and rear-cave regions (fig. 3.9; G = 50.510, df = 3, $p < 0.001$). The number of species producing interzooidal avicularia does not change significantly along the cave gradient (G = 2.366, df = 3, $0.50 < p < 0.10$), although the proportion of such species increases due to the decreased representation of species with adventitious avicularia. Similarly, the number of species producing no avicularia does not change significantly (G = 2.439, df = 3, $0.90 < p < 0.50$), although again the proportions increase due to the decrease in species with adventitious avicularia.

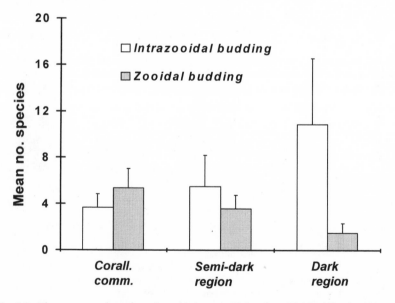

Fig. 3.8. The mean number of species with intrazooidal and zooidal budding in 400-cm² areas from the coralligenous community and from walls in semidark and dark regions. Regions and data for colonies as described for figure 3.7. Dark regions 1 and 2 in figure 3.7 have been pooled as dark region.

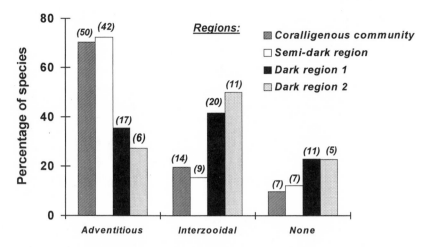

Distribution of avicularia in different regions

Fig. 3.9. The percentage of species (number of species in parentheses) with adventitious, interzooidal, and no avicularia from different regions. Regions and data for colonies as described for figure 3.7.

Predictions about Competitively Inferior Forms and Colony and Zooid Size

Other aspects of bryozoan distributions within the cave conform to the predictions of the food acquisition model but not of the spatial limitation model. Thus the total number of runner species remains similar throughout the cave rather than increasing with distance from the entrance as predicted by the adaptive strategy model in response to the availability of space (fig. 3.7). The small increase in proportional representation of runners in cave recesses reflects decreases in trees and sheets. These data are based on presence or absence of species with different colony morphologies, but they broadly conform with relative abundance. For example, runners also decrease in the frequency that they are encountered along transects into the cave (Harmelin 1986).

The total number and proportion of species forming small encrusting sheets greatly increase, reflecting the increased representation of both true spots and sheetlike encrusting species whose colonies clearly only reach small sizes before growth ceases (fig. 3.7; $G = 9.324$, df $= 3$, $0.025 < p < 0.05$). This trend is paralleled by a significant decrease in mean maximum colony sizes with distance from the cave entrance as measured for encrusting cheilostome species in different regions of the cave (fig. 3.10; $F_{2,14} = 18.468$, $p < 0.001$). Colonies at the rear of the caves are an order

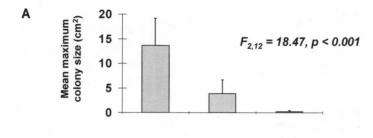

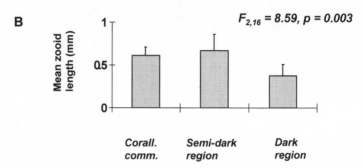

Fig. 3.10. The mean maximum sizes of colonies of encrusting cheilostomes *(A)* and the mean zooid lengths of these colonies *(B)* in different regions. The decrease in size in the dark regions of caves reflects the increased representation of true spots and of sheets that reach only a small, determinate size. Measurements from colonies just outside caves in the coralligenous community and from semidark regions (approximately 8–10 m from cave entrances) were taken from two series of twelve proxi-photographs (frame = 160 cm²). Measurements of colonies in dark regions (25–50 m from entrances) were taken from five wall samples of 400 cm² and from six limestone plates (area = 400 cm²) submerged for 9 years. Wall samples and photographs were from both Trémies and 3PP Caves. Maximum sizes of colonies represent the extreme of size ranges measured for colonies of each species encountered in photographs, wall samples, or limestone plates. In cases where the same species was present on each substrata, we included the larger, more conservative measurement for analysis.

of magnitude smaller than those toward the cave entrance and two orders of magnitude smaller than those in the coralligenous zone outside of the cave (fig. 3.10). Moreover, the small, sheetlike colonies in the rear of caves possess only a few functional zooids in distal regions while the rest of the colony is dead or senescent (Harmelin 2000). Thus these colonies effectively converge to spotlike forms, perhaps as a result of negative hydrodynamic interactions among lophophores (Grünbaum 1995) that constrain colonies to small size. In contrast, true spots do not exhibit such effects, and reproductive colonies of five spot species persist deep within caves (table 3.3).

Table 3.3. The number and associated percentage (in brackets) of reproductive and nonreproductive colonies of spot species

	Zone 2[a]		Zone 3[b]		Zone 4[c]	
	+R	−R	+R	−R	+R	−R
Tubulipora hemiphragmata	0	0	3	5	21	19
			(37.5)	(62.5)	(52.5)	(47.5)
Setosella cavernicola	0	0	11	3	30	8
			(78.6)	(21.4)	(78.9)	(21.1)
Puellina pedunculata	10	3	12	0	12	2
	(76.9)	(23.1)	(100)		(85.7)	(14.3)
Puellina cassidainsis	0	0	15	2	10	5
			(88.2)	(11.8)	(66.7)	(33.3)
Haplopoma sciaphilum	0	0	31	2	56	15
			(93.9)	(6.11)	(78.9)	(21.1)
Total of 5 species (zones 3 & 4)	—	—	72	12	129	49
			(85.7)	(14.3)	(72.5)	(27.5)

Note: These species were present in wall samples from different regions within 3PP Cave. Reproductive colonies (+R) were counted as those with ovicells, and nonreproductive colonies (−R) were those without ovicells. Note lack of spots apart from one species encountered in collections from zone 2, closer to entrance of cave.
[a] Zone 2 was 8–10 m from entrance (three wall samples of 400 cm^2).
[b] Zone 3 was 20–40 m from entrance (eight wall samples of 400 cm^2).
[c] Zone 4 was 50–100 m from entrance (twelve wall samples of 400 cm^2).

The mean zooid sizes of cheilostome species (measured by zooid length) are not significantly different in colonies near the entrance of the cave relative to just outside, but zooid length is significantly smaller in the rear ($F_{2,16} = 3.634$, $p = 0.0029$) (see fig. 3.10). Temperature effects on zooid size (Menon 1972; Hunter and Hughes 1994) do not explain these differences since temperature differences are minimal along the Trémies Cave gradient (see table 3.2) while cooler temperatures that develop with distance from the entrance of 3PP Cave should tend to increase zooid size.

It could be argued that variability in food supply (table 3.2) may explain the distributions and abundances of different colony forms. Thus colonies may have been established during periods when resources were temporarily increased, so that the patterns we report here reflect historical events. This is unlikely for three reasons. First, under such a scenario we would not expect that distribution and abundance patterns of different colony forms would be regularly repeated in different cave systems, but this is indeed the case worldwide (Harmelin 2000). Second, the exceedingly low colonization rates in caves now documented in several studies (Harmelin 2000) indicate that stands of colonies of particular morphologies are unlikely to develop during transient conditions of temporarily increased food resources. Third, it is unlikely that distribution due to historical events would conform precisely with predictions of the food acquisition model.

Tests from Other Environments

Small Cavities and Other Cryptic Environments

Bryozoans commonly occur on the undersurfaces of boulders and in other cryptic habitats such as small cavities within dense assemblages on hard substrata (McKinney and Jackson 1989). Small cavities range in size from interstices (centimeter scale or less) to larger holes (up to 1 m³) that result from geologic or biological processes (Harmelin 2000). Such cavity systems are prominent parts of coral reefs (Jackson 1977) and of Mediterranean coralligenous bioherms constructed by coralline algae (Hong 1982). The cumulative volume of cavity systems in coral reefs may exceed that of the reef framework (Garrett et al. 1971; Ginsburg 1983), and the same is likely in coralligenous reefs (Harmelin 2000). In both systems, encrusting bryozoans are significant secondary frame builders (Cuffey 1973, 1974; Hong 1982; Martindale 1992). Food availability within such small cryptic habitats has not been measured but is undoubtedly extremely reduced due to depletion of suspended particles by suspension feeders in the surrounding more open environment (e.g., Glynn 1973; Buss and Jackson 1981; Wildish and Kristmanson 1984; Fréchette and Bourget 1985) and reduction in the flow of water and thus the flux of food.

Mediterranean Cavities and Cobbles. In the French Mediterranean, such cryptic habitats along with cave environments shelter a large part of the bryozoan regional pool of species (Harmelin 1976, 2000). Some 70% of species found in the eurythermal coastal zone (0–80-m depth, $n = 200$ species) and 62% of species from the continental shelf and upper bathyal (0–500-m depth, $n = 227$ species) are residents of cryptic habitats and caves. The greatest number of species are cave residents, but 57.1% occur on the undersides of cobbles and 60.7% are found in cavities. Likewise, the greatest bryozoan diversity in the coastal zone of central Chile occurs on the undersides of boulders and in crevices (Juan Cancino, pers. comm.). Thus cryptic environments compose a significant habitat for bryozoans in many regions.

The relative proportions of species with different colony forms in Mediterranean cryptic environments is similar to the deep recesses of caves (fig. 3.11; Harmelin 2000). Both the proportion and number of species of trees is greatly decreased relative to that in the open coralligenous zone (goodness of fit test based on null hypothesis that the number of tree

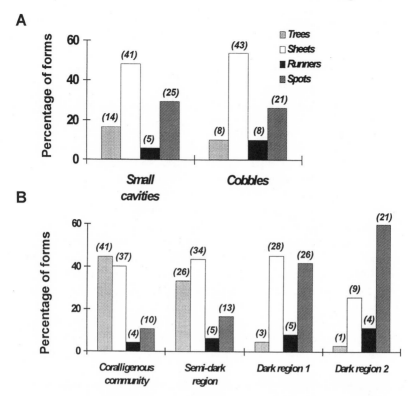

Fig. 3.11. The percentage by region of species of trees, runners, indeterminate sheets, and spots (including small, determinate sheets) in small cavities and under cobbles *(A)* and, for comparison, from the coralligenous community *(B)* just outside of caves and along the cave gradient.

species is equally distributed in open coralligenous regions, under cobbles, and in small cavities: G = 28.068, df = 2, *p* < 0.001). The scarcity of trees in cryptic habitats may be due to lack of room for upright growth, but occasional exceptional occurrence where room permits suggests that decreased food availability may also be involved (Harmelin 2000). In contrast, the proportion of spots increases in cryptic habitats (similar goodness-of-fit test: G = 7.051, df = 2, 0.025 < *p* < 0.05), but the frequency and proportion of runnerlike species is roughly similar in all habitats (goodness-of-fit test for distribution of runners in all locations: G = 1.877, df = 5, 0.90 < *p* < 0.50). Finally, the distribution of sheetlike species is not equivalent in all habitats (goodness of fit test for distribution of sheets in all locations: G = 30.287, df = 5, *p* < 0.001). Sheets are proportionately most abundant in cryptic habitats and least so in the deepest recesses of caves (fig. 3.11). The high diversity of sheet species in Mediterranean cryptic environments is similar to that on Caribbean coral reefs (Jackson 1984).

Disarticulated Bivalve Shells. Surfaces of disarticulated bivalves provide both cryptic and more exposed microhabitats for bryozoans (e.g., Bishop 1989; Ward and Thorpe 1991; McKinney 1996). The most hydrodynamically stable position is with the concave, interior surface facing downward. Bryozoans growing on these interior surfaces are effectively within a cavity or small cave, while those on the exterior have greater access to food-laden waters. The distributions and sizes of sheetlike and spotlike bryozoans differ between the inner and outer surfaces of disarticulated bivalve shells in the Adriatic (McKinney 2000). All nine species of small, spotlike colonies were significantly more common on interior valve surfaces than on exterior surfaces. However, five of the nine species of spots were significantly larger albeit rarer on external surfaces; and three of these five also had higher proportions of fertile colonies on exterior surfaces. In contrast, four of five species of larger, sheetlike colonies were significantly more common and larger on exterior valve surfaces while the remaining species showed a similar, but nonsignificant, pattern. Overall, the distributions of bryozoans on disarticulated bivalve shells are similar to distributions in caves and similarly broadly conform with the predictions of the food acquisition model.

Winston (1985) observed similar patterns for common species in early successional stages on artificial panels. The indeterminate, sheetlike species *Drepanophora tuberculatum* occurred more frequently on the outer parts of panel surfaces while the determinately growing, spotlike species *Disporella ovoidea* was found in greatest numbers on protected cave or tunnellike surfaces of panels.

Spots on Seagrass Blades

Spot colonies are common on blades of the seagrass *Posidonia oceanica*, where they may be particularly favored because the flexible and spatially restricted nature of the substratum precludes the development of more extensive, sheetlike colonies (Hayward 1974). Availability of food in open, seagrass environments should be very high compared to caves. Thus the presence of spots in Mediterranean seagrass beds allows a further test of the food acquisition model by considering zooid sizes of spots in *Posidonia* beds versus spots in the recesses of caves. Mean zooid length of spot species growing on *Posidonia* blades is greater than that of spot species found in the recesses of caves (fig. 3.12; two-tailed t-test; $t = 2.252$, df $= 9$, $p = 0.051$). This suggests that spots maintain advection ratios appropriate to food availability in different habitats, as predicted by the food acquisition model (fig. 3.5).

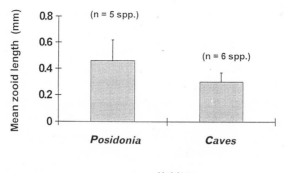

Fig. 3.12. Mean zooid length in spots on fronds of the seagrass *Posidonia oceanica* and in dark caves. Zooid length data for *Haplopoma impressum, Puellina gattyae, Electra posidoniae,* and *Fenestrulina joannae* on seagrass fronds taken as mean values of ranges of zooid lengths reported by Gautier (1961). Note that *E. posidoniae* possesses a ribbonlike colony morphology in which colonies develop directional growth as a number of narrow multiserial lobes. We infer that, as for spots, excurrent flow will be released along the edges of ribbonlike colonies. This, in combination with the noted limitation in sizes of *E. posidoniae* colonies (Hayward 1974), leads us to include the species in our treatment. Zooid length for *Rhamphostomellina posidoniae* on seagrass from Hayward 1975. Zooid length data for spot species in recesses of caves *(Puellina cassidainsis, P. innominata, P. pedunculata, P. radiata, Setosella cavernicola,* and *Haplopoma sciaphilum)* from Harmelin 1997 and unpublished data.

Spots in Interstitial Habitats

Many bryozoan species inhabit shell fragments and sand- and gravel-sized particles of interstitial environments (Winston and Håkansson 1986; McKinney and Jackson 1989). In such habitats a viscous flow regime prevails (Crenshaw 1980; Vogel 1981) and food availability is likely much lower than in open environments. While some species that form spots are specifically found interstitially, other species show wider distributions. Zooid lengths of three spot species in Florida are significantly greater in colonies collected from open habitats than from interstitial samples (fig. 3.13; two-way *t*-test for paired comparisons; $t = 4.799$, df = 2, $p = 0.041$). Winston and Håkansson (1986) similarly noted (but did not test statistically) that zooid sizes in colonies that occur interstitially are slightly smaller than in colonies of the same species that grow on larger substrata in more open habitats.

Of course, the small size of interstitial substrata may, in itself, enforce smaller zooid sizes and thereby explain these patterns. However, there is considerable unoccupied space on most interstitial shell fragments (e.g., see figs. 14, 19, 21, 40, 41, and 81 in Winston and Håkansson 1986). Thus data for colonies in interstitial habitats also agree with specific predictions of the food acquisition model.

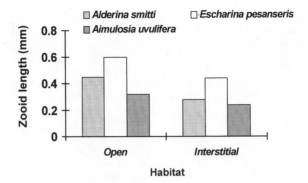

Fig. 3.13. Zooid lengths of three spot species collected from interstitial and more open habitats. Length data for interstitial colonies of *Alderina smitti, Escharina pesanseris,* and *Aimulosia uvulifera* from mean lengths reported in Winston and Håkansson 1986. Data for colonies of *Alderina smitti* (collected from grab samples at 15 and 27 m off Fort Walton Beach, Florida) and *Aimulosa uvulifera* (also collected from grab samples off Fort Walton Beach and from dredging rocky outcrop at 9 m off Dog Island, Florida) taken as median zooid lengths for length ranges reported by Schier (1964). Data for zooids of *Escharina pesanseris* from colonies collected in offshore environments on dead shells reported by Winston (1982).

Support for the Food Acquisition Model

The food acquisition model successfully predicts more aspects of the distribution and abundance of bryozoans than the spatial limitation model. Food acquisition has played an important role in determining colony morphology and strongly influences present-day patterns of distribution and abundance. Most predictions of the spatial limitation model are also well supported, but only when levels of food are sufficiently high. Both the food acquisition and spatial limitation models are relevant, and therefore complementary, to bryozoan macroevolution. However, the food acquisition model appears to provide a more comprehensive explanation of colony form and size, zooidal size, and patterns of distribution and abundance.

It is impossible to tell which of these factors were of primary importance for the evolution of bryozoan colony form without detailed analyses of early colony morphologies and modes of growth. Many spreading sheetlike colonies unequivocally adopt confrontational tactics, but early selection for sheetlike growth may have been based initially on the feeding benefit achieved by deploying tightly packed lophophores. Deployment of costly defenses in such confrontational forms should be promoted by the feeding benefit conferred by tightly packed lophophores. Likewise, upright growth may reflect primarily selection for exploitation of water levels above the bottom where replenishment of food to lophophores is greater. Alternatively, erect growth may have arisen as a means of over-topping and starving out encrusting neighbors below. Finally, the fact that runnerlike growth is primitive among both stenolaemates and gymnolaemates (Cheetham 1975; Taylor and Larwood 1990) suggests that the first bryozoans may have been adapted to locating spatial refuges to survive in competition with other encrusting organisms. However, an alternative hypothesis is that early runnerlike growth resulted from simple budding rules.

The Adaptive Significance of Spots

Bishop (1989) argued that spot colonies are adapted to particular configurations of substrata, such as the concave surfaces of disarticulated bivalve shells, and therefore do not appear to be relegated to spatial refuges from biotic enemies. This argument is similar to ours in identifying spot morphologies as specific adaptations that have resulted from strong selective forces associated with particular environmental conditions. Ideally, such hypotheses should be subject to comparative analysis to determine whether selective forces have produced similar shapes among multiple, independently derived taxa, but lack of sufficient data on the taxonomy and evolution of bryozoans precludes such formal analysis at present.

Nonetheless, it is certainly the case that, collectively, spots derived from multiple taxa are well suited to and occur disproportionately in conditions of low food availability. Furthermore, bryozoan genera are old (dating from the Paleogene or even Cretaceous), and many spots have close relatives (within the same genus or family) that do not grow as spots (e.g., Winston and Håkansson 1986). Thus the occurrence of spots in so many different genera, families, and even orders indicates that, despite the lack of a good phylogeny for the Bryozoa, it must be true that spots have evolved independently in many different clades. We therefore argue that there is strong evidence that low food availability has exerted considerable selection in shaping the evolutionary trajectories of a number of species in a variety of environments.

Key habitats in which such selection has operated on bryozoans are small cryptic environments and the recesses of caves. It is thus our prediction that historical and phylogenetic evidence will eventually support our contention that low food availability has resulted in strong selective forces that have produced small colony size. However, we should also point out that when spots occur on flexible substrata, such as algal and seagrass fronds, their distribution could be divorced from conditions of low food availability. In these situations, the physical nature of the substrata may have selected for colony form.

The Nature of Refugia and Refuge-Seeking Taxa

Many animals that live in cryptic habitats, caves, and the deep sea are regarded as primitive, anachronistic, refuge-dwelling species that have been relegated to marginal habitats through competition with superior biological enemies (Jackson et al. 1971; Vermeij 1987). However, the arguments presented here suggest that life in these marginal environments entails extreme conditions that require specific adaptations (see also Bishop 1989). In particular, spot colonies show a distinct suite of characters that specifically adapt them to low levels of food availability, including small, determinate size, early reproduction, and larval settlement preferences. Both the greater abundance of spots and the presence of reproductive colonies within the deep recesses of caves (table 3.3) confirm that these environments present important habitats for spots. Thus spot bryozoans do not conform to the view that specializations to cave dwelling are reductions and losses rather than specific adaptations (Vermeij 1987). Similar arguments can be made for the panoply of bizarre animal structures and behaviors characteristic of the deep sea, including tripod fishes perched on ooze, carnivorous sorberacean tunicates, swimming elasipodid holothurians, giant scavenging amphipods, and stalked glass sponges, as well as their astonishingly high diversity (Gage and Tyler 1991; Gage 1997).

Low food levels common to "refuge" habitats clearly limit the development of dense assemblages. Thus adaptations associated with evolutionary escalation in defense of space are not expected. But this does not preclude adaptation to limited food resources, as exemplified by spot bryozoans in caves and the elaborate mechanisms of prey capture employed by deep-sea carnivorous sponges (Vacelet and Boury-Esnault 1995), carnivorous tunicates, and various fishes (Gage and Tyler 1991). Indeed, competition for food may have resulted in evolutionary escalation for this extremely limited resource in "refuge" habitats. We conclude that the paradigm of food-poor environments as refugia for adaptively anachronistic species fails because it focuses on the wrong resource (space) and overlooks the fact that food-poor environments require specific and sophisticated adaptations for survival and reproduction.

Acknowledgments

We thank J. Bishop, F. K. McKinney, and M. LaBarbera for comments that helped us to clarify our arguments, even though they may not all agree with them, and to improve the manuscript. N. Knowlton suggested considering the implications for refuge species and provided a title. B. Okamura especially wishes to thank Alan Cheetham for his support, encouragement, and tuition in studying the ecology of bryozoans and Jim Eckman for his fundamental contribution in modeling bryozoan suspension feeding from boundary-layer flows.

References

Banta, W. C., F. K. McKinney, and R. L. Zimmer. 1974. Bryozoan monticules: Excurrent water outlets? *Science* 185:783–84.

Best, M. A., and J. P. Thorpe. 1986. Effects of food particle concentration on feeding current velocity in six species of marine Bryozoa. *Marine Biology* 93: 255–62.

Bishop, J. D. D. 1989. Colony form and the exploitation of spatial refuges by encrusting Bryozoa. *Biological Reviews* 64:197–218.

Buss, L. W. 1979. Habitat selection, directional growth, and spatial refuges: Why colonial animals have more hiding places. In *Biology and systematics of colonial organisms,* ed. G. Larwood and B. R. Rosen, 459–97. London: Academic Press.

———. 1981. Group living, competition, and the evolution of cooperation in a sessile invertebrate. *Science* 213:1012–14.

Buss, L. W., and J. B. C. Jackson. 1981. Planktonic food availability and suspension-feeder abundance: Evidence of in situ depletion. *Journal of Experimental Marine Biology and Ecology* 49:151–61.

Cheetham, A. H. 1975. Taxonomic significance of autozooid size and shape in some early multiserial cheilostomes from the Gulf Coast of the U.S.A. In *Bryozoa 1974,* ed. S. Pouyet, Documents des Laboratoires de Géologie de la Faculté des Sciences de Lyon, hors série 3, fasc. 2, 547–64.

———. 1986. Branching, biomechanics, and bryozoan evolution. *Proceedings of the Royal Society, London,* series B, 228:151–71.

Cheetham, A. H., and E. Thomsen. 1981. Functional morphology of arborescent animals: Strength and design of cheilostome bryozoan skeletons. *Paleobiology* 7:355–83.

Cook, P. L. 1977. Colony-wide water currents in living Bryozoa. *Cahiers de Biologie Marine* 18:31–47.

Crenshaw, D. G. 1980. How interstitial animals respond to viscous flows. Ph.D. diss., Duke University, Durham, NC.

Cuffey, R. J. 1973. Bryozoan distribution in the Modern reefs of Eniwetok Atoll and the Bermuda Platform. *Pacific Geology* 6:25–50.

———. 1974. Delineation of bryozoan constructional roles in reefs from comparison of fossil bioherms and living reefs. In *Proceedings of the Second International Coral Reef Symposium,* 1:357–64. Brisbane, Australia: Great Barrier Reef Committee.

Denny, M. W. 1988. *Biology and the mechanics of the wave-swept environment.* Princeton: Princeton University Press.

Dick, M. H. 1987. A proposed mechanism for chimney formation in encrusting bryozoan colonies. In *Bryozoa: Present and past,* ed. J. R. P. Ross, 73–80. Bellingham, WA: Western Washington University.

Eckman, J. E., and B. Okamura. 1998. A model of particle capture by bryozoans in turbulent flow: Significance of colony form. *American Naturalist* 152:861–80.

Fichez, R. 1989. Phénomènes d'oligotrophie en milieu aphotique: Étude des grottes sous-marine, comparaison avec les milieux profonds et bilans énergétiques. Ph.D. diss., Université d'Aix-Marseille II, Marseille, France.

———. 1990a. Decrease in allochthonous organic inputs in dark submarine caves: Connection with lowering in benthic community richness. *Hydrobiologia* 207:61–69.

———. 1990b. Les pigment chlorophylliens: Indices d'oligotrophie dans les grottes sous-marine. *Comptes-Rendus de l'Académie des Sciences, Paris* 310 (3):255–61.

———. 1991a. Benthic oxygen uptake and carbon cycling under aphotic and resource-limiting conditions in a submarine cave. *Marine Biology* 110:137–43.

———. 1991b. Suspended particular organic matter in a Mediterranean submarine cave. *Marine Biology* 108:167–74.

Fréchette, M., and E. Bourget. 1985. Food-limited growth of *Mytilus edulis* L. in relation to the benthic boundary layer. *Canadian Journal of Fisheries and Aquatic Sciences* 42:1166–70.

Gage, J. D. 1997. High species diversity in deep-sea sediments: The importance of hydrodynamics. In *Marine biodiversity: Patterns and processes,* ed. R. F. G. Ormond, J. D. Gage, and M. V. Angel, 148–77. Cambridge: Cambridge University Press.

Gage, J. D., and P. A. Tyler. 1991. Deep-sea biology: A natural history of organisms at the deep-sea floor. Cambridge: Cambridge University Press.

Garrett, P., D. L. Smith, A. O. Wilson, and D. Patriquin. 1971. Physiography, ecology, and sediments of two Bermuda patch reefs. *Journal of Geology* 79:647–68.

Gautier, Y. V. 1961. Recherches écologiques sur les Bryozoaire chilostomes en Méditerranée occidentale. *Recueil des Travaux de la station marine d'Endoume, Faculté des Sciences de Marseille* 38:1–434.

Ginsburg, R. N. 1983. Geological and biological roles of cavities in coral reefs. In *Perspectives on coral reefs,* ed. D. J. Barnes, 148–53. Townsville: Australian Institute of Marine Science Publications.

Glynn, P. W. 1973. Ecology of a Caribbean coral reef: The *Porites* reef flat biotope: Part 2, Plankton community with evidence for depletion. *Marine Biology* 22:1–21.

Grünbaum, D. 1995. A model of feeding currents in encrusting bryozoans shows interference between zooids within a colony. *Journal of Theoretical Biology* 174:409–25.

Harmelin, J.-G. 1976. Le sous-ordre des Tubuliporina (Bryozoaires cyclostomes) en Méditerranée: Écologie et systématique. *Mémoires. Institut Océanographique* (Monaco) 10:1–326.

———. 1986. Patterns in the distribution of bryozoans in the Mediterranean marine caves. *Stylogia* 2:10–25.

———. 1997. Diversity of bryozoans in a Mediterranean sublittoral cave with bathyal-like conditions: Role of dispersal processes and local factors. *Marine Ecology Progress Series* 153:139–52.

———. 2000. Ecology of cave and cavity dwelling bryozoans. In *Proceedings of the Eleventh International Bryozoology Association Conference,* ed. A. Herrera Cubilla and J. B. C. Jackson, 38–53. Balboa, Republic of Panamá: Smithsonian Tropical Research Institute.

Harvell, C. D., and M. LaBarbera. 1985. Flexibility: A mechanism for control of local velocities in hydroid colonies. *Biological Bulletin* 168:312–20.

Hayward, P. J. 1974. Observations on the bryozoan epiphytes of *Posidonia oceanica* from the island of Chios (Aegean Sea). In *Bryozoa 1974,* ed. S. Pouyet, Documents des Laboratoires de Géologie de la Faculté des Sciences de Lyon, hors série 3, fasc. 2, 347–56.

———. 1975. A new species of epiphytic bryozoan from the Aegean Sea. *Cahiers de Biologie Marine* 16:207–12.

Hong, J. S. 1982. Contribution à l'étude des peuplements d'un fond de concrétionnement coralligène dans la région marseillaise en Méditerranée nord-occidentale. *Bulletin Kordi* 4:27–51.

Hugues-Dit-Ciles, F. 1997. Environmental controls on recruitment of epibenthos on artificial panels in a submarine cave. B.S. diss., University of Plymouth, Plymouth.

Hunter, E., and R. Hughes. 1994. The influence of temperature, food ration, and genotype on zooid size in *Celleporella hyalina* (L.). In *Biology and palaeobiology of bryozoans,* ed. P. J. Hayward, J. S. Ryland, and P. D. Taylor, 83–86. Fredensborg, Denmark: Olsen and Olsen.

Jackson, J. B. C. 1977. Competition on marine hard substrata: The adaptive significance of solitary and colonial strategies. *American Naturalist* 111:743–67.

———. 1979. Morphological strategies of sessile animals. In *Biology and systematics of colonial organisms,* ed. G. Larwood and B. R. Rosen, 499–555. London: Academic Press.

———. 1983. Biological determinants of present and past sessile animal distributions. In *Biotic interactions in Recent and fossil benthic communities,* ed. M. J. S. Tevesz and P. L. McCall, 39–120. New York: Plenum Publishing.

———. 1984. Ecology of cryptic coral reef communities: 3, Abundance and aggregation of encrusting bryozoans with particular reference to cheilostome Bryozoa. *Journal of Experimental Marine Biology and Ecology* 75:37–57.

Jackson, J. B. C., T. F. Goreau, and W. D. Hartman. 1971. Recent brachiopod-coralline sponge communities and their paleoecological significance. *Science* 173:623–25.

Jackson, J. B. C., and F. K. McKinney. 1990. Ecological processes and progressive macroevolution in marine clonal benthos. In *Causes of evolution: A paleontological perspective,* ed. R. M. Ross and W. D. Allmon, 173–209. Chicago: University of Chicago Press.

Jackson, J. B. C., and J. E. Winston. 1982. Ecology of cryptic coral reef communities: 1, Distribution and abundance of major groups of encrusting organisms. *Journal of Experimental Marine Biology and Ecology* 57:135–47.

Kay, A. M., and M. J. Keough. 1981. Occupation of patches in the epifaunal communities on pier pilings and the bivalve *Pinna bicolor* at Edithburgh, South Australia. *Oecologia* 48:123–30.

Keough, M. J. 1984. Dynamics of the epifauna of the bivalve *Pinna bicolor:* Interactions among recruitment, predation, and competition. *Ecology* 65:677–68.

Lidgard, S. 1981. Water flow, feeding, and colony form in an encrusting cheilostome. In *Recent and fossil Bryozoa,* ed. G. P. Larwood and C. Nielsen, 135–42. Fredensborg, Denmark: Olsen and Olsen.

———. 1985. Zooid and colony growth in encrusting cheilostome bryozoans. *Palaeontology* 28:255–91.

Martindale, W. 1992. Calcified epibionts as palaeoecological tools: Examples from the Recent and Pleistocene reefs of Barbados. *Coral Reefs* 11:167–17.

McKinney, F. K. 1990. Feeding and associated colonial morphology in marine bryozoans. *Aquatic Science* 2:255–80.

———. 1996. Encrusting organisms on co-occurring disarticulated valves of two marine bivalves: Comparison of living assemblages and skeletal residues. *Paleobiology* 22:543–67.

———. 2000. Colony sizes and occurrence patterns among Bryozoa encrusting disarticulated bivalves in the northeastern Adriatic Sea. In *Proceedings of the Eleventh International Bryozoology Association Conference,* ed. A. Herrera Cubilla and J. B. C. Jackson, 282–90. Balboa, Republic of Panamá: Smithsonian Tropical Research Institute.

McKinney, F. K., and J. B. C. Jackson. 1989. *Bryozoan evolution.* Boston: Unwin Hyman.

McKinney, F. K., and P. D. Taylor. 1997. Life histories of some Mesozoic encrusting cyclostome bryozoans. *Palaeontology* 40:515–56.

Menon, N. R. 1972. Heat tolerance, growth, and regeneration on three North Sea bryozoans exposed to different constant temperatures. *Marine Biology* 15:1–11.

Nowell, A. R. M., and P. A. Jumars. 1984. Flow environments of aquatic benthos. *Annual Review of Ecology and Systematics* 15:303–28.

Okamura, B. 1984. The effects of ambient flow velocity, colony size, and upstream colonies on the feeding success of Bryozoa: 1, *Bugula stolonifera* Ryland, an arborescent species. *Journal of Experimental Marine Biology and Ecology* 83:179–93.

———. 1985. The effects of ambient flow velocity, colony size, and upstream colonies on the feeding success of Bryozoa: 2, *Conopeum reticulum* (Linnaeus), an encrusting species. *Journal of Experimental Marine Biology and Ecology* 889:69–80.

———. 1988. The influence of neighbors on the feeding of an epifaunal bryozoan. *Journal of Experimental Marine Biology and Ecology* 120:105–23.

———. 1992. Microhabitat variation and patterns of colony growth and feeding in a marine bryozoan. *Ecology* 73:1502–13.

Okamura, B., and J. E. Eckman. 1997. Modelling particle capture rates by encrusting bryozoans: Adaptive significance of colony form. In *Proceedings of the Eighth International Coral Reef Symposium,* ed. H. A. Lessios and I. G. Macintyre, 2:1077–80. Balboa, Republic of Panama: Smithsonian Tropical Research Institute.

Passelaigue, F. 1989. Les migrations journalières du mysidace marin cavernicole *Hemimysis speluncola:* Comparison avec les migrations verticales du plancton. Ph.D. diss., Université d'Aix-Marseille II, Marseille, France.

Riisgård, H. U., and P. Manriquez. 1997. Filter-feeding in fifteen marine ectoprocts (Bryozoa): Particle capture and water pumping. *Marine Ecology Progress Series* 154:223–39.

Schier, D. E. 1964. Marine Bryozoa from northwest Florida. *Bulletin of Marine Science of the Gulf and Caribbean* 14:603–62.

Taylor, P. D., and G. P. Larwood. 1990. Major evolutionary radiations in the Bryozoa. In *Major evolutionary radiations,* ed. P. D. Taylor and G. P. Larwood, 209–33. Oxford: Clarendon Press.

Vacelet, J., and N. Boury-Esnault. 1995. Carnivorous sponges. *Nature* 373:333–35.

Vermeij, G. J. 1987. *Evolution and escalation.* Princeton: Princeton University Press.

Vogel, S. 1981. *Life in moving fluids: The physical biology of flow.* Princeton: Princeton University Press.

Ward, M. A., and J. P. Thorpe. 1991. Distribution of encrusting bryozoans and other epifauna on the subtidal bivalve *Chlamys opercularis. Marine Biology* 110:253–59.

Wildish, D. J., and D. D. Kristmanson. 1984. Importance to mussels of the benthic boundary layer. *Canadian Journal of Fisheries and Aquatic Sciences* 41: 1618–25.

Winston, J. E. 1978. Polypide morphology and feeding behavior in marine ectoprocts. *Bulletin of Marine Science* 28:1–31.

———. 1979. Current-related morphology and behavior in some Pacific coast bryozoans. In *Advances in bryozoology,* ed. G. P. Larwood and M. B. Abott, 247–67. New York: Academic Press.

———. 1982. Marine bryozoans (Ectoprocta) of the Indian River area (Florida). *Bulletin of the American Museum of Natural History* 173:99–176.

———. 1985. Life history studies of *Disporella* and *Drepanophora* in Jamaica. In *Bryozoa: Ordovician to Recent,* ed. C. Nielsen and G. P. Larwood, 350. Fredensborg, Denmark: Olsen and Olsen.

Winston, J. E., and E. Håkansson. 1986. The interstitial bryozoan fauna from Capron Shoal, Florida. *America Museum Novitates* 2865:1–50.

Part 2

Recognition of Species, and the
Tempo of Speciation and Extinction

Recognizing Coral Species Present and Past 4

NANCY KNOWLTON AND ANN F. BUDD

Introduction

Species are a fundamental unit of biodiversity because they represent distinct and independent lineages, often with characteristic ecological requirements, life histories, and physiologies. In some groups they are also morphologically well defined, so that recognizing species is relatively easy. Mayr (1963), for example, estimated that only 5% of birds belong to problematic complexes of sibling species. Although more recent analyses have increased this figure somewhat, it remains the case that a marine biologist or paleontologist with a pair of binoculars and a field guide can often correctly identify the hummingbirds of Panama. The corals of Panama are, however, a different story, as there is little agreement on species boundaries in many of the important genera, or their relationships to similar taxa elsewhere.

While the problematic nature of scleractinian coral species is widely recognized (Budd 1990; Veron 1995; Wallace and Willis 1994; Willis et al. 1997), there is no consensus as to why this problem exists. Veron (1995) has argued that species boundaries in corals may not be clear-cut because of extensive and complex patterns of hybridization. The relatively simple morphology of corals may make developmental catastrophes less likely in hybrids, and the fact that many different coral species spawn at the same time ("mass spawning") may give their gametes numerous opportunities for interspecific fertilizations (Veron 1995). Alternatively, species may be quite discrete in terms of their breeding biology, but exhibit much overlap in the characters traditionally used to tell them apart. Phenotypic plasticity (Willis 1985), morphological stasis (Potts et al. 1993; Budd, Johnson, and Potts 1994), slow rates of molecular evolution (Romano and Palumbi 1996, 1997; van Oppen et al. 1999), and relatively recent origins (Palumbi 1994; Budd, Stemann, and Johnson 1994) combined with long generation times (Potts 1984) could all contribute to the difficulty of identifying reproductively discrete groups. In any case, the problem of recognizing species gets even worse as one moves from sympatric morphotypes to allopatric populations (e.g., Johnson 1991) because there is no commonly accepted criterion for species status in allopatry (Cracraft 1989; Knowlton and Weigt 1997; Knowlton 2000).

Although enthusiasm for Veron's reticulate view is widespread, the data needed to test it are largely lacking. This is worrisome, because the plant literature is rife with supposed examples of ancient or ongoing hybridization that have turned out to be false or overstated upon closer investigation (Howard et al. 1997; Rieseberg 1997). The best support comes from the west Pacific. Several studies (Willis et al. 1997; Hatta et al. 1999) have shown that fertilization success is often high in the laboratory between conventionally defined species in the genera *Acropora*, *Montipora*, and *Platygyra*. In *Acropora*, patterns of chromosome numbers (Kenyon 1997), multiple, highly distinctive ITS sequences within nominal species (Odorico and Miller 1997), and phylogenetic analyses (Hatta et al. 1999) suggest a possible history of past hybridization. In *Platygyra*, distinct morphotypes show little or no evidence for either genetic differentiation or barriers to interbreeding (Miller 1994; Miller and Babcock 1997; Miller and Benzie 1997). On the other hand, despite the suggestive nature of the data for these corals, no genetic evidence for routine hybridization in the field exists (e.g., abundant F_1s), the viability and fertility of F_2s remain to be demonstrated, and shared ancestral polymorphisms are an alternative explanation for some of the genetic patterns. For Caribbean corals, there are almost no data of the type needed to examine this problem.

Montastraea annularis as a Model System

Montastraea annularis sensu lato provides an important example for exploring species boundaries in Caribbean corals. This coral is the dominant reef builder of the region (Goreau 1959) and has been so for millions of years (Budd, Stemann, and Johnson 1994). As such, it has been a model organism for a variety of topics, including phenotypic plasticity, coral bleaching, stable isotopes, and symbiosis (e.g., Graus and Macintyre 1976; Fairbanks and Dodge 1979; Dustan 1982; Porter et al. 1989; Szmant and Gassman 1990; Fitt et al. 1993; Dunbar and Cole 1993; Gleason and Wellington 1993; Rowan and Knowlton 1995; Rowan et al. 1997).

For decades, *M. annularis* sensu lato was considered the archetypal generalist coral (Connell 1978) with a distribution ranging from the intertidal to over 80 m (Goreau and Wells 1967). The extensive variability in colony morphology (heads, columns, and plates) exhibited over this depth range was believed to be an adaptive response to differing light levels (Graus and Macintyre 1976), despite the apparent absence of intermediates (Graus 1977). More recently, however, a number of features have been found to covary with the different types of colony morphologies, including allozymes, aggressive behavior, ecology, growth rate, life history, corallite morphometrics, and stable isotopes (Tomascik 1990; Knowlton et al. 1992; Van Veghel and Bak 1993).

Weil and Knowlton (1994) consequently resurrected two previously synonymized species (*M. faveolata* and *M. franksi*) making, together with *M. annularis* sensu stricto, a total of three species in the complex. This decision was based on the widely accepted principle of concordance (Avise and Ball 1990), namely that a broad array of traits would not consistently covary if reproductive barriers between taxa were absent. Van Veghel and Bak (1993) also found comparable differences among the three morphotypes in Curaçao, whose reefs lie about 1,000 km from those of Panama, but argued that species-level designation was inappropriate without fixed, diagnostic differences. Szmant et al. (1997) also pointed out that many colonies in the northern Caribbean do not easily fit into the categories defined by Weil and Knowlton (1994), although no genetic data or morphological analyses are available to determine whether such colonies are likely to be hybrids or undescribed taxa.

Nonmorphological studies have often provided important early clues to the existence of unrecognized sibling or cryptic species (Lang 1984). Ultimately, however, morphological characters will be needed to recognize these taxa in the fossil record and to reconstruct their origins spatially and temporally. Below we review the nonmorphological evidence supporting the species described by Weil and Knowlton (1994), and then present results of our recent attempts to distinguish them morphologically.

Nonmorphological Approaches

Reproductive Biology

Differences in reproductive biology provide important characters in their own right, as well as possible clues to the underlying cause of genetic barriers between cryptic species. Such evidence falls into two broad classes: differences in the timing of reproduction, and fertilization or early developmental failures (which are separate in mechanism but often difficult to distinguish in practice).

Members of the *M. annularis* complex, like many other major reef builders (Harrison et al. 1984), spawn on just a few nights of the year. At many sites, spawning occurs 7–8 days after the full moon in August, or in September if the August full moon is very early (Gittings et al. 1992;

Knowlton et al. 1997; Szmant et al. 1997). *Montastraea* spawns 1 month earlier in Bermuda (Wyers et al. 1991) and 1 month later in Curaçao (Van Veghel 1994), perhaps due to local differences in the annual temperature cycle (Van Veghel 1994). Nevertheless, at a variety of sites (e.g., Panama, Curaçao, and Florida), the three taxa show considerable overlap in the dates of peak spawning.

Across the Caribbean, spawning generally occurs 1.5–4.5 hours after local sunset time. However, in Panama (Knowlton et al. 1997 and unpublished data), the Florida Keys (Szmant et al. 1997), and the Texas Flower Gardens (Hagman, Gittings, and Deslarzes 1998), there is a consistent 1–2-hour difference in the time of spawning between *M. franksi* and the other two taxa. Although Van Veghel (1994) reported simultaneous spawning by the three taxa in Curaçao, recent observations (Van Veghel and Knowlton, unpublished data) suggest that *M. franksi* spawns earlier than the other two taxa at this site as well. The time of spawning can be accelerated in all three taxa by providing an artificially early sunset in the laboratory (Knowlton et al. 1997 and unpublished data from Panama), a phenomenon also reported for some Pacific corals (Babcock 1984; Hunter 1989; Harrison 1989).

A difference of 1–2 hours in spawning time should dramatically reduce the potential for gene exchange in the field, despite long sperm life spans under laboratory conditions. For example, sperm of *M. franksi* could either move off the reef or become too dilute to be effective by the time eggs of the other species are released. In support of this, Oliver and Babcock (1992) found that fertilization of previously collected eggs exposed to water taken from over the reef was high immediately after conspecifics spawned in the field but dropped sharply for water collected from the same site 2 hours later. Dilution can also decrease the longevity of sperm (Levitan et al. 1991), so that laboratory estimates of longevity at high concentrations may not be relevant to naturally occurring situations.

Synchronous spawning, such as that exhibited by *M. annularis* and *M. faveolata*, does not guarantee cross-fertilization. The importance of chemical blocks to fusion between eggs and sperm of different species has been documented in both echinoderms (Metz et al. 1994) and mollusks (Swanson and Vacquier 1998), and may play a major role in the evolution of new species in many groups that spawn their gametes into the water column (Palumbi 1994). Chemicals that attract sperm to eggs also have the potential to diverge and promote speciation (Coll et al. 1994).

Preliminary studies of *Montastraea* fertilization under laboratory conditions in Panama (Knowlton et al. 1997; Levitan and Knowlton, unpublished data) suggest that *M. annularis* and *M. franksi* cross-fertilize readily under laboratory conditions, but that *M. faveolata* may not cross well with the other two taxa, particularly *M. annularis*. Thus the least successful crosses are between the two taxa that spawn simultaneously. As with many broadcast spawners (Heyward and Babcock 1986), crosses using gamete bundles from the same colony (selfing) yielded few successful fertilizations (Knowlton et al. 1997; Levitan and Knowlton, unpublished data).

Reproductive barriers between *M. faveolata* and *M. franksi* were also found by Hagman, Gittings, and Vize (1998) on the Texas Flower Garden reefs; thus studies from at least two widely separated locations indicate that *M. faveolata* may be limited in its potential to cross-fertilize with the other two taxa. In contrast, Szmant et al. (1997) reported considerable cross-fertilization in all possible combinations (as well as failures to fertilize within taxa) in corals from the Florida Keys. It is difficult to know whether the differences among studies stem from actual biological differences between regions or differences in technique.

Symbiotic Associations

Like all reef-building corals, *Montastraea* hosts dinoflagellate symbionts (termed "zooxanthellae") of the genus *Symbiodinium*. The discovery that these and other corals can host different types of zooxanthellae (Rowan 1998) complicates interpretation of other differences among coral taxa, because the characteristics of the coral might be influenced by the type of symbiont being hosted. In the Caribbean, there are at least four major clades of zooxanthellae, and their distribution in *Montastraea* appears to be determined by ambient light levels and other less well understood factors, both across the reef and within individual colonies (Rowan and Knowlton 1995; Rowan et al. 1997; Toller et al. in press a, in press b). There are subtle differences among the coral taxa in the patterns of symbiont association, but the four taxa of zooxanthellae have broadly similar depth distributions in all three members of the *M. annularis* complex (Rowan and Knowlton 1995; Rowan et al. 1997; Toller et al. in press b). This suggests that the substantial differences in colony morphology exhibited by the three coral taxa at depths where they co-occur cannot be readily attributed to differences in their symbionts.

Aggressive Behavior

Lang (1971) was the first to document that a certain form of aggressive behavior, termed "extracoelenteric digestion," was associated with species boundaries. When colonies of different species are placed next to each other, the dominant species everts its stomach and digests away the adjacent portions of its neighbor within 24–48 hours (Lang 1973). In the case of *Montastraea*, the three taxa show a linear dominance hierarchy, with *M. franksi* dominant over both other taxa and *M. faveolata* dominant over *M. annularis* (Knowlton et al. 1992; Van Veghel and Bak

1993; Weil and Knowlton 1994). This reaction is highly consistent and provides a quick and dirty field assay for recognizing genetically distinct forms in sympatry.

Biochemical and Genetic Analyses

Genetic analyses have considerable potential to reveal cryptic species, particularly in sympatry, because such differences are difficult to explain in the absence of barriers to gene flow between taxa (Avise and Ball 1990; Avise 1994; Thorpe and Solé-Cava 1994). When several independent genetic markers are consistently associated with particular morphologies, the evidence for cryptic species is compelling.

Earlier work on nine polymorphic allozyme loci from Panama and Curaçao showed that *M. franksi* and *M. annularis* are more closely related to each other (Nei's D of 0.06–0.07) than either are to *M. faveolata* (Nei's D of 0.13–0.26); the latter is distinguished by a nearly fixed difference at one locus (Knowlton et al. 1992; Van Veghel and Bak 1993; Weil and Knowlton 1994). Cluster analysis including populations from both sites generates groupings consistent with morphologically based characterizations (e.g., *M. franksi* from Curaçao is more similar genetically to *M. franksi* from Panama than to other Curaçao taxa; Van Veghel, Weil, and Knowlton, unpublished data).

More recently, DNA-based analyses have been tried (Lopez and Knowlton 1997; Lopez et al. 1999; Medina et al. 1999), typically using gametes as a source of DNA because they are free of symbionts (Szmant 1991). Attempts to find diagnostic sequence differences by targeting specific genes (ITS-1 and ITS-2 of ribosomal DNA, a beta-tubulin intron, and the mtDNA COI gene) have been unsuccessful, although there appear to be minor frequency differences in a few nucleotide positions (Lopez and Knowlton 1997). The failure to find diagnostic differences may be related to very slow rates of molecular evolution. For example, Medina et al. (1999) found only 2.4% divergence in 658 base pairs of COI sequence between *Montastraea cavernosa* and the members of the *M. annularis* complex, with all but one of thirteen individuals of the latter being genetically identical. Given that *M. cavernosa* probably diverged from the lineage leading to *M. annularis* sensu lato approximately 50 million years ago (see below), these data imply an exceptionally slow rate of molecular evolution of less than 0.05% per million years. It is thus not surprising that the members of the *M. annularis* complex, which separated less than 10 million years ago (Ma; see below), show essentially no divergence.

The generally low level of variability in ITS, beta-tubulin, and COI sequences suggests that approaches that examine a broader portion of the genome might be more successful. Analyses based on amplified fragment length polymorphisms (AFLPs; Vos et al. 1995; Mueller et al. 1996) revealed two markers able to distinguish *M. faveolata* from *M. franksi* and *M. annularis* (Lopez and Knowlton 1997; Lopez et al. 1999). For one of these, the DNA from diagnostic bands was sequenced and primers were designed for the region, allowing the insertions/deletions that distinguish *M. faveolata* from the other two taxa to be identified (Lopez et al. 1999). The differences revealed by analysis of gametes can also be seen when somatic tissues are analyzed. As with allozymes, there were no fixed or nearly fixed differences between *M. annularis* and *M. franksi;* however only twelve of many hundred possible primers have been screened.

Morphology and the Fossil Record

Analyses of Recent Material

Morphology is the traditional tool of systematics, and it is the only tool for identifying species in fossil deposits. Thus studies of the origination and extinction of species and their stability through geologic time require morphological characters. Nevertheless, use of morphology in scleractinian systematics has been hampered by (1) a general shortage of morphological characters, (2) a lack of discrete morphological characters (many consist of quantitative measurements or counts), and (3) high phenotypic plasticity, both documented (Foster 1979; Willis 1985; Bruno and Edmunds 1997) and inferred. Most of the classic monographs have distinguished species using a few imprecisely defined measurements or counts (e.g., in *Montastraea;* see Vaughan 1919; Veron et al. 1977), and many workers are unwilling to accept species that overlap morphologically, either in single characters or in character combinations (e.g., Best et al. 1984; Riegel and Piller 1995). This perspective dates back at least to Wood-Jones in the early 1900s, who stated that "from the study of the life of the colony in different surroundings, and from repair of injury, and death, in unsuitable habitats, I think it will be seen that the number of the true species of corals is by no means so great as is at present supposed . . . In very many cases one single colony could be found to provide several types of growth, that if presented as fragments, would be deemed to merit individual description of species" (Wood-Jones 1907, 554–55). The end result in some cases has been lumping of clearly distinct taxa (e.g., Zlatarski and Estalella 1982).

In the southern Caribbean, *Montastraea annularis, M. faveolata,* and *M. franksi* can be visually distinguished in the field using characters related to overall colony shape and the colony growing edge. However, initial morphological analyses of corallite measures revealed no single diagnostic difference among the three species in three key characters that have traditionally been used to distinguish species of *Montastraea:* number of septa per corallite, calice diameter, and calice spacing (Knowlton et al. 1992; Weil and Knowlton 1994). Univariate analyses of variance and canonical discriminant analysis of traditional measurements and counts did show statistically significant differences among the species in both Panama (Weil and Knowlton 1994) and Curaçao (Van Veghel and Bak 1993). Nevertheless, because the distributions of these data overlap among species, specimens cannot be identified with complete confidence using these methods.

Inspired by the pioneering analyses of Cheetham (1986, 1987), we are beginning to develop statistical protocols that use new, more refined and biologically meaningful morphological characters that are evenly sampled across each colony. Our initial efforts have focused on capturing three-dimensional landmark data on corallite surfaces using a Reflex microscope, and exploring a variety of different geometric approaches to analyze the data (Budd, Johnson, and Potts 1994; Budd and Johnson 1996; Johnson and Budd 1996). In these analyses, three-dimensional Cartesian coordinates have been obtained for 34 landmarks (fig. 4.1A) on calices of 21 colonies (5 *M. annularis*, 7 *M. faveolata*, and 9 *M. franksi*). Six calices were digitized on samples from the top, middle, and edge of each colony. Size and shape coordinates (Bookstein 1991) were calculated for selected triplets of landmarks using a program for three-dimensional landmark analysis written by K. G. Johnson (Budd, Johnson, and Potts 1994). Centroid size was used to estimate calice size and spacing; shape coordinates were determined by studying triangles formed by selected triplets of landmarks (see Budd, Johnson, and Potts 1994 for details).

The size and shape coordinate data were then analyzed using two multivariate statistical procedures, cluster analysis and canonical discriminant analysis, and the digitized colonies were grouped into species by splitting to the highest levels of statistical significance. This approach follows Jackson and Cheetham (1990, 1994), who used breeding experiments and protein electrophoresis to confirm that morphospecies of bryozoans defined by splitting to the highest significance levels are genetically distinct. To select the size and shape coordinates that were included in cluster analyses, we explored the data distributions for outliers and correlations among characters. The size and shape coordinate data were then used to calculate Mahalanobis distances (Klecka 1980; Marcus 1993a) between samples, and the resulting distance matrix was analyzed using average linkage cluster analysis.

The resulting cluster dendrogram (fig. 4.2A) clearly shows three distinct groups that generally match independent field identifications of the three species based on overall colony shape. Canonical discriminant analysis indicates that the most important variables in discriminating species consist of characters related to the elevation and thickness of the septa

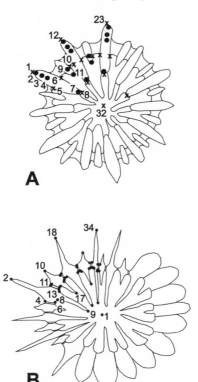

Fig. 4.1. Schematic diagrams showing the locations of landmarks whose Cartesian coordinates have been digitized. The landmarks consist of spatially homologous points selected to capture information about the structure and relief of the wall, septa, and columella (see Budd and Johnson 1996 for further details). *A*, 14 of 34 landmarks digitized on calices in three dimensions using a Reflex microscope. Centroid size was calculated by summing the squared differences between landmarks 1, 32, and 33 to a common centroid, and between landmarks 32, 33, and 34 to a common centroid. Shape coordinates were calculated for seven triplets of landmarks: primary septum elevation, 1-3-32; calical platform shape, 1-7-32; primary septum length, 1-8-32; costa height, 1-2-32; tertiary septum development, 9-10-11; tertiary costa development, 3-11-14; wall development, 4-5-6. *B*, 12 of 45 landmarks digitized on corallites in two dimensions using thin sections. Size and shape coordinates were calculated for 12 selected landmarks (1, 2, 4, 6, 8, 9–11, 13, 17, 18, 34) with landmarks 2 and 9 designated as the baseline.

and the shape of the septal margin (fig. 4.3A–C). The groups corresponding with *M. annularis* and *M. franksi* are the most similar, as was found in the genetic analyses described above. Interestingly, corallites from *M. franksi* consistently cluster with corallites from the sides of *M. annularis* columns. This pattern may reflect the fact that these corallites have similar, slow growth rates: *M. franksi* grows slowly throughout the colony,

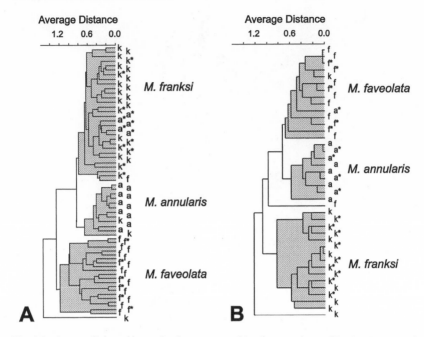

Fig. 4.2. Average linkage cluster dendrograms resulting from analyses of Bookstein size and shape coordinates. Each branch of the two dendrograms represents one sample consisting of six corallites on the top, middle, or edge of 21 colonies identified using field observations (5 colonies of *M. annularis,* 7 colonies of *M. faveolata,* and 9 colonies of *M. franksi*). Asterisks indicate samples of colony edges; *a, k,* and *f* refer to field identifications made respectively for *M. annularis, M. franksi,* and *M. faveolata*. *A,* Results based on three-dimensional analyses of the 14 landmarks shown in figure 4.1A (after Budd and Johnson 1996). *B,* Results based on two-dimensional analyses of the 12 landmarks shown in figure 4.1B.

while the sides of colonies of *M. annularis* columns are nearly senescent. This would suggest that, in these genetically similar corals, corallite morphology may converge when growth rates are similar.

Applying morphometric analyses of three-dimensional landmark data is difficult in fossil material, because many key diagnostic features in three dimensions are worn or recrystallized. Therefore, we conducted a pilot study in which 45 landmarks (fig. 4.1B) were digitized in two dimensions on thin sections of six corallites from the tops and edges of the same 21 colonies of the *M. annularis* complex as above. Size and shape coordinates were calculated for 11 selected landmarks (fig. 4.1B) using the Unigraph 3 computer program (Marcus 1993b). As in the three-dimensional analyses, Mahalanobis distances were calculated using centroid size and nine pairs of shape coordinates, and entered into an average linkage cluster analysis. Like the three-dimensional analyses, three clearly distinct groups were detected (fig. 4.2B); however, *M. faveolata* and *M. annularis*

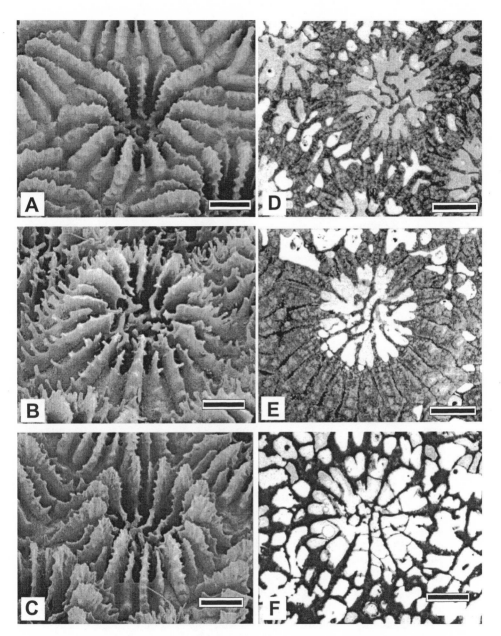

Fig. 4.3. Calical surfaces and transverse thin sections of the three species. SEM photos of *A, Montastraea annularis* (NK465); *B, M. franksi* (NK95-3); *C, M. faveolata* (NK428). Thin sections of *D, M. annularis* (NK464); *E, M. franksi* (NK426); *F, M. faveolata* (NK490). Septa of *M. annularis* and *M. franksi* slope gently from the wall to the columella, and their corallite walls are formed by coalescing septa and costae ("septothecal"). The septa and wall are thinner and more regular in *M. annularis* than in *M. franksi*. In contrast, septa of *M. faveolata* are noticeably more elevated and drop steeply from the wall to the columella. The corallite walls of *M. faveolata* are formed by dissepiments ("parathecal"). *Scale bar* = 1 mm.

appear more closely related to one another than to *M. franksi*. Thus, the new thin-section characters are effective at distinguishing species in fossil material, but perhaps less so at revealing their fine-scale evolutionary relationships.

Landmark analyses will never reveal evolutionary relationships accurately, however, if the landmarks are nonhomologous. Recent analyses of wall structure in the *M. annularis* complex (fig. 4.3D–F) suggest that there are important differences in the components making up the wall between *M. faveolata* and the other two taxa. In *M. faveolata,* the septa usually do not thicken significantly, and the wall is formed primarily by dissepiments. In contrast, in *M. franksi,* the wall is usually formed by septal thickening, and thick costae extend variably between adjacent corallites. In *M. annularis,* the wall structure is similar to *M. franksi,* except the costae are less variable and not as well developed. These observations are preliminary (data from appropriate outgroups are lacking), but they suggest that evolutionary relationships assessed from thin-section characters may match the results of molecular and three-dimensional morphological analyses once the landmarks are homologous.

The Fossil Record of Caribbean *Montastraea*

Comparably detailed analyses have not yet been applied to fossil material. However, cluster and canonical discriminant analysis of linear measurements and septal counts for Miocene to Pleistocene *Montastraea* (Budd 1990; Budd 1991; Budd et al. 1992; Budd, Johnson, and Potts 1994; Budd and Johnson 1997) reveal at least 16 apparent species of *Montastraea* from Caribbean localities over the past 50 million years. Preliminary phylogenetic reconstructions suggest that the genus consists of two distinct evolutionary clades that diverged during Eocene time. Clade I contains at least ten species with <36 septa per corallite, and clade II at least six species with >36 septa per corallite (Budd 1991; http://porites.geology.uiowa.edu). The former includes the *M. annularis* complex (~5.6–0 Ma) and the *M. limbata* complex (16.2–1.5 Ma), and the latter includes the *M. cavernosa* complex (8.3–0 Ma) (fig. 4.4).

The oldest occurrences of the *M. annularis* complex are of *M. faveolata* and *M. franksi* in the Seroe Domi Formation of Curaçao in horizons whose age dates range from 5.6 to 3 Ma. The next oldest occurrences of the complex are for these two species in the Pinecrest Sandstone of Florida (3.5–3 Ma) and the Buenos Aires reef trend of the Limón Group of Costa Rica (3.2–2.9 Ma). The oldest occurrences of the third species in the complex, *M. annularis,* are also in the Seroe Domi Formation but are higher in the section and have age dates of 2.6–2 Ma. The next oldest occurrences of *M. annularis* are in the Bahamas Drilling Project cores at

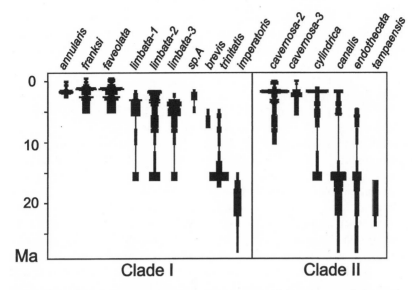

Fig. 4.4. Spindle diagram showing the stratigraphic ranges of the 16 species of *Montastraea* known from the late Oligocene to Recent. The width of each line is proportional to the number of localities (total = 112) in which the species occurred. Occurrence data are based on Budd and Johnson 1997 and Budd 1991. The *M. annularis* complex arose at the onset of turnover between 5.6 and 3 Ma and overlapped with the *M. limbata* complex.

intervals with age dates of 1.8–1.6 Ma. These data suggest that *M. faveolata* and *M. franksi* arose early during the evolution of the complex, whereas *M. annularis* appeared a million or more years later (fig. 4.4). This temporal pattern of divergence agrees with the genetic results above, and indicates that the distinctiveness of these morphologies has been maintained for millions of years. Furthermore, in Curaçao, the three taxa can be recognized (together with an additional, now extinct form) from 125,000-year-old terraces. Their local distribution patterns broadly resemble those found today (Pandolfi, Jackson, and Geister, this volume; Pandolfi and Jackson 2001), suggesting long-term constancy of ecological roles as well.

Conclusions

As the previous summary indicates, many years have been devoted to resolving the species controversy in the *M. annularis* complex. In the southern Caribbean (i.e., Panama and Curaçao) evidence for the three taxa is relatively strong, although we still have no unambiguous genetic tools for distinguishing *M. annularis* and *M. franksi*. In the northern Caribbean, numerous forms that are difficult to place make the picture less clear-cut: are they hybrids or unique taxa that do not occur farther south? The persistence of such uncertainties is unsettling, considering the effort that has been invested in trying to resolve the nature of these taxa and their ecological importance.

Species boundaries in many other Caribbean taxa are even less clear, in part because they are largely unstudied. Table 4.1 summarizes various species and forms whose status is debatable. Taxonomic controversies swirl around many of the dominant reef-building genera of the Caribbean, but in the absence of a "solution" to the *Montastraea* problem, it is no wonder that scientists and funding agencies are reluctant to become involved. Many of the forms listed in table 4.1 may turn out to be good, conventional species, and Veron (1995) himself argued that the limited geographic scale of the Caribbean would be less likely to promote evolutionary reticulation and its attendant problems for species recognition.

In the Pacific, Veron's idea (1995) that species are rather arbitrary units due to widespread hybridization has changed the focus of discussion and research, and mismatches between morphological, reproductive, and genetic species are often interpreted in light of his hypothesis (e.g., Willis

Table 4.1. Modern Caribbean reef coral species whose taxonomic status is uncertain

Species	Reference
Stephanocoenia intersepta (2 morphs)	Goreau and Wells 1967; Foster 1987
Acropora prolifera	Vaughan 1901; Knowlton, pers. observation
"Agaricia" agaricites (>3 morphs), *"A." crassa*	Wells and Lang 1973; Van Moorsel 1983; Stemann 1991; Morse 1993
Agaricia fragilis (2 morphs)	Wells and Lang 1973
Siderastrea siderea (2 morphs)	Budd and Guzman 1994; Knowlton and Budd, pers. observation
Porites astreoides (2 morphs)	Weil 1993; Potts et al. 1993; Budd, Johnson, and Potts 1994
Porites furcata, P. porites	Potts et al. 1993; Budd, Johnson, and Potts 1994
Porites divaricata (2 morphs)	Potts et al. 1993; Budd, Johnson, and Potts 1994
Diploria labyrinthiformis (2 morphs)	Matthai 1928; Johnson, pers. observation
Manicina areolata (>3 morphs), *M. mayori*	Wells 1936; Johnson 1991
Colpophyllia natans (2 morphs), *C. amaranthus*	Matthai 1928; Budd and Johnson 1999
Montastraea cavernosa (>2 morphs)	Lasker 1981; Budd 1993
Meandrina meandrites (2 morphs)	Goreau and Wells 1967
Dichocoenia stokesi (2 morphs)	Matthai 1928; Wells 1973
Eusmilia fastigiata (2 morphs)	Wells 1973

Notes: Some species have fuzzy species boundaries; others may consist of species complexes. Species of *Mycetophyllia* are also debated (Wells 1973).

et al. 1997; Odorico and Miller 1997; Hatta et al. 1999). Because of the enormous diversity that researchers on Pacific corals must confront, however, genetic studies remain preliminary, morphological homologies have not been scrutinized in detail, morphometric analyses such as those pioneered by Cheetham are generally lacking, and species from fossil deposits have not been carefully analyzed using modern morphological techniques. It thus remains unclear whether coral species in the Pacific are fundamentally different in character from their Caribbean counterparts.

Resolving species boundaries in corals is an important task for marine ecology (see discussion in Knowlton 1993; Knowlton and Jackson 1994), evolutionary biology, and paleontology (Budd and Coates 1992). Poorly defined species compromise phylogenetic analyses (Jackson and Cheetham 1994), estimation of rates of molecular evolution depends on recognizing species in fossil deposits, and management of coral reefs requires knowing which taxa are capable of interbreeding. Answering evolutionary questions of fundamental importance, such as whether most morphological change is associated with speciation ("punctuated equilibrium") and whether morphological changes during speciation are random or caused by selection (Cheetham et al. 1993, 1994; Cheetham and Jackson 1995), requires that we be able to show whether morphological and genetic divergence are concordant in living forms (Jackson and Cheetham 1990).

Throughout the tropics, solution of the coral species problem will require a multifaceted approach, drawing on insights from observations on living colonies, combined with genetic analyses and sophisticated morphological studies of the type pioneered and used with such success by Alan Cheetham. Our ongoing studies of *Montastraea* suggest that species boundaries may be difficult to detect both morphologically and genetically, requiring that we split to the highest level of statistical significance in morphological analyses and search widely for diagnostic genetic markers. In this light, negative evidence for differentiation (e.g., Medina et al. 1999) may be a particularly unreliable guide for making decisions about species boundaries in corals.

Acknowledgments

We are especially grateful to K. G. Johnson for discussion of morphometric protocols, to J. Jackson, D. Levitan, J. Lopez, R. Kersanach, H. Guzmán, and R. Rowan for sharing data and ideas, and to J. Jara and J. Maté for help collecting corals. We also thank P. Thorson-Work for collecting three-dimensional landmark data using the Reflex microscope, and C. Colton for digitizing thin sections. This research was supported by grants from the Carver Scientific Research Program at the University of Iowa (to AFB), the U.S. National Science Foundation (to AFB), and the Scholarly Studies Program of the Smithsonian Institution (to NK). Permission was granted by the government of Panamá (Autoridad Nacional del Ambiente, Departamento de Cuarentena Agropecuaria del Ministerio de Desarrollo Agropecuario and Recursos Marinos) and the Kuna Nation for fieldwork, collecting, and exportation.

References

Avise, J. C. 1994. *Molecular markers, natural history, and evolution.* New York: Chapman Hall.

Avise, J. C., and R. M. Ball, Jr. 1990. Principles of genealogical concordance in species concepts and biological taxonomy. In *Oxford surveys in evolutionary biology,* ed. D. Futuyma and J. Antonovics, 7:45–67. Oxford: Oxford University Press.

Babcock, R. C. 1984. Reproduction and distribution of two species of *Goniastrea* Scleractinia from the Great Barrier Reef Province. *Coral Reefs* 2:187–95.

Best, M. B., G. J. Boekschoten, and A. Oosterbaan. 1984. Species concept and ecomorph variation in living and fossil Scleractinia. *Palaeontographica Americana* 54:58–69.

Bookstein, F. L. 1991. *Morphometric tools for landmark data.* Cambridge: Cambridge University Press.

Bruno, J. F., and P. J. Edmunds. 1997. Clonal variation for phenotypic plasticity in the coral *Madracis mirabilis. Ecology* 78:2177–90.

Budd, A. F. 1990. Longterm patterns of morphological variation within and among species of reef-corals and their relationship to sexual reproduction. *Systematic Botany* 15:150–65.

———. 1991. Neogene paleontology in the northern Dominican Republic: 11, The family Faviidae (Anthozoa: Scleractinia): Part 1. *Bulletins of American Paleontology* 101:5–83.

———. 1993. Variation within and among morphospecies of *Montastraea. Courier Forschungs-Institut Senckenberg* 164:241–54.

Budd, A. F., and A. G. Coates. 1992. Non-progressive evolution in a clade of Cretaceous *Montastraea*-like corals. *Paleobiology* 18:425–46.

Budd, A. F., and H. Guzmán. 1994. *Siderastrea glynni,* a new species of scleractinian coral (Cnidaria: Anthozoa) from the eastern Pacific. *Proceedings of the Biological Society of Washington* 107:591–99.

Budd, A. F., and K. G. Johnson. 1996. Recognizing species of late Cenozoic Scleractinia and their evolutionary patterns. *Paleontological Society Papers* 1:59–79.

———. 1997. Coral reef community dynamics over 8 million years of evolutionary time: Stasis and turnover. In *Proceedings of the Eighth International Coral Reef Symposium,* ed. H. A. Lessios and I. G. Macintyre, 1:423–28. Balboa, Republic of Panamá: Smithsonian Tropical Research Institute.

———. 1999. Neogene paleontology in the northern Dominican Republic: The family Faviidae (Anthozoa: Scleractinia): 2, The genera *Caulastraea, Favia,*

Diploria, Manicina, Hadrophyllia, Thysanus, and *Colpophyllia. Bulletins of American Paleontology* 109:5–83.

Budd, A. F., K. G. Johnson, and D. C. Potts. 1994. Recognizing morphospecies in colonial reef corals: 1, Landmark-based methods. *Paleobiology* 20:484–505.

Budd, A. F., T. A. Stemann, and K. G. Johnson. 1994. Stratigraphic distributions of genera and species of Neogene to Recent Caribbean reef corals. *Journal of Paleontology* 68:951–77.

Budd, A. F., T. A. Stemann, and R. H. Stewart. 1992. Eocene Caribbean reef corals: A unique fauna from the Gatuncillo Formation of Panama. *Journal of Paleontology* 66:570–94.

Cheetham, A. H. 1986. Tempo of evolution in a Neogene bryozoan: Rates of morphologic change within and across species boundaries. *Paleobiology* 12:190–202.

————. 1987. Tempo of evolution in a Neogene bryozoan: Are trends in single morphologic characters misleading? *Paleobiology* 13:286–96.

Cheetham, A. H., and J. B. C. Jackson. 1995. Process from pattern: Tests for selection versus random change in punctuated bryozoan speciation. In *New approaches to speciation in the fossil record,* ed. D. H. Erwin and R. L. Anstey, 184–207. New York: Columbia University Press.

Cheetham, A. H., J. B. C. Jackson, and L. A. Hayek. 1993. Quantitative genetics of bryozoan phenotypic evolution: 1, Rate tests for random change versus selection in differentiation of living species. *Evolution* 47:1526–38.

————. 1994. Quantitative genetics of bryozoan phenotypic evolution: 2, Analysis of selection and random change in fossil species using reconstructed genetic parameters. *Evolution* 48:360–75.

Coll, J. C., B. F. Bowden, G. V. Meehan, G. M. Konig, A. R. Carroll, D. M. Tapiolas, P. M. Aliño, A. Heaton, R. De Nys, P. A. Leone, M. Maida, T. L. Aceret, R. H. Willis, R. C. Babcock, B. L. Willis, Z. Florian, M. N. Clayton, and R. L. Miller. 1994. Chemical aspects of mass spawning in corals: 1, Sperm attractant molecules in the eggs of the scleractinian coral *Montipora digitata. Marine Biology* 118:177–82.

Connell, J. H. 1978. Diversity in tropical rain forests and coral reefs. *Science* 199:1302–10.

Cracraft, J. 1989. Speciation and its ontology: The empirical consequences of alternative species concepts for understanding patterns and processes of differentiation. In *Speciation and its consequences,* ed. D. Otte and J. A. Endler, 28–59. Sunderland, MA: Sinauer.

Dunbar, R. B., and J. E. Cole. 1993. *Coral records of ocean-atmosphere variability.* NOAA Climate and Global Change Program Special Report no. 10. Boulder, CO: University Corporation for Atmospheric Research.

Dustan, P. 1982. Depth-dependent photoadaptation by zooxanthellae of the reef coral *Montastrea annularis. Marine Biology* 68:253–64.

Fairbanks, R. G., and R. E. Dodge. 1979. Annual periodicity of the $^{18}O/^{16}O$ and $^{13}C/^{12}C$ ratios in the coral *Montastrea annularis. Geochimica Cosmochimica Acta* 43:1009–20.

Fitt, W. K., H. J. Spero, J. Halas, M. W. White, and J. W. Porter. 1993. Recovery of the coral *Montastrea annularis* in the Florida Keys after the 1987 Caribbean "bleaching event." *Coral Reefs* 12:57–64.

Foster, A. B. 1979. Phenotypic plasticity in the reef corals *Montastraea annularis* and *Siderastrea siderea*. *Journal of Experimental Marine Biology and Ecology* 39:25–54.

———. 1987. Neogene paleontology in the northern Dominican Republic: 4, The genus *Stephanocoenia* (Anthozoa: Scleractinia: Astrocoeniidae). *Bulletins of American Paleontology* 93:5–22.

Gittings, S. R., G. S. Boland, K. J. P. Deslarzes, C. L. Combs, B. S. Holland, and T. J. Bright. 1992. Mass spawning and reproductive viability of reef corals at the East Flower Garden bank, northwest Gulf of Mexico. *Bulletin of Marine Science* 51:420–28.

Gleason, D. F., and G. M. Wellington. 1993. Ultraviolet radiation and coral bleaching. *Nature* 365:836–38.

Goreau, T. F. 1959. The ecology of Jamaican coral reefs: 1, Species composition and zonation. *Ecology* 40:67–90.

Goreau, T. F., and J. W. Wells 1967. The shallow-water Scleractinia of Jamaica: Revised list of species and their vertical distribution range. *Bulletin of Marine Science* 17:442–53.

Graus, R. R. 1977. Investigation of coral growth adaptations using computer modelling. In *Proceedings of the Third International Coral Reef Symposium*, 2:463–69. Miami: Rosenstiel School of Marine and Atmospheric Science, University of Miami.

Graus, R. R., and I. G. Macintyre. 1976. Light control of growth form in colonial reef corals: Computer simulation. *Science* 193:895–97.

Hagman, D. K., S. R. Gittings, and K. J. P. Deslarzes. 1998. Timing, species participation, and environmental factors influencing annual mass spawning at the Flower Garden Banks (northwest Gulf of Mexico). *Gulf of Mexico Science* 16:170–79.

Hagman, D. K., S. R. Gittings, and P. D. Vize. 1998. Fertilization in broadcast spawning corals of the Flower Garden Banks National Marine Sanctuary. *Gulf of Mexico Science* 16:180–87.

Harrison, P. L. 1989. Pseudo-gynodioecy: An unusual breeding system in the scleractinian coral *Galaxea fascicularis*. In *Proceedings of the Sixth International Coral Reef Symposium*, ed. J. H. Choat et al., 2:699–705. Townsville, Australia: Symposium Executive Committee.

Harrison, P. L., R. C. Babcock, G. D. Bull, J. K. Oliver, C. C. Wallace, and B. L. Willis. 1984. Mass spawning in tropical reef corals. *Science* 223:1186–89.

Hatta, M., H. Fukami, W. Wang, M. Omori, K. Shimoike, T. Hayashibara, Y. Ina, and T. Sugiyama. 1999. Reproductive and genetic evidence for a reticulate evolutionary history of mass-spawning corals. *Molecular Biology and Evolution* 16:1607–13.

Heyward, A. J., and R. C. Babcock. 1986. Self- and cross-fertilization in scleractinian corals. *Marine Biology* 90:191–95.

Howard, D. J., R. W. Preszler, J. Williams, S. Fenchel, and W. J. Boecklen. 1997. How discrete are oak species? Insights from a hybrid zone between *Quercus grisea* and *Quercus gambelii*. *Evolution* 51:747–55.

Hunter, C. L. 1989. Environmental cues controlling spawning in two Hawaiian corals, *Montipora verrucosa* and *M. dilatata*. In *Proceedings of the Sixth International Coral Reef Symposium*, ed. J. H. Choat et al., 2:727–32. Townsville, Australia: Symposium Executive Committee.

Jackson, J. B. C., and A. H. Cheetham. 1990. Evolutionary significance of morphospecies: A test with cheilostome Bryozoa. *Science* 248:579–83.

———. 1994. Phylogeny reconstruction and the tempo of speciation in cheilostome Bryozoa. *Paleobiology* 20:407–23.

Johnson, K. G. 1991. Population ecology of a free-living coral: Reproduction, population dynamics, and morphology of *Manicina areolata* (Linnaeus). Ph.D. thesis, University of Iowa, Iowa City.

Johnson, K. G., and A. F. Budd. 1996. Three-dimensional landmark techniques for the recognition of reef coral species. In *Advances in morphometrics*, NATO ASI Series A, ed. L. F. Marcus, M. Corti, A. Loy, G. J. P. Naylor, and D. Slice, 284:345–53. New York: Plenum.

Kenyon, J. C. 1997. Models of reticulate evolution in the coral genus *Acropora* based on chromosome numbers: Parallels with plants. *Evolution* 51:756–67.

Klecka, W. R. 1980. *Discriminant analysis*. Beverly Hills, CA: Sage.

Knowlton, N. 1993. Sibling species in the sea. *Annual Review of Ecology and Systematics* 24:189–216.

———. 2000. Molecular genetic analyses of species boundaries in the sea. *Hydrobiologia* 420:73–90.

Knowlton, N., and J. B. C. Jackson. 1994. New taxonomy and niche partitioning on coral reefs: Jack of all trades or master of some? *Trends in Ecology and Evolution* 9:7–9.

Knowlton, N., J. L. Maté, H. M. Guzmán, R. Rowan, and J. Jara. 1997. Direct evidence for reproductive isolation among the three species of the *Montastraea annularis* complex in Central America (Panama and Honduras). *Marine Biology* 127:705–11.

Knowlton, N., and L. A. Weigt. 1997. Species of marine invertebrates: A comparison of the biological and phylogenetic species concepts. In *Species: The units of biodiversity*, ed. M. F. Claridge, H. A. Dawah, and M. R. Wilson, 199–219. London: Chapman and Hall.

Knowlton, N., E. Weil, L. A. Weigt, and H. M. Guzmán. 1992. Sibling species in *Montastraea annularis*, coral bleaching, and the coral climate record. *Science* 255:330–33.

Lang, J. 1971. Interspecific aggression by scleractinian corals: 1, The rediscovery of *Scolemia cubensis* (Milne Edwards & Haime). *Bulletin of Marine Science* 21:952–59.

———. 1973. Interspecific aggression by scleractinian corals: 2, Why the race is not only to the swift. *Bulletin of Marine Science* 23:260–79.

———. 1984. Whatever works: The variable importance of skeletal and nonskeletal characters in scleractinian taxonomy. *Palaeontographica Americana* 54:18–44.

Lasker, H. R. 1981. Phenotypic variation in the coral *Montastrea cavernosa* and its effects on colony energetics. *Biological Bulletin* 160:292–302.

Levitan, D. R., M. A. Sewell, and F. S. Chia. 1991. Kinetics of fertilization in the sea urchin *Strongylocentrotus franciscanus:* Interaction of gamete dilution, age, and contact time. *Biological Bulletin* 181:371–78.

Lopez, J. V., R. Kersanach, S. A. Rehner, and N. Knowlton. 1999. Molecular determination of species boundaries in corals: Genetic analysis of the *Montastraea annularis* complex using amplified fragment length polymorphisms and a microsatellite marker. *Biological Bulletin* 196:80–93.

Lopez, J. V., and N. Knowlton. 1997. Discrimination of species in the *Montastraea annularis* complex using multiple genetic loci. In *Proceedings of the Eighth International Coral Reef Symposium,* ed. H. A. Lessios and I. G. Macintyre, 2:1613–18. Balboa, Republic of Panamá: Smithsonian Tropical Research Institute.

Marcus, L. F. 1993a. Some aspects of multivariate statistics for morphometrics. In *Contributions to morphometrics,* ed. L. F. Marcus, E. Bello, and A. Garcia-Valdecasas, 95–130. Madrid: Museo Nacional de Ciencias Naturales.

———. 1993b. UNIGRAPH3: Estimation of uniform components. Software available at http://life.bio.sunysb.edu/morph.

Matthai, G. 1928. *A monograph of the recent meandroid Astraeidae.* Vol. 7 of *Catalogue of the Madreporarian corals in the British Museum (Natural History).* London: British Museum (Natural History).

Mayr, E. 1963. *Animal species and evolution.* Cambridge: Harvard University Press.

Medina, M., E. Weil, and A. M. Szmant. 1999. Examination of the *Montastraea annularis* species complex (Cnidaria: Scleractinia) using ITS and COI sequences. *Marine Biotechnology* 1:89–97.

Metz, E. C., R. E. Kane, H. Yanagamachi, and S. R. Palumbi. 1994. Fertilization between closely related sea urchins is blocked by incompatibilities during sperm-egg attachment and early stages of fusion. *Biological Bulletin* 187:23–34.

Miller, K. J. 1994. Morphological species boundaries in the coral genus *Platygyra:* Environmental variation and taxonomic implications. *Marine Ecology Progress Series* 110:19–28.

Miller, K. J., and R. Babcock. 1997. Conflicting morphological and reproductive species boundaries in the coral genus *Platygyra. Biological Bulletin* 192:98–110.

Miller, K. J., and J. A. H. Benzie. 1997. No clear genetic distinction between morphological species within the coral genus *Platygyra. Bulletin of Marine Science* 61:907–17.

Morse, A. N. C. 1993. Unique patterns of substratum selection by distinct populations of *Agaricia humilis* contribute to opportunistic distribution within the Caribbean. In *Proceedings of the Seventh International Coral Reef Symposium,* ed. R. H. Richmond, 1:501–2. Mangilao, Guam: University of Guam Press.

Mueller, U. G., S. E. Lipari, and M. G. Milgroom. 1996. Amplified fragment length polymorphism (AFLP) fingerprinting of symbiotic fungi cultured by the fungus-growing ant *Cyphomyrmex minutus. Molecular Ecology* 5:119–22.

Odorico, D. M., and D. J. Miller. 1997. Variation in the ribosomal internal transcribed spacers and 5.8s rDNA among five species of *Acropora* (Cnidaria:

Scleractinia): Patterns of variation consistent with reticulate evolution. *Molecular Biology and Evolution* 14:465–73.

Oliver, J., and R. Babcock. 1992. Aspects of the fertilization ecology of broadcast spawning corals: Sperm dilution effects and in situ measurements of fertilization. *Biological Bulletin* 183:409–18.

Palumbi, S. R. 1994. Genetic divergence, reproductive isolation, and marine speciation. *Annual Review of Ecology and Systematics* 25:547–72.

Pandolfi, J. M., and J. B. C. Jackson. 2001. Community structure of Pleistocene coral reefs of Curaçao, Netherlands Antilles. *Ecological Monographs* 71:49–67.

Porter, J. W., W. K. Fitt, H. J. Spero, C. S. Rogers, and M. W. White. 1989. Bleaching in reef corals: Physiological and stable isotopic responses. *Proceedings of the National Academy of Sciences, USA* 86:9342–46.

Potts, D. C. 1984. Generation times and the Quaternary evolution of reef-building corals. *Paleobiology* 10:48–58.

Potts, D. C., A. F. Budd, and R. L. Garthwaite. 1993. Soft tissue vs. skeletal approaches to species recognition and phylogeny reconstruction in corals. *Courier Forschungs-Institut Senckenberg* 164:221–31.

Riegel, B., and W. E. Piller. 1995. Variability of calyx characteristics in Recent *Porites:* Implications for fossil identification. In *VII International Symposium on Fossil Cnidaria and Porifera* (Madrid), ed. S. Rodríguez, abstract 72. Madrid: Universidad Complutense.

Rieseberg, L. H. 1997. Hybrid origins of plant species. *Annual Review of Ecology and Systematics* 28:359–89.

Romano, S. L., and S. R. Palumbi. 1996. Evolution of scleractinian corals inferred from molecular systematics. *Science* 231:1528–33.

———. 1997. Molecular evolution of a portion of the mitochondrial 16S ribosomal gene region in scleractinian corals. *Journal of Molecular Evolution* 45:397–411.

Rowan, R. 1998. Diversity and ecology of zooxanthellae on coral reefs. *Journal of Phycology* 34:407–17.

Rowan, R., and N. Knowlton. 1995. Intraspecific diversity and ecological zonation in coral-algal symbiosis. *Proceedings of the National Academy of Sciences, USA* 92:2850–53.

Rowan, R., N. Knowlton, A. Baker, and J. Jara. 1997. Landscape ecology of algal symbionts creates variation in episodes of coral bleaching. *Nature* 388:265–69.

Stemann, T. A. 1991. Evolution of the reef-coral family Agariciidae (Anthozoa: Scleractinia) in the Neogene through Recent of the Caribbean. Ph.D. thesis, University of Iowa, Iowa City.

Swanson, W. J., and V. D. Vacquier. 1998. Concerted evolution in an egg receptor for a rapidly evolving abalone sperm protein. *Science* 281:710–12.

Szmant, A. M. 1991. Sexual reproduction by the Caribbean reef corals *Montastrea annularis* and *M. cavernosa. Marine Ecology Progress Series* 74:13–25.

Szmant, A. M., and N. J. Gassman. 1990. The effects of prolonged "bleaching" on the tissue biomass and reproduction of the reef coral *Montastrea annularis. Coral Reefs* 8:217–24.

Szmant, A. M., E. Weil, M. W. Miller, and D. E. Colon. 1997. Hybridization within the species complex of the scleractinian coral *Montastraea annularis.* *Marine Biology* 129:561–72.

Thorpe, J. P., and A. M. Solé-Cava. 1994. The use of allozyme electrophoresis in invertebrate systematics. *Zoological Scripta* 23:3–18.

Toller, W. W., R. G. Rowan, and N. Knowlton. In press a. Repopulation of zooxanthellae in the Caribbean corals *Montastraea annularis* and *M. faveolata. Biological Bulletin.*

———. In press b. Zooxanthellae of the *Montastraea annularis* species complex: Patterns of distribution of four taxa of *Symbiodinium* on different reefs across depths. *Biological Bulletin.*

Tomascik, T. 1990. Growth rates of two morphotypes of *Montastrea annularis* along a eutrophication gradient, Barbados, W.I. *Marine Pollution Bulletin* 21: 376–81.

Van Moorsel, G. W. N. M. 1983. Reproductive strategies in two closely related stony corals (*Agaricia,* Scleractinia). *Marine Ecology Progress Series* 13:273–83.

Van Oppen, M. J. H., B. L. Willis, and D. J. Miller. 1999. Atypically low rate of cytochrome b evolution in the scleractinian coral genus *Acropora. Proceedings of the Royal Society, London,* series B, 266:179–83.

Van Veghel, M. L. J. 1994. Reproductive characteristics of the polymorphic Caribbean reef building coral *Montastrea annularis:* 1, Gametogenesis and spawning behavior. *Marine Ecology Progress Series* 109:209–19.

Van Veghel, M. L. J., and R. P. M. Bak. 1993. Intraspecific variation of a dominant Caribbean reef building coral, *Montastrea annularis:* Genetic, behavioral, and morphometric aspects. *Marine Ecology Progress Series* 92:255–65.

Vaughan, T. W. 1901. The stony corals of the Porto Rico waters. *Bulletin of the U.S. Fish Commission for 1900* 20:290–320.

———. 1919. Fossil corals from Central America, Cuba, and Porto Rico with an account of the American Tertiary, Pleistocene, and Recent coral reefs. *U.S. National Museum Bulletin* 103:189–524.

Veron, J. E. N. 1995. *Corals in space and time: The biogeography and evolution of the Scleractinia.* Sydney, Australia: UNSW Press.

Veron, J. E. N., M. Pichon, and M. Wijman-Best. 1977. Scleractinia of eastern Australia: 2, Families Faviidae and Trachyphyllidae. *Australian Institute of Marine Sciences Monograph Series* 3:1–233.

Vos, P., R. Hogers, M. Bleeker, M. Reijans, T. V. de Lee, M. Hornes, A. Frijters, J. Pot, J. Peleman, M. Kuiper, and M. Zabeau. 1995. AFLP: A new technique for DNA fingerprinting. *Nucleic Acids Research* 23:4407–14.

Wallace, C. C., and B. L. Willis. 1994. Systematics of the coral genus *Acropora:* Implications of new biological findings for species concepts. *Annual Review of Ecology and Systematics* 25:237–62.

Weil, E. 1993. Genetic and morphological variation in Caribbean and eastern Pacific *Porites* (Anthozoa: Scleractinia): Preliminary results. In *Proceedings of the Seventh International Coral Reef Symposium,* ed. R. H. Richmond, 2:643–56. Mangilao, Guam: University of Guam Press.

Weil, E., and N. Knowlton. 1994. A multi-character analysis of the Caribbean coral *Montastraea annularis* (Ellis and Solander, 1786) and its two sibling species, *M. faveolata* (Ellis and Solander, 1786) and *M. franksi* (Gregory, 1895). *Bulletin of Marine Science* 55:151–75.

Wells, J. W. 1936. The nomenclature and type species of some genera of recent and fossil corals. *American Journal of Science* 31:97–134.

———. 1973. New and old scleractinian corals from Jamaica. *Bulletin of Marine Science* 23:16–55.

Wells, J. W., and J. C. Lang. 1973. Systematic list of Jamaican shallow-water Scleractinia. *Bulletin of Marine Science* 23:55–58.

Willis, B. L. 1985. Phenotypic plasticity versus phenotypic stability in the reef corals *Turbinaria mesenterina* and *Pavona cactus*. In *Proceedings of the Fifth International Coral Reef Congress*, 4:107–12. Moorea, French Polynesia: Antenne Museum–EPHE.

Willis, B. L., R. C. Babcock, P. L. Harrison, and C. C. Wallace. 1997. Experimental hybridization and breeding incompatibilities within the mating systems of mass spawning reef corals. *Coral Reefs* 16, supp.: S53–S65.

Wood-Jones, F. 1907. On the growth-forms and supposed species in corals. *Proceedings of the Zoological Society, London* 518–56.

Wyers, S. C., H. S. Barnes, and S. R. Smith. 1991. Spawning of hermatypic corals in Bermuda: A pilot study. *Hydrobiologia* 216/217:109–16.

Zlatarski, V. N., and N. M. Estalella. 1982. *Les Scléractiniaires de Cuba avec des données sur les organismes associés*. Sofia: Editions de l'Académie bulgare des Sciences.

5 Geologically Sudden Extinction of Two Widespread Late Pleistocene Caribbean Reef Corals

John M. Pandolfi, Jeremy B. C. Jackson, and Jörn Geister

Introduction

In the face of accelerating loss of biological habitats and diversity due to human impacts, we urgently need to understand how and why species become extinct and the effects of extinction on surviving lineages. This is necessarily a mostly historical endeavor because loss of diversity, like the quantity of diversity itself, has proven remarkably difficult to document among recent biota despite enormous loss of habitats across the globe (May 1988, 1990; May et al. 1995). In contrast, the fossil record exhibits much extinction that varies widely in intensity over time (Raup 1991b, 1996; Jablonski 1995).

Paleontological interest in extinction increased enormously in the 1980s, largely due to the bold hypothesis of asteroid impact as the cause of mass extinction at the end of the Cretaceous period (Alvarez et al. 1980). Mass extinctions were relatively unselective ecologically (Jablonski 1995; Smith and Jeffrey 1998), apparently due to the overwhelming magnitude of the forces involved, so that survival or extinction may have been largely a matter of chance (Raup 1991a, 1991b). However, only about 4% of all species extinction occurred during the five major mass extinctions (Valentine et al. 1978; Raup 1991b; Valentine in Rosenzweig 1995, 149). Thus over 95% of all extinctions, including that of the closest extinct relatives of species alive today, occurred during relatively quiet periods of earth history when basic differences in ecology appear to have made a significant difference for survival (Jackson 1974; Jablonski 1995; Johnson et al. 1995; Cheetham and Jackson 1996, 1999). It is this so-called background extinction (Raup and Sepkoski 1982; Jablonski 1986) that is most likely to be informative for understanding the present situation.

Simberloff (1986) differentiated "ultimate" from "proximate" causes of extinction. Ultimate causes are those responsible for a decline to small population size, such as a drop in sea level or the introduction of a superior competitor. In contrast, proximate causes are those that lead to the final killing of the few remaining individuals, including any process that can adversely affect species with very low abundance and restricted geographic distribution (Diamond 1984; Simberloff 1986). Only ultimate causes may be investigated paleontologically, because temporal and spatial resolution are inevitably inadequate to examine population-level processes in small, local populations. McLaren's "ultimate causes" and "immediate causes" (1983) are broader in temporal and geologic scale than Simberloff's causes (1986), and are usefully applied to mass extinctions in the fossil record. Raup (1991a) summarized the search for causes of extinction in the fossil record and concluded that "for none of the thousands of well-documented extinctions in the geologic past do we have a solid explanation of why the extinction occurred" (17). Here we look at species extinction on Pleistocene coral reefs, stressing pre- and postextinction species distribution patterns, rather than the possible causes.

Recent extinction has not been observed on coral reefs despite very extensive global habitat degradation and loss of coral cover. These problems are especially bad in the Caribbean where once enormous stands of *Acropora* and other framework-building corals have virtually disappeared over vast areas (Lewis 1984; Ginsburg 1993; Hughes 1994; Jackson 1991, 1992, 1997). Glynn and De Weerdt (1991) reported the extinction of the fire coral *Millepora boschmai* in the tropical eastern Pacific following an extreme El Niño event, but this species was soon rediscovered (Glynn and Feingold 1992). Perhaps the metapopulation structure of many coral-reef organisms helps to buffer them against habitat destruction up to some unknown threshold value (Jackson et al. 1996; Roberts 1997; Pandolfi 1999). Of course, extinctions of other taxa may have occurred unobserved because we are still very ignorant of the great majority of smaller species (Reaka-Kudla 1996).

The Quaternary history of Caribbean coral reefs was foreshadowed by the turnover, toward the end of the Pliocene, of shallow reefs dominated by small, branching pocilloporid finger corals to reefs dominated by large, rapidly growing elkhorn and staghorn *Acropora* (Johnson et al. 1995; Budd et al. 1996; Budd and Johnson 1997, 1999). The main extinction of the older fauna occurred in an intense pulse between 2 and 1.5 million years ago (Ma), but this was preceded by a more drawn-out burst of coral species origination beginning 3.5 Ma. Altogether about three-quarters of all the Caribbean reef coral species turned over during this period. Since then the composition of Pleistocene coral assemblages has changed little (Budd et al. 1994, 1996; Mesolella 1967, 1968; Jackson 1992), and only five species have become extinct in the Caribbean during the past 1.5 million years (Budd et al. 1994). In this paper we document the geologically rapid extinction of two of these coral species that occurred between 82 and 3 thousand years ago (Ka). The cases are interesting because both species were geographically widespread, and common to abundant through to their final occurrences. Moreover, extinction was clearly selective and in one case may be responsible for striking changes in the morphology and distribution of surviving species.

Pocillopora cf. *palmata*

Pocillopora cf. *palmata* has a distinctive, slender, branching growth form that is characterized by corallites that partly group into verrucae (warts) on the colony surface (Geister 1975, 1977b) (fig. 5.1). Geister's original observations from the 125 Ka terraces at San Andrés and Barbados included colonies up to 40 cm in height and 1 m in diameter, but our subsequent investigations have revealed much larger colonies from the Lesser and Greater Antilles, Curaçao, and Barbados that are up to 10 m in width and over 2 m in height (table 5.1; Pandolfi and Jackson 2001). *Pocillopora* cf. *palmata* ranged throughout most of the oceanic tropical

Fig. 5.1. Large colony of *Pocillopora* cf. *palmata* from Curaçao, illustrating the typical slender branching growth form. This colony is from the 125 Ka Pleistocene sheltered leeward reef-crest environment, where it probably lived in around 7 m water depth. It extends for at least 10 m and is up to 1.5 m high. Hammer is 32 cm high.

Table 5.1. Abundance and distribution of *Pocillopora* cf. *palmata* throughout its biogeographic and ecological range in the Caribbean

Country	Island	Locality	Formation	Age (Ka)	Environment[a]	Abundance	Reference	Comments
Bahamas	San Salvador		Cockburn Town	125	Reef tract	Rare	White and Curran 1995; Greenstein pers. comm.	Rare except for two m-scale stands at N end of quarry
			Cockburn Town	125+	Bank-barrier reef (Ap framework)	Rare	Curran pers. comm.	"2 or 3 large 'heads' present and v. distinctive"
Colombia	San Andrés	Schooner Bight	San Luis	125	Few to <10 m	Single patch of colonies	Geister 1977b	
		Southwest Cove	San Luis	125	Leeward—marginal to lagoonal	Especially frequent	Geister 1977b	"Small monospecific isolated patches"
		NW coast	San Luis	125	Leeward M	Present	Pandolfi unpub. data 1995	
		NW coast	San Luis	125	Leeward Ac	Present	Pandolfi unpub. data 1995	
Netherlands Antilles	Aruba	Cape California	Lower terrace	125–175	Relatively protected	Part of 1 colony	Geister 1977b	Not in situ
	Curaçao	NE coast	Lower terrace	125	Windward reef crest	0.31% of 40-m transect	Pandolfi and Jackson 2001	
		NE coast	Lower terrace	125	Windward reef crest	3.15% of 40-m transect		
		NE coast	Lower terrace	125	Windward reef crest	0.71% of 40-m transect		
		NE coast	Lower terrace	125	Windward back reef	Present		
		NE coast	Lower terrace	125	Windward back reef	Present		
		SE coast	Lower terrace	125	Windward reef crest	27% of 40-m transect		
		NW coast	Lower terrace	125	Leeward reef crest	Present		
		NW coast	Lower terrace	125	Leeward reef crest	0.26% of 40-m transect		
		NW coast—central	Lower terrace	125	Leeward reef crest	Present		

Continued on next page

Table 5.1. (continued)

Country	Island	Locality	Formation	Age (Ka)	Environment[a]	Abundance	Reference	Comments
Barbados	NW coast	Maycock's Bay		82	Poorly bedded carbonate sand facies (see James et al. 1971:2012–13)	Numerous colonies	Geister 1977b	Probably no colony deposited in situ; possibly reworked
	NW coast	Hangman's Bay		82	Protected/leeward; semilagoonal	Numerous colonies/small thickets	Geister 1977b	Frequent forms almost monospecific patches and small thickets
	NE coast	Pie Corner		82	Quiet back-reef	16 broken pieces	Geister 1977b	
	NW coast	Maycock Fort		104	Lagoonal sands—back-reef of leeward reef complex	Numerous fragments in quarry	Geister 1977b	No in situ colonies
	NW coast	Hangman's Bay	Highest terrace	104	Lagoonal sands—back-reef of leeward reef complex	Few broken pieces in weathered surface rubble; 2 in situ colonies	Geister 1977b	
	SE coast	Bentom Bottom		125 (111)	Protected back-reef; semilagoonal	1 colony in situ	Geister 1977b	
		Clermont Nose		125	Rear zone	0.30% of 40-m transect	Pandolfi unpub. data 1996.	
		Clermont Nose		125	Ap	0.41% of 40-m transect	Pandolfi unpub. data 1996	
		Clermont Nose		125	Ac	0.29% of 40-m transect	Pandolfi unpub. data 1996	
		Christ Church		125	Ap	0.36% of 40-m transect	Pandolfi unpub. data 1996	
Cayman Islands	Grand Cayman		Ironshore	125	Protected areas behind reef crest; 2–6 m depth	Very common—up to 10-m-diameter colonies	Hunter & Jones 1996	*Pocillapora-Dendrogyra* association (not present in modern)
			Ironshore	125	Protected areas behind reef crest; 2–6 m depth	Very rare		*Montastraea annularis–Isophyllastrea rigida* association

Region	Locality	Formation	Age	Environment	Description	Reference	*Porites porites–I. rigida* association
Dominican Republic	La Romana	Ironshore	125	Protected areas behind reef crest; 2–6 m depth	Scarce; formed large thickets in the 125 Ka reefs	Geister 1982	
Cuba	La Galeta		125	Back-reef/lagoon	Fragments		
	E of Boca Chica		125	Rear of barrier reef	Few fragments		
	Guantanamo Bay		Late Pleistocene	?	Colonies ?	J. W. Wells, pers. comm. to Geister 1976	
Antilles							
Guadeloupe	Grande Terre		125	Adjacent to *Acropora cervicornis* and *Montastraea annularis* facies	Several thickets, each several meters across and 1 m high	Geister unpub. data	
	Pointe d'Antigues		125	Surrounded by *A. cervicornis* and *M. annularis* facies	3-m patch		
	Pointe d'Antigues		125	*M. annularis* zone adjacent to lagoonal facies	Thickets up to 7-m diameter		
	Anse Colas		125	Ap	Isolated branches and several colonies 0.5–1.0-m diameter		
Marie-Galante	Pointe de Tali		125	M	1 patch of several colonies over several square meters		No other *Pocillopora* along several kilometers of coast
Le Désirade	South Point		125	Lagoonal facies	Thickets at least 15-m diameter		Absent in *Acropora palmata* and *Montastraea annularis* zones along rest of south coast
Barbuda Island	Hog Point		125	Back reef	Several large patches, each more than 10-m diameter		Absent along long coastal stretch south of Hog Point

a Environments from Geister 1977a. M = "*annularis*" zone; Ac = *Acropora cervicornis* zone; Ap = *Acropora palmata* zone; WFR = windward fringing reef.

western Atlantic from the Bahamas in the north to Curaçao and Aruba in the south, and from San Andrés and Grand Cayman in the west to Barbados in the east (table 5.1, fig. 5.2). In contrast, it has not been found anywhere on the continent despite extensive and detailed surveys (see fig. 5.2).

There are no other species of *Pocillopora* in the Quaternary fossil record of the Caribbean, and the nearest Caribbean relative to *P.* cf. *palmata* is probably *Pocillopora crassoramosa* from the late Miocene to Pliocene Buff Bay Formation of Jamaica (Budd et al. 1994). The Pleistocene record of Caribbean reefs is largely incomplete from the early Pleistocene about 1.4 Ma until 125 Ka (Budd et al. 1996; Budd and Johnson 1997, 1999), except for subsurface cores in the Bahamas (Budd and Kievman 1993); little-studied outcrops in Curaçao, Cuba, Haiti, and the Dominican Republic; and Barbados, where late Pleistocene terraces extend back to about 600 Ka (Mesolella et al. 1970). The fossil record of *P.* cf. *palmata* ranges from only 125 to 82 Ka, and it is surprising that *P.* cf. *palmata* has no fossil record in Barbados before 125 Ka given its abundance there afterward. It is improbable that an incursion from the eastern Pacific occurred during the Pleistocene since, by all current knowledge, the isthmian barrier was complete long before (Coates and Obando 1996; Cronin and Dowsett 1996). Likewise, we know of no other *Pocillopora* from anywhere in the Atlantic during the Pleistocene to present (Budd et al. 1994). After *P.* cf. *palmata* first appeared in Barbados on the 125 Ka terrace, it also occurred on each of the two subsequent terraces (104 and 82 Ka).

Most occurrences of *P.* cf. *palmata* are in protected shallow water behind wave-resistant reef crests, but we have also found it in high-energy reef crest zones (table 5.1). *P.* cf. *palmata* may be dominant in monospecific patches greater than 10 m across, and we have found large populations on the 125 Ka leeward reef crest of Curaçao (table 5.1; Pandolfi and Jackson 1997 2001), and in Guadeloupe, Marie-Galante, Le Désirade, and Barbuda. Moreover, it characteristically occurs as extremely large colonies on the 125 Ka terrace at Curaçao and the 104 Ka terrace at Barbados (Pandolfi, unpublished data).

In summary, *Pocillopora* cf. *palmata* was common in a variety of shallow-water environments throughout the oceanic Caribbean from 125 to 82 Ka and is absent thereafter, including Holocene deposits. The oldest Holocene reef surface outcrops sampled in the Caribbean are the Rosario Islands and adjacent Colombian coast, and the Lake Enriquillo area of the Dominican Republic, both around 3 Ka. Extinction therefore occurred within no more than about 80 Ka. There is no archeological evidence for human occupation on tropical American islands at this time (Meltzer 1997), so that the extinction was entirely natural.

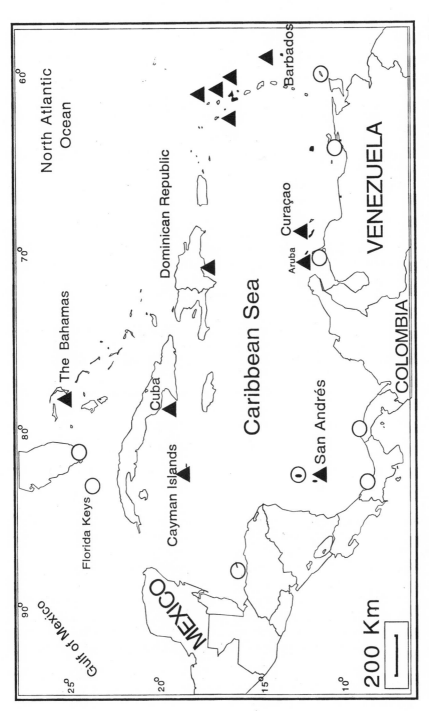

Fig. 5.2. Pleistocene distribution of extinct *Pocillopora* cf. *palmata* throughout the Caribbean Sea. The broad distribution of this species was confined to insular settings. See table 5.1 for abundance patterns at each site. *Black triangles* = sites where *Pocillopora* cf. *palmata* has been found; *open circles* = sites where *Pocillopora* cf. *palmata* has been searched for but not found.

Organ-Pipe *Montastraea*

The extinct organ-pipe *Montastraea* is yet another member of the large species complex of *Montastraea "annularis"* (Knowlton et al. 1992). Three living species within this complex have been recognized so far from depths less than 20 m in Panama, Honduras, and Curaçao (Knowlton et al. 1992; Weil and Knowlton 1994; Knowlton et al. 1997; Knowlton and Budd, this volume), although Van Veghel and Bak (1994) considered the differences to be intraspecific variation. *Montastraea annularis* (sensu stricto) is broadly columnar, *M. faveolata* is massive with a smooth colony surface, and *M. franksi* is massive to platelike with large superficial "bumps" and possesses a more heavily calcified skeleton than the other two species.

There has been much controversy and confusion about the *Montastraea "annularis"* species complex, in part caused by inconsistent criteria for defining species (reviewed in Knowlton and Budd, this volume). Moreover, southern and northern Caribbean corals of approximately similar colony morphology need close examination to determine whether they represent different species. Support for the distinctiveness of the three southern species as described by Weil and Knowlton (1994) includes differences in genetics (Knowlton et al. 1992; Lopez and Knowlton 1997), habitat (Van Veghel and Bak 1994; Rowan and Knowlton 1995), morphology of the calyx (Knowlton et al. 1992; Budd 1993; Budd et al. 1994), and isotopic composition of the skeleton (Knowlton et al. 1992). In addition, the three species support strikingly different proportions of symbiotic dinoflagellate taxa in their tissues (Rowan and Knowlton 1995; Rowan et al. 1997) and spawn at different times of the night during the annual mass-spawning event (Knowlton et al. 1997). These three species have been documented from shallow-water habitats from Panamá and Curaçao. There are at least six additional fossil species in the clade (Ann Budd, pers. comm.), and it seems almost inevitable that more Recent and fossil species will be added as populations from greater depths and from wider geographic areas are studied consistently at the same level of detail. We have found four species of the *M. "annularis"* species complex in our studies of Pleistocene reefs from Curaçao and Barbados (fig. 5.3).

Pleistocene organ-pipe *Montastraea* has a very distinct growth form composed of slender branches, or "pipes," that rise from a common massive base. Both "pipes" and colonies may eventually reach a height of 2–3 m, and colonies are up to 8 m wide (fig. 5.3C). Hoffmeister and Multer (1968, 1491) noted that the mature colonies of the organ-pipe *Montastraea* "can be likened to a huge pincushion from which protrude numerous closely set club-shaped pins." The pipes are thin, with a mean width of 7.8 cm in Barbados (n = 426 branches, s.d. = 2.7 cm).

The extinct organ-pipe form has a very different morphology from other species of the *Montastraea "annularis"* species complex. Recent columnar *M. annularis* usually have much broader columns that diverge from one another throughout colony development (fig. 5.4). These colonies commonly attain diameters and heights of up to 3 m and have

Fig. 5.3. Photographs of colony growth form of the four species of the *Montastraea "annularis"* complex recognized in this study, all from the 125 Ka Pleistocene of Curaçao. *A*, Columnar *M. annularis* from near San Pedro, windward back-reef. *Scale bar* = 3 cm. *B*, Massive *M. faveolata* from near San Pedro, windward back-reef. *Scale bar* = 4 cm. *C*, Undescribed, extinct organ-pipe *Montastraea* from Boca Mansaliña, windward back-reef (3–5-m water depth). This colony illustrates the typical branching growth form. *Scale bar* = 5 cm. *D*, Undescribed sheet *Montastraea* from Punte Halvedag, leeward reef crest. *Scale bar* = 3 cm.

Fig. 5.4. Photographs of modern columnar *Montastraea annularis* from Barbados showing variability in column width. Colonies with broad columns *(top)* and thin columns *(bottom)* characterize populations of this species today, but in the Pleistocene only broad columns predominated. The thinner columns may have developed as the columnar form became more abundant in shallow water after the extinction of the organ-pipe form. Both photos taken in 2-m water depth. Large colony in upper photo is approximately 1.5 m wide. Field of view in lower photo is approximately 1.5 m.

been described throughout the Caribbean, including the Cayman Islands (Hunter and Jones 1996), Jamaica (Dustan 1975; T. Goreau, pers. comm.), Honduras (Knowlton et al. 1997), Panama (Knowlton et al. 1992), Bonaire (Scatterday 1974), and Barbados (Lewis 1960; Pandolfi, unpublished data). Narrow columns more suggestive of the fossil organ-pipe species than typical *M. annularis* can also develop, but their mode of formation is usually different. Such narrow branches usually form as a result of dying back of originally large columns, and they do not extend very far back toward the base of the colony (fig. 5.4). Scatterday (1974) described this process in *M. annularis* from Bonaire. We have received sporadic reports of living organ-pipe colonies since this work began, but so far all of these have turned out to be somewhat narrower than normal columns of *M. annularis*. Of course, it is always possible given the widespread taxonomic confusion in the genus that the organ-pipe may still be alive somewhere in the Caribbean. If so, however, it is vastly rarer than it was for the preceding half million years.

Organ-pipe *Montastraea* had a broad and oceanic Caribbean distribution (fig. 5.5) similar to *Pocillopora* cf. *palmata,* but its true geographic range may have been greater owing to its very recent recognition. For the same reason, we hesitate to declare that the species was exclusively oceanic, although we believe that will turn out to be the case. Organ-pipe *Montastraea* ranges back to the oldest terraces on Barbados, at least 600 Ka, and extends forward to 82 Ka. Its closest relatives are undoubtedly other members of the *M. "annularis"* complex which extend back to the middle Miocene, including three thin-columned species of *Montastraea "limbata"* (Ann Budd, pers. comm.). Phylogenetic analyses have yet to be made, but there is clearly no dearth of possible Caribbean ancestors.

During the past several years we have made quantitative censuses of Pleistocene reef coral communities from many localities around the Caribbean, including Key Largo, Florida; San Andrés, Curaçao; and Barbados (Pandolfi and Jackson 1997, 2001; Greenstein et al. 1998). With the exception of Barbados, all censuses have been taken from the 125 Ka terrace when sea level was 2–6 m higher than present. From these censuses it is clear that organ-pipe *Montastraea* had a very broad ecological range in shallow water, from protected back-reef areas behind barrier reef crests to leeward and windward reef-crest habitats (fig. 5.5, table 5.2). Organ-pipe populations often dominated these different reef habitats, especially in areas of low wave energy. It was the dominant species in the Pleistocene back-reef environment of Key Largo and Curaçao and was codominant with the elkhorn coral *Acropora palmata* in the leeward reef crest of Curaçao and San Andrés. Organ-pipe *Montastraea* was also locally abundant throughout the 500+ Ka Pleistocene history of reef growth in Barbados as described below.

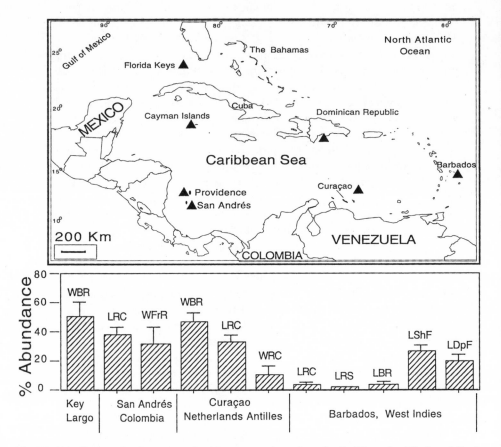

Fig. 5.5. Distribution of extinct organ-pipe *Montastraea* throughout the Caribbean Sea. The broad distribution of this species is so far confined to insular settings. The low number of reported occurrences probably reflects its recent recognition as a separate species, and a broader range is expected as workers become aware of the species. Below the map are estimates of quantitative abundance of organ-pipe *Montastraea* along 40-m transects from the Pleistocene (125 Ka) of Key Largo (Florida, USA), San Andrés and Providencia Islands (Colombia), Curaçao (Netherlands Antilles), and Barbados (West Indies). Note the high abundance of this species relative to the rest of the community in environments ranging from protected back-reef to high-energy windward reef-crest zones. *Black triangles* = sites where extinct organ-pipe *Montastraea* has been found. WBR = windward back-reef; LRC = leeward reef crest (both *annularis* zone and *A. cervicornis* zone of Geister 1977a); WFrR = windward fringing reef; WRC = windward reef crest; LRS = leeward reef slope; LBR = leeward back-reef; LShF = leeward shallow fore-reef; and LDpF = leeward deep fore-reef. See table 5.2 for more information on abundance patterns at each site.

Table 5.2. Abundance and distribution of organ-pipe *Montastraea* throughout its biogeographic and ecological range in the Caribbean

Country	Island	Locality	Formation	Age (Ka)	Environment	Abundance[a]	Reference
Netherlands An-tilles	Curaçao	NE coast 1	Lower Terrace	125	Windward back-reef	7.6	Pandolfi & Jackson 2001
		NE coast 2			Windward back-reef	47.2	
		NE coast 3			Windward back-reef	61.8	
		NE coast 4			Windward back-reef	21.2	
		NE coast 5			Windward reef crest	4.81	
		NE coast 6			Windward back-reef	2.1	
		NE coast 7			Windward back-reef	34.3	
		NE coast 8			Windward back-reef	29.7	
		NE coast 9			Windward reef crest	Present	
		NE coast 10			Windward reef crest	Present	
		NE coast 11			Windward reef crest	Present	
		East coast 1			Windward reef crest	Present	
		East coast 2			Windward reef crest	0.75	
		East coast 3			Windward reef crest	Present	
		East coast 4			Windward back-reef	43.3	
		East coast 5			Windward back-reef	85.1	
		East coast 6			Windward back-reef	71.7	
		East coast 7			Windward back-reef	51.6	
		East coast 8			Windward back-reef	30.3	
		East coast 9			Windward back-reef	80.5	
		East coast 10			Windward back-reef	27.6	
		SE coast 1			Windward reef crest	Present	
		SE coast 2			Windward reef crest	69.2	
		SE coast 3			Windward reef crest	10.6	
		SE coast 4			Windward reef crest	43	
		SE coast 5			Windward reef crest	2.7	

Continued on next page

Table 5.2. (continued)

Country	Island	Locality	Formation	Age (Ka)	Environment	Abundance[a]	Reference
		SE coast 6			Windward reef crest	Present	
		SE coast 7			Windward back-reef	53.1	
		SE coast 8			Windward back-reef	57.7	
		SE coast 9			Windward back-reef	85.3	
		NW coast 1			Leeward reef crest	51.8	
		NW coast 2			Leeward reef crest	40.6	
		NW coast 3			Leeward reef crest	31.2	
		NW coast 4			Leeward reef crest	38.6	
		NW coast 5			Leeward reef crest	20.5	
		NW coast 6			Leeward reef crest	15.3	
		NW coast 7			Leeward reef crest	42.7	
		West coast 1			Leeward reef crest	27.2	
		West coast 2			Leeward reef crest	9.8	
		West coast 3			Leeward reef crest	90.3	
		West coast 4			Leeward reef crest	21.4	
		West coast 5			Leeward reef crest	15.5	
		West coast 6			Leeward reef crest	7.7	
		West coast 7			Leeward reef crest	25.5	
		SW coast 1			Leeward reef crest	36.3	
		SW coast 2			Leeward reef crest	46.2	
United States	Florida[b]	Key Largo 1	Key Largo	125	Patch reef	26.7	Pandolfi & Greenstein, unpub. data 1995
		Key Largo 2				45.7	
		Key Largo 3				71.4	
		Key Largo 4				77.6	
		Key Largo 5				43.1	
		Key Largo 6				0.00	
		Key Largo 7				80.4	
		Key Largo 8				57.9	

Cayman Islands	Grand Cayman		Ironshore	125	"Upper shoreface"	?	Hunter & Jones 1996, fig. 5D
Dominican Republic		E of Boca Chica		125	Montastraea "annularis" zone	?	Geister 1982, fig. 13
Colombia	San Andrés[c]	SW coast 1	San Luis	125	M[c]	34.3	Pandolfi, unpub. data 1995
		SW coast 2			M	51.8	
		SW coast 3			M	43.8	
		SW coast 4			M	48.4	
		SW coast 5			M	69.4	
		SW coast 6			M	61.4	
		SW coast 7			M	62.5	
		SW coast 8			M	64.4	
		SW coast 9			Ac	23.0	
		NW coast 1			M	43.0	
		NW coast 2			M	0.2	
		NW coast 3			M	48.0	
		NW coast 4			M	53.5	
		NW coast 5			M	50.8	
		NW coast 6			M	53.7	
		NW coast 7			M	73.9	
		NW coast 8			Ac	4.8	
		NW coast 9			Ac	8.9	
		NW coast 10			Ac	7.9	
		NW coast 11			Ac	10.5	
		NW coast 12			Ac	0.5	
		SE coast 1			WFR	39.5	
		SE coast 2			WFR	67.2	
		East coast 1			WFR	39.9	
		East coast 2			WFR	3.4	
		East coast 3			WFR	6.6	
	Providencia	South coast	+4 m terrace	125	Seaward of Porites zone	17.0	Pandolfi, unpub. data 1995

Continued on next page

Table 5.2. (continued)

Country	Island	Locality	Formation	Age (Ka)	Environment	Abundance[a]	Reference
Barbados[d]		Clermont Nose		125	Rear zone	5.4	Pandolfi, unpub. data 1996
		Clermont Nose		125	Ap	1.2	
		Clermont Nose		125	Ap	2.0	
		Clermont Nose		125	Ap	6.3	
		Clermont Nose		125	Ac	3.6	
		Clermont Nose		125	Buttress	6.7	
		Clermont Nose		125	Buttress	1.2	
		Clermont Nose		125	Ap	1.3	
		Clermont Nose		125	Buttress	1.8	
		Clermont Nose		125	Ac	1.1	
		Christ Church		125	Ac	3.2	
		Christ Church		125	Ap	0.7	
		Christ Church		125	Buttress	1.8	
		Christ Church		125	Rear zone	1.2	
		Christ Church		125	Ac	0.2	
		Christ Church		125	Coral head	1.1	
		Cobbler's Reef		125?	Ap	2.0	
		Cobbler's Reef		125?	Ac	1.0	
		Cobbler's Reef		125?	Buttress	7.5	
		Cobbler's Reef		125?	Buttress, Ap/buttress	15.2	
		NW Isle		125	Coral head	3.0	
		NW Isle		125	Ap	0.6	
		NW Isle		125	Ap	3.0	
		NW Isle		125	Buttress/Ac	13.9	
		NW Isle		125	Coral head	0.8	

[a] Percentage of corals along 40-m transect.

[b] Reported extensively throughout the Key Largo Formation by Stanley (1966) and Hoffmeister and Multer (1968).

[c] Environments from Geister 1977a. M = "annularis" zone; Ac = Acropora cervicornis zone; Ap = Acropora palmata zone; WFR = windward fringing reef.

[d] Reported extensively throughout the terraces of Barbados by Mesollela (1967) and this paper. Environments correspond to those of James et al. (1977). Buttress = shallow fore-reef; coral head = deep fore-reef.

Temporal Changes in the *Montastraea "Annularis"* Species Complex from Barbados

We chose Barbados to examine trends in late Pleistocene abundance of the four species of the *Montastraea "annularis"* species complex over the past 600 Ka because the record is more complete and better studied there than anywhere else in the Caribbean where late Pleistocene reefs older than 125 Ka can be observed (Mesolella 1967, 1968; Mesolella et al. 1970; Bender et al. 1979; Radtke et al. 1988; Bard et al. 1990; Ku et al. 1990; Taylor and Mann 1991; Gallup et al. 1994). Radiometric age dating reveals reef growth from more than 600 to 82 Ka around more than half of the island, and abundant corals are well preserved throughout so that replicate ecological sampling is possible. Detailed geologic and stratigraphic studies provide a general framework for the relative age and correlation of reef terraces around the island.

Most of Barbados is covered by a series of raised reef tracts represented as geomorphological terraces (fig. 5.6). The majority of the reef-building episodes preserve transgressive reefs that were deposited around the margin of the island during glacio-eustatic sea-level rise. In general, the reefs are youngest at the periphery of the island and progressively older toward the interior. The terraces preserve the original fossil reef geomorphology as well as the original zonation patterns of the corals with respect to water depth (fig. 5.7).

Mesolella (1967) used the distribution and abundance of corals living in Jamaica (Goreau 1959) as a guide to describing zonation patterns in the Pleistocene of Barbados, and in general found excellent correspondence between the Recent and fossil communities (see also Jackson 1992). Interpretations of paleoenvironments and paleodepths were made quasi-independently of the corals using the terrace geomorphology (tops of reef terraces as a datum), physical characteristics of the sediments, and facies relationships (Mesolella 1967, 1968; Mesolella et al. 1970; see Pandolfi et al. 1999 for discussion of paleoenvironmental interpretation in Pleistocene reefs of the Caribbean). Subsequent workers have refined and corroborated these environmental interpretations based on carbonate diagenesis (James 1974), refined age dating (Bender et al. 1979), sequence stratigraphy (Humphrey and Kimbell 1990), and epibionts (Martindale 1992).

The typical coral zonation with increasing depth is as follows (fig. 5.7): shallow reef crest (0–5 m) dominated by *Acropora palmata,* shallow to mid-fore-reef slope (5–20 m) characterized by *Acropora cervicornis,* and deep fore-reef slope characterized by a diverse assemblage of "head corals." Commonly a buttress zone is developed between the shallow reef crest and shallow to mid-fore-reef zones. In 1996 we made a quantitative census of the relative abundance of coral species in each of these environ-

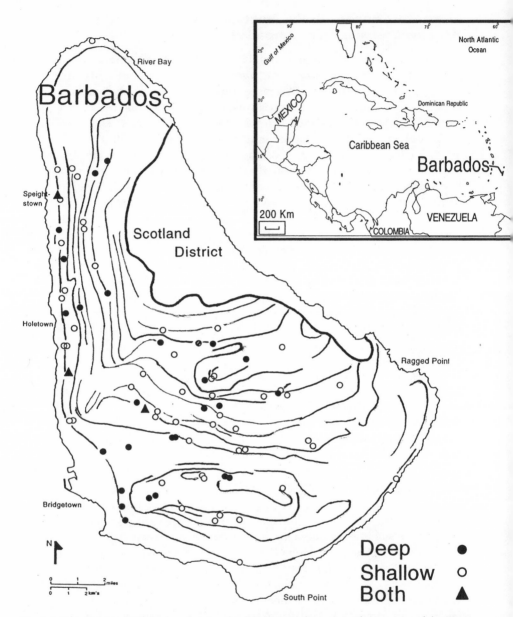

Fig. 5.6. Locality map of Barbados, showing the distribution of transects where species of the *Monta-straea "annularis"* species complex were censused. Corals were censused from 72 transects, each 40 m in length, from two reef environments, the shallow (49 transects) and deep fore-reef (23 transects). *Lines* on larger map are topographic contours.

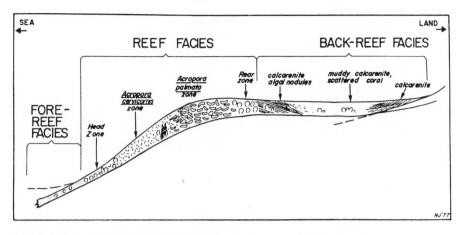

Fig. 5.7. Schematic section showing the environments, water depth, and coral zonation patterns that are characteristic of the raised reef terraces of Barbados (from James et al. 1977).

ments on the 125 Ka terrace at 58 localities around the island (J. Pandolfi, unpublished data). The results supported Mesolella's zonation (1967) based on qualitative observations, which we therefore used as a general framework for our censuses of the relative abundance of the different species in the *Montastraea "annularis"* complex.

We found four late Pleistocene species of the *Montastraea "annularis"* species complex in Barbados (fig. 5.3), all of which are common to abundant throughout the island from roughly 600 to 82 Ka. These are the organ-pipe species, columnar *M. annularis,* massive *M. faveolata,* and a sheetlike species that resembles but is apparently distinct from *M. franksi.* Whether this fourth species is extinct or survives cannot be determined until recent colonies of similar morphology from elsewhere around the Caribbean (e.g., Dustan 1975) have been examined as carefully as those from Panamá and Curaçao.

Methods

We made a quantitative census of all four species of the *Montastraea "annularis"* complex (fig. 5.3) along 72 40-m-long transects. Forty-nine of these transects came from the shallow fore-reef "buttress zone" and twenty-three from the deep fore-reef environment "head coral zone" (terminology of Mesolella 1967) (circles and triangles in fig. 5.6). We chose localities for these transects by three strict criteria to assure that we were sampling reasonably intact communities with minimal transport. (1) We sampled only transgressive reefs. A transgressive reef is built during rising sea level over older sediments, and usually only contains corals that lived during the same reef-building event. Regressive reefs are built during sea-level fall over a previous transgressive reef, and therefore may contain corals that are stratigraphically indistinguishable from the underlying or adjacent transgressive reef. By sampling only transgressive reefs, we minimized the danger of including corals from different time periods in a single assemblage. (2) We sampled only reefs where the top of the geomorphological terrace was exposed to provide a clear sea-level datum with the entire reef zonation below. Exceptions were made only when incomplete sections could be stratigraphically correlated unambiguously with adjacent complete sections. (3) We sampled only where the majority of corals were in upright and growth position, and usually whole. The overwhelming majority of corals in the transects were upright and whole (fig. 5.8).

The age of Pleistocene reefs at Barbados increases with elevation for most reefs that have been dated by radiometric methods. We therefore used elevation as a proxy for age in our analyses of temporal abundance. This allowed us to sample a much greater spatial and temporal range of localities than would have been possible if we had only sampled places where radiometric age dates were available. The assumption of a positive linear relationship between terrace age and elevation inevitably introduces some error due to the complex terrain and geologic history of Barbados, and the dense human settlement on the island that may obscure faults and other geologic complexities. However, because these errors were confined to narrow temporal ranges relative to the total range of the study and involved very few of the transects, we judged them to be small for our purposes of analyses of overall temporal trends.

Censuses were made by laying a 40-m tape across the outcrop and recording the position and length along the transect of all colonies of the four species of the M. "annularis" complex using standard line-intercept transect procedures (Loya 1972, 1978). If there was no colony directly intercepting the transect tape, colonies were recorded if they were within 50 cm above or below the tape. We analyzed our data on a "per meter" basis. We plotted raw (untransformed) abundance per meter against time for the four species combined in each of the two reef environments. Because of the high frequency of zero occurrences, individual species abundance per meter was transformed using the arc sine (square root $(\log(\text{abundance}/m + 1)))$, which provided the closest fit to a normal distribution. The transformed variable for each species in each environment was plotted against time, and a regression line was fitted. We used Spearman's rank correlation coefficient to assess the significance of association between time and raw (untransformed) species abundance per meter of transect for each species and for all species combined for both shallow and deep fore-reef environments.

We also used Spearman's rank correlation coefficient to assess the significance of association of each species pair in each environment, using

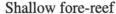

Shallow fore-reef

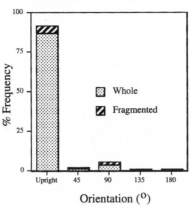

Orientation (^{O})

Deep fore-reef

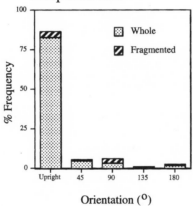

Orientation (^{O})

Fig. 5.8. Histograms of preservational aspects of the corals censused along the 72 transects in the shallow fore-reef *(top)* and deep fore-reef *(bottom)* environments of Barbados. In both environments, the overwhelming majority of censused corals were upright and whole.

the raw (untransformed) species abundance per meter of transect. Results are presented in tabular form.

Results of Censuses

The entire *Montastraea "annularis"* species complex remained abundant and constant (Spearman's $r_s = 0.10$, $p > 0.05$) in shallow fore-reef environments of Barbados for over half a million years, but declined significantly (Spearman's $r_s = 0.73$, $p < 0.0005$) over the same period in deep fore-reef environments (fig. 5.9). In shallow depths, the organ-pipe

Shallow fore-reef

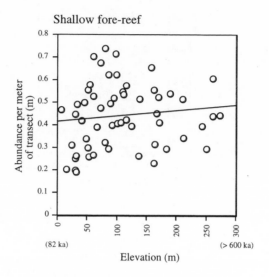

Deep fore-reef

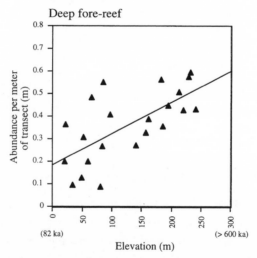

Fig. 5.9. The abundance per meter of all species combined from the *Montastraea* "*annularis*" species complex from the Pleistocene terraces of Barbados. Overall abundance remained the same within the shallow fore-reef environment *(top)*, but declined steadily in the deep fore-reef environment *(bottom)*. Terrace elevation is used as a proxy for time, with lower elevations denoting younger reef-building events.

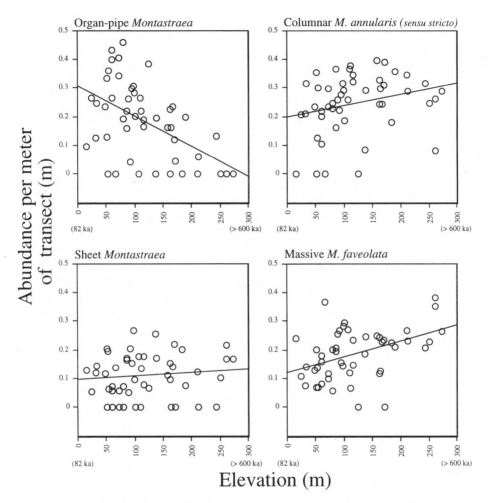

Fig. 5.10. The transformed abundance per meter (arcsin (log $(x + 1))^{1/2}$, where $x =$ abundance per meter of transect) in relation to elevation for each individual species of the *Montastraea "annularis"* species complex. Abundances were measured from the shallow fore-reef environment of the Pleistocene terraces of Barbados. The extinct organ-pipe coral gained in abundance throughout the history of the island, though it may have slightly declined near the end of the record. Columnar *Montastraea annularis* and massive *M. faveolata* decreased through time. Sheet *M. "annularis"* was relatively constant through time, with low abundance compared to the other species. Terrace elevation is used as a proxy for time, with lower elevations denoting younger reef-building events.

Table 5.3. Spearman's rank correlation of species abundance between species pairs in the *Montastraea "annularis"* species complex from the Pleistocene shallow fore-reef of Barbados

	Columnar	Massive	Sheet
Massive	−0.02	—	
	ns		
Sheet	−0.03	0.38	—
	ns	$0.0025 < p < 0.005$	
Organ-pipe	0.06	−0.52	−0.49
	ns	$p < 0.0005$	$p < 0.0005$

Note: ns = not significant.

coral gained significantly in abundance throughout the majority of the interval (Spearman's $r_s = -0.48$, $p < 0.0005$), with a slight fall below 50–100 m (fig. 5.10). In contrast, massive *M. faveolata* decreased steadily in abundance (Spearman's $r_s = 0.41$, $0.001 < p < 0.0025$); columnar *M. annularis* showed an uneven but significant decrease (Spearman's $r_s = 0.33$, $0.01 < p < 0.025$); and the sheetlike species showed consistently low abundance throughout (Spearman's $r_s = 0.13$, $p > 0.05$) (fig. 5.10). Significantly negative pairwise correlations in abundance occur between the organ-pipe species versus the massive and sheetlike species, whereas a significantly positive pairwise correlation in abundance occurs between the massive and sheetlike species (table 5.3).

On the deep fore-reef, abundance of the organ-pipe *Montastraea* (Spearman's $r_s = -0.03$, $p > 0.05$) and massive *M. faveolata* (Spearman's $r_s = 0.28$, $p > 0.05$) remained relatively unchanged throughout the entire interval, whereas columnar *M. annularis* (Spearman's $r_s = 0.46$, $0.01 < p < 0.025$) and the sheetlike species (Spearman's $r_s = 0.40$, $0.025 < p < 0.05$) decreased significantly (fig. 5.11). There was also a significantly negative pairwise correlation between the organ-pipe species and the sheetlike species, and significantly positive correlations between columnar *M. annularis* and both the organ-pipe species and massive *M. faveolata* (table 5.4).

In summary so far, the organ-pipe species was consistently as abundant or more so than any of the other species of the *M. "annularis"* species complex in Barbados for the half million years before its extinction, and the same was true geographically throughout the Caribbean 125 Ka (table 5.2; Pandolfi and Jackson 2001). Just like *Pocillopora* cf. *palmata*, extinction occurred during the roughly 80,000 years between 82 Ka and late Holocene surface outcrops, and was natural. Nothing about the distribution or abundance of organ-pipe *Montastraea* foretold its demise.

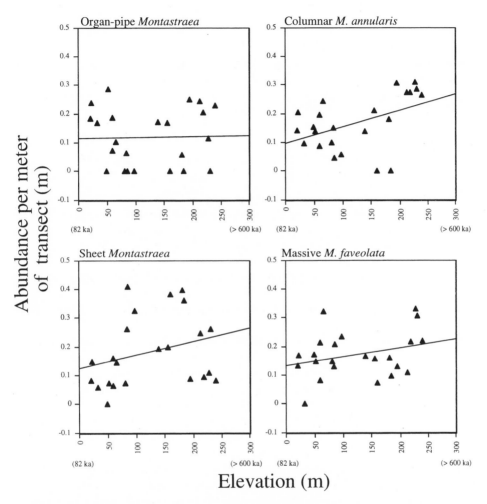

Fig. 5.11. The transformed abundance per meter (arcsin (sqrt(log $(x + 1)$))$^{1/2}$, where x = abundance per meter of transect) in relation to elevation for each individual species of the *Montastraea* *"annularis"* species complex. Abundances were measured from the deep fore-reef environment of the Pleistocene terraces of Barbados. Both the extinct organ-pipe *Montastraea* and the massive *M. faveolata* maintained approximately constant abundance throughout the island's history. However, both columnar *M. annularis* and sheet *M. "annularis"* increased significantly throughout the 600-Ka (300-m) interval. Terrace elevation is used as a proxy for time, with lower elevations denoting younger reef-building events.

Table 5.4. Spearman's rank correlation of species abundance between species pairs in the *Montastraea "annularis"* species complex from the Pleistocene deep fore-reef of Barbados

	Columnar	Massive	Sheet
Massive	0.37	—	
	$0.025 < p < 0.05$		
Sheet	−0.26	0.08	—
	ns	ns	
Organ-pipe	0.51	−0.16	−0.45
	$0.005 < p < 0.01$	ns	$0.01 < p < 0.025$

Note: ns = not significant.

What Happened after the Extinction of Organ-Pipe *Montastraea?*

The next data point after 82 Ka is the late Holocene, so extinction occurred sometime between 82 and around 3 Ka. Two interesting changes occurred among the surviving species of the *M. "annularis"* complex after the extinction of the organ-pipe species. First, shifts in abundance patterns among different reef habitats occurred for all three surviving species. Second, the columns of *M. annularis* decreased significantly in diameter toward the dimensions of the extinct organ-pipe species.

Shallow coral communities in the Pleistocene leeward reef-crest environment in Curaçao were dominated by *Acropora palmata* and the organ-pipe *Montastraea,* whereas columnar *M. annularis* was much less abundant (fig. 5.12). The organ-pipe species was also more abundant than all the other species of the *M."annularis"* complex in the Pleistocene windward reef-crest and windward back-reef environments at Curaçao (Pandolfi and Jackson in press). In contrast, recent reefs at Curaçao exhibit a strikingly different pattern (fig. 5.12; Van Veghel and Bak 1994). Although the data were collected differently, columnar *M. annularis* and massive *M. faveolata* are clearly much more abundant in Recent shallow-water environments now than they were 125 Ka in the Pleistocene. Likewise, the sheetlike species of the *M."annularis"* complex in Barbados declined dramatically in abundance in the deep fore-reef environment for 500 Ka (fig. 5.11), but sheetlike species are the dominant form of *M. "annularis"* today in deeper water throughout the Caribbean (Goreau 1959; Graus and Macintyre 1976; Dustan 1975; Bak and Luckhurst 1980). Thus both columnar and massive species appear to have filled the ecological space left vacant in shallow water by the extinction of the organ-pipe species, and platelike species have correspondingly increased in deeper water.

We measured the diameter of the columns of Pleistocene organ-pipe *Montastraea* and living *M. annularis* from Barbados, and living

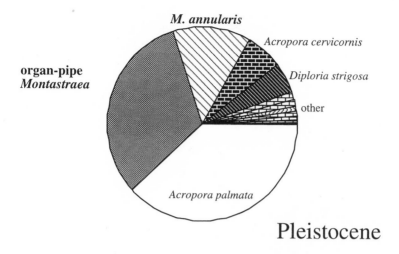

Pleistocene

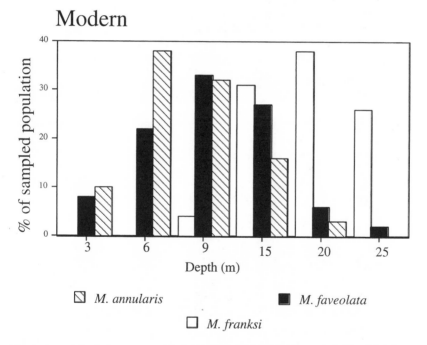

Fig. 5.12. Contrast between Pleistocene and modern species abundances from the shallow leeward reef of Curaçao. In the Pleistocene *(top)*, the leeward reef crest is dominated by both *Acropora palmata* and the extinct organ-pipe *Montastraea*. Columnar *Montastraea annularis* is much less abundant than the organ-pipe form, and the massive *M. faveolata*, while present, is not abundant. In contrast, living reefs of Curaçao are characterized by abundant columnar and massive forms of *Montastraea* in similar shallow-water environments *(bottom)*.

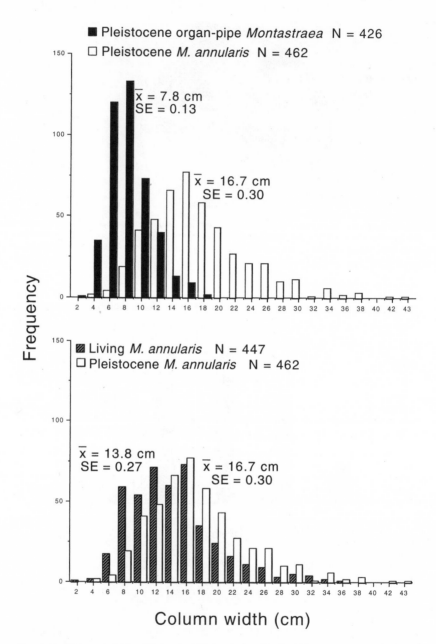

Fig. 5.13. Column width of living and Pleistocene columnar *Montastraea annularis* and Pleistocene organ-pipe *Montastraea* from Barbados. Following the extinction of the organ-pipe form, the living columnar form displays more slender column widths, but not as slender as the extinct organ-pipe form (see text for discussion). *SE* = standard error on the mean.

Table 5.5. Student's *t*-test of mean column width among fossil columnar, living columnar, and fossil organ-pipe colonies of *Montastraea "annularis"*

	Pleistocene organ-pipe	Pleistocene columnar	Living columnar
Pleistocene organ-pipe	—		
Pleistocene columnar	−31.36*	—	
Living columnar	−21.76*	−7.78*	—

Note: Tests were completed on log-transformed data.
* $p < 0.05$.

M. annularis from Barbados and St. Vincent island. The Pleistocene organ-pipe species had highly significantly thinner columns than both Pleistocene and living *M. annularis* (fig. 5.13; table 5.5). However, living *M. annularis* also had significantly thinner columns than Pleistocene *M. annularis*. Thus column diameter of columnar *M. annularis* decreased dramatically in the direction of the organ-pipe species after the organ-pipe species became extinct. Moreover, the column width of Pleistocene *M. annularis* when it occurred in sympatry with the organ-pipe *Montastraea* (mean = 17.16 cm, N = 366) was significantly greater ($t = -2.762$, df = 102, $p = 0.007$) than when the two species were not in each other's presence (mean = 15.20 cm, N = 62) (as measured along the 40-m transects).

Discussion

Sudden Extinction of Widespread Species

We commonly associate extinction with rarity, but when last observed neither *Pocillopora* cf. *palmata* nor organ-pipe *Montastraea* was rare by any of the commonly accepted criteria of narrow geographic distribution, extreme habitat specificity, or small local population size everywhere within its range (Rabinowitz 1981). While it is true that many modern coral species have widespread distributions (Veron 1993) regardless of whether they are common or rare, it is equally true that most coral species are rare, and Caribbean communities tend to be dominated by only a handful of coral species (Goreau 1959; Goreau and Wells 1967; Mesollela 1967; Geister 1977a; Pandolfi and Jackson 1997, 2001). Moreover, roughly 90% of all late Pleistocene Caribbean coral species survived the same interval, including several rare species. Thus population size at 82 Ka is a poor predictor of extinction for these corals. This is all the more important because the extinctions must have been natural, occurring no later than about the same time as the first human occupation of the Caribbean islands (Meltzer 1997), and certainly well before such occupation could have made any significant ecological difference in the sea.

Pocillopora cf. *palmata* and the organ-pipe *Montastraea* are only very distantly related, but both became extinct during the same geologically brief interval between 82 Ka and roughly 3 Ka. The last record for each of the two corals was 82 Ka from Barbados. There are no reports of *Pocillopora* cf. *palmata* or the organ-pipe *Montastraea* from any of the numerous Recent reef localities throughout the Caribbean that have been surveyed in detail. The few emergent outcrop areas of late Holocene reef rocks in the Caribbean (Colombian coast; Lake Enriquillo, Dominican Republic) need further examination, but have not yet revealed the two extinct species. Thus extinction probably occurred within no more than about 80,000 years. With generation times and longevities of reef corals measured in terms of many decades to centuries (Potts 1984; Potts et al. 1985; Babcock 1991), this works out to roughly only 1,000 generations for the elimination of common to abundant, geographically widespread species! Moreover, the great reduction in habitat during the last glacial maximum about 18 Ka (Kleypas 1997), and the fact that predominantly oceanic species went extinct, while other, rare species survived, led Pandolfi (1999) to raise the hypothesis that extinction occurred during only a few thousand years when habitat area for coral reefs in the Caribbean was near its Pleistocene low. A more detailed discussion of the possible causes of the extinctions is presented elsewhere (Pandolfi 1999).

Eighty thousand years (or considerably less) is a geological instant compared with the 5 million years' duration of the great majority of Neogene Caribbean reef coral species, including other species of *Pocillopora* and *Montastraea* (Budd et al. 1994). In contrast, the fossil records of *P.* cf. *palmata* and the organ-pipe *Montastraea* are only 37 and 500 Ka, respectively. These extremely short ranges are difficult to interpret because both species were discovered so recently that they are probably commonly overlooked in most general surveys of Pleistocene reef terraces. The widespread acceptance of extreme phenotypic plasticity of *Montastraea "annularis"* also has diminished interest in documenting the geologic ranges of the different "growth forms." There seems little doubt that the range of the organ-pipe species will be extended backward through ongoing studies of the fossil record of speciation and extinction in the *Montastraea "annularis"* clade (Ann Budd, pers. comm.). The absence of *Pocillopora* cf. *palmata* from before 125 Ka at Barbados may be due simply to the extreme geographic and oceanographic isolation of Barbados as the easternmost island of the Caribbean, and therefore detailed surveys are needed from places like Cuba, Curaçao, Haiti, and the Dominican Republic where late Pleistocene reefs older than 125 Ka are abundantly preserved.

In conclusion, the geologically sudden extinction of two such widespread and abundant coral species demonstrates that natural extinction may be just as punctuated as the origin of species (e.g., Jackson and Cheetham 1994). Such rapid extinction of widespread and abundant species emphasizes the vulnerability of increasingly threatened reef coral species in the face of rapid environmental and climatic change (Pandolfi 1999). A similar conclusion was reached for the late Quaternary extinction of an eastern North American tree species (Jackson and Weng 1999).

Effects on Surviving Lineages

Comparison of patterns before and after extinction of single species or entire biotas can reveal much about underlying causes of distribution and abundance on many scales (Hansen 1988; Jablonski 1998). Schluter and McPhail (1992, 89) defined character displacement as "the influence of one species on the evolution of resource use in another species as a consequence of resource competition." A rigorous evaluation of character release is presently being undertaken and will be the subject of a future contribution. However, we raise the hypothesis that the ecological and morphological changes observed in the three surviving species of the *M. "annularis"* complex occurred by character release when the organ-pipe *Montastraea* went extinct.

The acquisition of the primary resources, food and light, depends upon growth form in corals that depend upon symbiosis for growth (Barnes 1973; Sebens 1997). Because growth form is directly related to growth rate, we can think of resource use in terms of growth rate. Growth rates of branching corals are much higher than massive corals, which in turn grow slightly faster than sheet- or platelike species (Jackson 1991); and the same is true for the columnar, massive, and platelike "growth forms" of living colonies of the *M. "annularis"* species complex (Dustan 1975; Hubbard and Scaturo 1985; Tomascik 1990; Van Veghel 1994). Higher light levels in shallower water result in thinner branches with greater emphasis on vertical extension (i.e., higher growth rates). Thus as light attenuates with increasing water depth, first columns, then massive, and finally sheetlike forms predominate (Barnes 1973; Graus and Macintyre 1976, 1982; Dustan 1975). Perhaps self-shading in branching forms limits their distribution in deeper water, so they are confined to shallow water where they may also shade out and exclude massive and sheetlike forms. By this scenario, extinction of the competitively superior organ-pipe species opened up shallow habitat for columnar *M. annularis* and massive *M. faveolata,* which had been at least partially excluded by faster and taller growing, narrower branches of the organ-pipe species. This is consistent with the generally increasing abundance of the organ-pipe relative to the other *M. "annularis"* species over 500 Ka in Barbados, and with the columns of living *M. annularis* being narrower than Pleistocene columnar *M. annularis* (fig. 5.13; table 5.5). Growth rate, as measured by annual bands, is higher in fossil organ-pipe than columnar M. *"annularis"* (Pandolfi et al., unpublished data).

Not only does resource use appear different before and after the extinction, but also the different Recent species compete aggressively for space by causing tissue lesions (Knowlton et al. 1992; Van Veghel and Bak 1994), as well as possible shading. Moreover, various genetic studies suggest that species differences within the M. *"annularis"* complex, and therefore colony forms, are genetic (Knowlton et al. 1992; Van Veghel 1994; Lopez and Knowlton 1997). If it turns out that character release after extinction led to the ecological and morphological changes observed in the surviving M. *"annularis"* species, then biological interactions may have been important determinants of their distribution and abundance both before and after the Pleistocene extinction.

Acknowledgments

C. Lovelock first suggested measuring the widths of columns of modern and fossil *Montastraea "annularis."* She, A. Budd, N. Knowlton, and G. Llewellyn made many other useful suggestions.

References

Alvarez, L. W., W. Alvarez, F. Asaro, and H. V. Michel. 1980. Extraterrestrial cause for the Cretaceous-Tertiary extinction. *Science* 208:1095–98.

Babcock, R. C. 1991. Comparative demography of three species of scleractinian corals using age-dependent and size-dependent classifications. *Ecological Monographs* 61:225–44.

Bak, R. P. M., and B. E. Luckhurst. 1980. Constancy and change in coral reef habitats along depth gradients at Curaçao. *Oecologia* 47:145–55.

Bard, E., B. J. Hamelin, and R. G. Fairbanks. 1990. U-Th ages obtained by mass spectrometry in corals from Barbados: Sea level during the past 130,000 years. *Nature* 346:456–58.

Barnes, D. J. 1973. Growth in colonial scleractinians. *Bulletin of Marine Science* 23:280–98.

Bender, M. J., R. G. Fairbanks, F. W. Taylor, R. K. Matthews, J. G. Goddard, and W. S. Broecker. 1979. Uranium-series dating of the Pleistocene reef tracts of West Indies. *Geological Society of America Bulletin* 90:577–94.

Budd, A. F. 1993. Variation within and among morphospecies of *Montastraea*. *Courier Forschungs-Institut Senckenberg* 164:241–54.

Budd, A. F., and K. G. Johnson. 1997. Coral reef community dynamics over 8 million years of evolutionary time: Stasis and turnover. In *Proceedings of the Eighth International Coral Reef Symposium,* ed. H. A. Lessios and I. G. Macintyre, 1:423–28. Balboa, Republic of Panamá: Smithsonian Tropical Research Institute.

————. 1999. Origination preceding extinction during late Cenozoic turnover of Caribbean reefs. *Paleobiology* 25:188–200.

Budd, A. F., K. G. Johnson, and T. A. Stemann. 1996. Plio-Pleistocene turnover and extinctions in the Caribbean reef-coral fauna. In *Evolution and environments in tropical America,* ed. J. B. C. Jackson, A. F. Budd, and A. G. Coates, 168–204. Chicago: University of Chicago Press.

Budd, A. F., and C. M. Kievman. 1993. Coral assemblages and reef environments in the Bahamas Drilling Project cores. Part 3 of *Final draft report of the Bahamas Drilling Project.* Coral Gables, FL: Rosenstiel School of Marine and Atmospheric Science, University of Miami.

Budd, A. F., T. A. Stemann, and K. G. Johnson. 1994. Stratigraphic distributions of genera and species of Neogene to Recent Caribbean reef corals. *Journal of Paleontology* 68:951–77.

Cheetham, A. H., and J. B. C. Jackson. 1996. Speciation, extinction, and the decline of arborescent growth in Neogene and Quaternary cheilostome Bryozoa of tropical America. In *Evolution and environment in tropical America,* ed. J. B. C. Jackson, A. F. Budd, and A. G. Coates, 205–33. Chicago: University of Chicago Press.

————. 1999. Neogene history of cheilostome Bryozoa in tropical America. In *Advances in bryozoan research*, ed. J. B. C. Jackson and A. Herrera. Washington, DC: Smithsonian Institution Press.

Coates, A. G., and J. A. Obando. 1996. The geological evolution of the Central American isthmus. In *Evolution and environment in tropical America,* ed. J. B. C. Jackson, A. F. Budd, and A. G. Coates, 21–56. Chicago: University of Chicago Press.

Cronin, T. M., and H. J. Dowsett. 1996. Biotic and oceanographic response to the Pliocene closing of the Central American isthmus. In *Evolution and environment in tropical America,* ed. J. B. C. Jackson, A. F. Budd, and A. G. Coates, 76–104. Chicago: University of Chicago Press.

Diamond, J. M. 1984. Historic extinctions: A Rosetta Stone for understanding prehistoric extinctions. In *Quaternary extinctions: A prehistoric evolution,* ed. P. S. Martin and R. G. Klein, 824–62. Tucson: University of Arizona Press.

Dustan, P. 1975. Growth and form in the reef-building coral *Montastraea annularis. Marine Biology* 33:101–7.

Gallup, C. D., R. L. Edwards, and R. G. Johnson. 1994. The timing of high sea levels over the past 200,000 years. *Science* 263:796–800.

Geister, J. 1975. Riffbau und geologische Entwicklungsgeschichte der Insel San Andrés (westl. Karibisches Meer, Kolumbien). *Stuttgarter Beiträge zur Naturkunde, Geologie, und Paläontologie* 15:1–203.

————. 1977a. The influence of wave exposure on the ecological zonation of Caribbean coral reefs. In *Proceedings of the Third International Coral Reef Symposium,* 1:23–29. Miami: Rosenstiel School of Marine and Atmospheric Science, University of Miami.

————. 1977b. Occurrence of *Pocillopora* in late Pleistocene Caribbean coral reefs. *Mémoires du Bureau de Recherches Géologiques et Minières* 89:378–88.

————. 1982. Pleistocene reef terraces and coral environments at Santo Domingo and near Boca Chica, southern coast of the Dominican Republic. In *Transactions of the 9th Caribbean Geological Conference,* ed. W. Snow, N. Gil, R. Llinas, R. Rodriguez-Torres, M. Seaward, and L. Tavares, 2:689–703. Santa Domingo, Dominican Republic.

Ginsburg, R. N., ed. 1993. *Proceedings of the colloquium on global aspects of coral reefs: Health, hazards, and history.* Miami: Rosenstiel School of Marine and Atmospheric Science, University of Miami.

Glynn, P. W., and W. H. De Weerdt. 1991. Elimination of two reef-building hydrocorals following the 1982–83 El Niño warming event. *Science* 253:69–71.

Glynn, P. W., and J. S. Feingold. 1992. Hydrocoral species not extinct. *Science* 257:1845.

Goreau, T. F. 1959. The ecology of Jamaican coral reefs: I. Species composition and zonation. *Ecology* 40:67–90.

Goreau, T. F., and J. W. Wells. 1967. The shallow-water Scleractinia of Jamaica: Revised list of species and their vertical distribution range. *Bulletin of Marine Science* 17:442–53.

Graus, R. R., and I. G. Macintyre. 1976. Light control of growth form in colonial reef corals: Computer simulation. *Science* 193:895–97.

———. 1982. Variations in the growth forms of the reef coral *Montastraea annularis* (Ellis and Solander): A quantitative evaluation of growth response to light distribution using computer simulation. In *The Atlantic barrier reef ecosystem at Carrie Bow Cay, Belize: 1, Structure and communities,* ed. K. Rützler and I. G. Macintyre, Smithsonian Contributions to Marine Science 12:441–64. Washington, DC: Smithsonian Institution Press.

Greenstein, B. J., H. A. Curran, and J. M. Pandolfi. 1998. Shifting ecological baselines and the demise of *Acropora cervicornis* in the western North Atlantic and Caribbean Province: A Pleistocene perspective. *Coral Reefs* 17:249–61.

Hansen, T. A. 1988. Early Tertiary radiation of marine molluscs and the longterm effects of the Cretaceous-Tertiary extinction. *Paleobiology* 14:37–51.

Hoffmeister, J. E., and H. G. Multer. 1968. Geology and origin of the Florida Keys. *Geological Society of America Bulletin* 79:1487–1502.

Hubbard, D. K., and D. Scaturo. 1985. Growth rates of seven species of scleractinian corals from Cane Bay and Salt River, St. Croix, USVI. *Bulletin of Marine Science* 36:325–38.

Hughes, T. P. 1994. Catastrophes, phase shifts, and large-scale degradation of a Caribbean coral reef. *Science* 265:1547–51.

Humphrey, J. D., and T. N. Kimbell. 1990. Sedimentology and sequence stratigraphy of upper Pleistocene carbonates of southeastern Barbados, West Indies. *American Association of Petroleum Geologists Bulletin* 74:1671–84.

Hunter, I., and B. Jones. 1996. Coral associations of the Pleistocene Ironshore Formation, Grand Cayman. *Coral Reefs* 15:249–67.

Jablonski, D. 1986. Background and mass extinctions: The alternation of macroevolutionary regimes. *Science* 231:129–33.

———. 1995. Extinctions in the fossil record. In *Extinction rates,* ed. J. H. Lawton and R. M. May, 25–44. Oxford: Oxford University Press.

————. 1998. Geographic variation in the molluscan recovery from the end-Cretaceous extinction. *Science* 279:1327–30.

Jackson, J. B. C. 1974. Biogeographic consequences of eurytopy and stenotopy among marine bivalves and their evolutionary significance. *American Naturalist* 108:541–60.

————. 1991. Adaptation and diversity of reef corals. *Bioscience* 41:475–82.

————. 1992. Pleistocene perspectives on coral reef community structure. *American Zoologist* 32:719–31.

————. 1997. Reefs since Colombus. *Coral Reefs* 16, supp.: S23–S32.

Jackson, J. B. C., A. Budd, and J. M. Pandolfi. 1996. The shifting balance of natural communities? In *Evolutionary paleobiology: Essays in honor of James W. Valentine,* ed. D. Jablonski, D. H. Erwin, and J. H. Lipps, 89–122. Chicago: University of Chicago Press.

Jackson, J. B. C., and A. H. Cheetham. 1994. Phylogeny reconstruction and the tempo of speciation in cheilostome Bryozoa. *Paleobiology* 20:407–23.

Jackson, S. T., and C. Weng. 1999. Late Quaternary extinction of a tree species in eastern North America. *Proceedings of the National Academy of Sciences, USA* 96:13847–52.

James, N. 1974. Diagenesis of scleractinian corals in the subaerial vadose environment. *Journal of Paleontology* 48:785–99.

James, N. P., E. W. Mountjoy, and A. Omura. 1971. An early Wisconsin reef terrace at Barbados, West Indies, and its climatic implications. *Geological Society of America Bulletin* 82:2011–18.

James, N. P., C. W. Stearn, and R. S. Harrison. 1977. Field guide book to modern and Pleistocene reef carbonates, Barbados, W.I. Third International Symposium on Coral Reefs, Miami Beach, FL.

Johnson, K. G., A. F. Budd, and T. S. Stemann. 1995. Extinction selectivity and ecology of Neogene Caribbean reef corals. *Paleobiology* 21:52–73.

Kleypas, J. A. 1997. Modeled estimates of global reef habitat and carbonate production since the last glacial maximum. *Paleoceanography* 12:533–45.

Knowlton, N., J. L. Maté, H. M. Guzmán, R. Rowan, and J. Jara. 1997. Direct evidence for reproductive isolation among the three species of the *Montastraea annularis* complex in Central America (Panamá and Honduras). *Marine Biology* 127:705–11.

Knowlton, N., E. Weil, L. A. Weigt, and H. M. Guzmán. 1992. Sibling species in *Montastraea annularis,* coral bleaching, and the coral climate record. *Science* 255:330–33.

Ku, T. L., M. Ivanovich, and S. Luo. 1990. U-series dating of last interglacial high sea stands: Barbados revisited. *Quaternary Research* 33:129–47.

Lewis, J. B. 1960. The coral reefs and coral communities of Barbados, W.I. *Canadian Journal of Zoology* 38:1133–45.

————. 1984. The *Acropora* inheritance: A reinterpretation of the development of fringing reefs in Barbados, West Indies. *Coral Reefs* 3:117–22.

Lopez, J. V., and N. Knowlton. 1997. Discrimination of species in the *Montastraea annularis* complex using multiple genetic loci. In *Proceedings of the Eighth International Coral Reef Symposium,* ed. H. A. Lessios and I. G. Macintyre, 2:1613–18. Balboa, Republic of Panamá: Smithsonian Tropical Research Institute.

Loya, Y. 1972. Community structure and species diversity of hermatypic corals at Eilat, Red Sea. *Marine Biology* 13:100–23.

———. 1978. Plotless and transect methods. In *Coral reefs: Research methods,* ed. D. R. Stoddart and R. E. Johannes, UNESCO Monographs on Oceanographic Methodology 5, 197–217. Paris: UNESCO.

Martindale, W. 1992. Calcified epibionts as palaeoecological tools: Examples from the Recent and Pleistocene reefs of Barbados. *Coral Reefs* 11:167–77.

May, R. M. 1988. How many species are there on Earth? *Science* 241:1441–49.

———. 1990. How many species? *Philosophical Transactions of the Royal Society of London,* series B, 330:293–304.

May, R. M., J. H. Lawton, and N. E. Stork. 1995. Assessing extinction rates. In *Extinction rates,* ed. J. H. Lawton and R. M. May, 1–24. Oxford: Oxford University Press.

McLaren, D. J. 1983. Bolides and biostratigraphy. *Geological Society of America Bulletin* 94:313–24.

Meltzer, D. J. 1997. Monte Verde and the Pleistocene peopling of the Americas. *Science* 276:754–55.

Mesolella, K. J. 1967. Zonation of uplifted Pleistocene coral reefs on Barbados, West Indies. *Science* 156:638–40.

———. 1968. The uplifted reefs of Barbados: Physical stratigraphy, facies relationships, and absolute chronology. Ph.D. diss., Brown University, Providence, RI.

Mesolella, K. J., H. A. Sealy, and R. K. Matthews. 1970. Facies geometries within Pleistocene reefs of Barbados, West Indies. *American Association of Petroleum Geologists Bulletin* 54:1899–1917.

Pandolfi, J. M. 1999. Response of Pleistocene coral reefs to environmental change over long temporal scales. *American Zoologist* 39:113–30.

Pandolfi, J. M., and J. B. C. Jackson. 1997. The maintenance of diversity on coral reefs: Examples from the fossil record. In *Proceedings of the Eighth International Coral Reef Symposium,* ed. H. A. Lessios and I. G. Macintyre, 1:397–404. Balboa, Republic of Panamá: Smithsonian Tropical Research Institute.

———. 2001. Community structure of Pleistocene coral reefs of Curaçao, Netherlands Antilles. *Ecological Monographs* 71:49–67.

Pandolfi, J. M., G. Llewellyn, and J. B. C. Jackson. 1999. Interpretation of ancient reef environments in paleoecological studies of community structure: Curaçao, Netherlands Antilles, Caribbean Sea. *Coral Reefs* 18:107–22.

Pandolfi, J. M., C. Lovelock, and A. F. Budd. 2000. Character release following extinction in a Caribbean reef coral species complex. Manuscript.

Potts, D. C. 1984. Generation times and the Quaternary evolution of reef-building corals. *Paleobiology* 10:48–58.

Potts, D. C., T. J. Done, P. J. Isdale, and D. A. Fisk. 1985. Dominance of a coral community by the genus *Porites* (Scleractinia). *Marine Ecology Progress Series* 23:79–84.

Rabinowitz, D. 1981. Seven forms of rarity. In *The biological aspects of rare plant conservation,* ed. H. Synge, 205–17. Chichester, NY: John Wiley and Sons.

Radtke, U., R. Gruen, and H. P. Schwarcz. 1988. Electron spin resonance dating of the Pleistocene coral reef tracts of Barbados. *Quaternary Research* 29:197–215.

Raup, D. M. 1991a. *Extinction: Bad genes or bad luck?* New York: W. W. Norton.

———. 1991b. A kill curve for Phanerozoic marine species. *Paleobiology* 17:37–48.

———. 1996. Extinction models. In *Evolutionary paleobiology: Essays in honor of James W. Valentine,* ed. D. Jablonski, D. H. Erwin, and J. H. Lipps, 419–33. Chicago: University of Chicago Press.

Raup, D. M., and J. J. Sepkoski. 1982. Mass extinctions in the marine fossil record. *Science* 215:1501–3.

Reaka-Kudla, M. L. 1996. The global biodiversity of coral reefs: A comparison with rain forests. In *Biodiversity II: Understanding and Protecting Our Natural Resources,* ed. M. L. Reaka-Kudla, D. E. Wilson, and E. O. Wilson. Washington, DC: Joseph Henry/National Academy Press.

Roberts, C. M. 1997. Connectivity and management of Caribbean coral reefs. *Science* 278:1454–57.

Rosenzweig, M. L. 1995. *Species diversity in space and time.* Cambridge: Cambridge University Press.

Rowan, R., and N. Knowlton. 1995. Intraspecific diversity and ecological zonation in coral-algal symbiosis. *Proceedings of the National Academy of Sciences, USA* 92:2850–53.

Rowan, R., N. Knowlton, A. Baker, and J. Jara. 1997. Landscape ecology of algal symbionts creates variation in episodes of coral bleaching. *Nature* 388:265–69.

Scatterday, J. W. 1974. Reefs and associated coral assemblages off Bonaire, Netherlands Antilles, and their bearing on Pleistocene and Recent reef models. In *Proceedings of the Second International Coral Reef Symposium,* 2:85–106. Brisbane: Great Barrier Reef Committee.

Schluter, D., and J. D. McPhail. 1992. Ecological character displacement and speciation in sticklebacks. *American Naturalist* 140:85–108.

Sebens, K. P. 1997. Adaptive responses to water flow: Morphology, energetics, and distribution of reef corals. In *Proceedings of the Eighth International Coral Reef Symposium,* ed. H. A. Lessios and I. G. Macintyre, 2:1053–58. Balboa, Republic of Panamá: Smithsonian Tropical Research Institute.

Simberloff, D. S. 1986. The proximate causes of extinction. In *Patterns and processes in the history of life,* ed. D. M. Raup and D. Jablonski, 259–76. Berlin: Springer-Verlag.

Smith, A. B., and C. H. Jeffrey. 1998. Selectivity of extinction among sea urchins at the end of the Cretaceous period. *Nature* 392:69–71.

Stanley, S. M. 1966. Paleoecology and diagenesis of Key Largo limestone, Florida. *American Association of Petroleum Geologists Bulletin* 50:1927–47.

Taylor, F. W., and P. Mann. 1991. Late Quaternary folding of coral reef terraces, Barbados. *Geology* 19:103–6.

Tomascik, T. 1990. Growth rates of two morphotypes of *Montastrea annularis* along a eutrophication gradient, Barbados, W.I. *Marine Pollution Bulletin* 21: 376–80.

Valentine, J. W., T. C. Foin, and D. Peart. 1978. A provisional model of Phanerozoic marine diversity. *Paleobiology* 4:55–66.

Van Veghel, M. L. J. 1994. Polymorphism in the Caribbean reef building coral *Montastrea annularis*. Ph.D. diss., University of Amsterdam.

Van Veghel, M. L. J., and R. P. M. Bak. 1994. Intraspecific variation of a dominant Caribbean reef building coral, *Montastrea annularis:* Genetic, behavioral, and morphometric aspects. *Marine Ecology Progress Series* 92:255–65.

Veron, J. E. N. 1993. *A biogeographic database of hermatypic corals.* Australian Institute of Marine Science Monograph Series 10. Townsville, Queensland: Australian Institute of Marine Science.

Weil, E., and N. Knowlton. 1994. A multi-character analysis of the Caribbean coral *Montastraea annularis* (Ellis and Solander, 1786) and its two sibling species, *M. faveolata* (Ellis and Solander, 1786), and *M. franksi* (Gregory, 1895). *Bulletin of Marine Science* 55:151–75.

White, B., and H. A. Curran. 1995. Entombment and preservation of Sangamonian coral reefs during glacioeustatic sea-level fall, Great Inagua Island, Bahamas. In *Terrestrial and shallow marine geology of the Bahamas and Bermuda,* ed. H. A. Curran and B. White, Geological Society of America Special Paper 300, 51–61. Boulder: Geological Society of America.

Linking Macroevolutionary Pattern and Developmental Process in Marginellid Gastropods

6

Ross H. Nehm

Introduction

What processes generate, maintain, and constrain the production of morphological disparity through geologic time? Studies of morphological disparity have primarily focused on temporal patterns of the differential filling of morphospace within and among clades (Foote 1991, 1993a, 1993b, 1994; Foote and Gould 1992; Willis et al. 1994; Wagner 1995; Willis 1998; Smith and Bunje 1999). These patterns of morphological disparity have triggered considerable debate regarding the role of developmental processes in macroevolutionary change (Valentine and Erwin 1991; Lee 1992; Briggs et al. 1992; Gerhart and Kirschner 1997; Raff 1996; Wray and Bely 1994; Wray et al. 1996; Davidson et al. 1995). Nevertheless, paleobiological studies have not examined patterns of developmental variation or developmental evolution and their relationship to patterns of disparity. In order to understand the role of developmental processes in patterns of disparity, it is necessary to perform detailed phylogenetic, morphogenetic, and morphometric studies of well-preserved and abundant fossil clades (Gould 1977; McKinney 1988; McKinney and McNamara 1991; McNamara 1995; Williamson 1987; Hall 1999; Thomson 1988; Rachootin and Thomson 1981). This study employs a model taxon in a new research approach designed to analyze the role of developmental processes in patterns of morphological disparity.

Paleontologists have much to contribute to our understanding of the role of developmental processes in the generation of large-scale morphological patterns, but must limit such research to groups with good fossil records, well-established phylogenies, a complete preservation of ontogeny, clear morphological markers of sexual maturity, and morphologies amenable to quantitative analysis and morphogenetic study. Fossil gastropods are one group that satisfies these requirements.

Although gastropods have figured prominently in studies of large-scale morphological patterns in the fossil record (Wagner 1995; Vermeij 1978, 1987; Gould 1969, 1977; Tissot 1988a, 1988b; Geary 1988, 1990b; Lindberg 1988; Allmon 1994), the role of development as a generating or constraining force in these patterns has been fully explored in only a few clades, including terrestrial *Cerion* (Gould 1969, 1977, 1984) and freshwater *Melanopsis* (Geary 1988, 1990a, 1990b). Fossil marine neogastropods, one of the largest and most diverse marine gastropod clades (Ponder and Lindberg 1997), represent a major lacuna in our knowledge of developmental evolution. As a clade that has undergone extensive evolutionary radiation in the Cenozoic, it would provide useful phylogenetic and ecological comparisons with heterobranch gastropods.

Two genera of fossil (and living) marine neogastropods, *Prunum* and *Volvarina,* from the family Marginellidae compose an excellent research system for the investigation of patterns and processes of morphological and developmental evolution in the fossil record. First, *Prunum* and *Volvarina* have determinate growth, which permits the recognition of juvenile and adult shells. Second, adult shells preserve a complete record of ontogeny, which is easily examined quantitatively using X-ray images, and compositionally using scanning electron microscopy (SEM) and hard-tissue histology. Third, stratophenetic and maximum-parsimony phylogenetic analyses of morphological and molecular data have established the evolutionary relationships of the Marginellidae in general and the Prunini in particular (Nehm 1996; Coovert 1987, 1989; Coovert and Coovert 1990, 1995; Nehm and Tran 1997; Nehm 1998). Fourth, a large amount of information has been assembled about the geographic, stratigraphic, and bathymetric distributions of *Prunum* and *Volvarina* species, and their relationship to ontogenetic and morphological variation (Nehm and Geary 1994; Nehm and Hickman 1995; Nehm 1998). Fifth, *Prunum* and *Volvarina* fossils are remarkably rich throughout the American tropics because the shells are taphonomically durable (as a result of small size, great thickness, and extensive body-whorl callusing).

This study investigates how developmental processes contribute to patterns of morphological disparity in a clade of marginellid gastropods (*Prunum* + *Volvarina*) in tropical America. This study addresses three questions: (1) Are the direction, magnitude, and environmental contexts of morphological evolution similar within and among clades? (2) Is proximity in morphospace a result of similar developmental processes? (3) How do developmental processes facilitate and constrain patterns of morphological disparity in marginellid gastropods?

Systematics and Phylogenetic Relationships of Marginellid Gastropods

Monophyly of the Marginellidae and the Prunini

Robust phylogenetic hypotheses are prerequisite to investigations of the polarity of morphological and developmental change, and nearly every

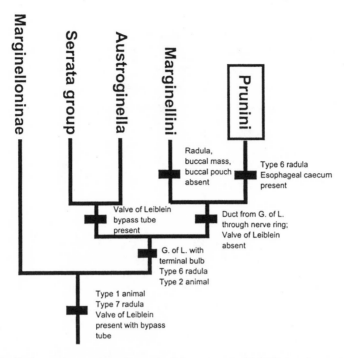

Fig. 6.1. Phylogenetic overview of marginellid gastropods, highlighting the position of the Prunini (containing *Prunum* and *Volvarina*). See text for a discussion of synapomorphies.

other aspect of comparative biology (Harvey and Pagel 1991). Several different approaches contribute to a very good understanding of marginellid gastropod phylogeny. Phylogenetic analyses based on morphological (Nehm 1996; Coovert 1987, 1989; Coovert and Coovert 1990, 1995), molecular (Nehm and Tran 1997; Nehm and Simison, unpublished data), and stratophenetic data (Nehm 1994, 1998) establish the monophyly of marginellid gastropods, and several clades therein (fig. 6.1). Within the Marginellidae, the monophyletic tribe Prunini is diagnosed by the type 6 radula (Coovert 1989) and the presence of an esophageal caecum (Coovert and Coovert 1995; Coovert 1989; Strong and Nehm, unpublished data). The Prunini include the genera *Prunum, Volvarina, Bullata, Rivomarginella, Hyalina,* and *Cryptospira.*

Monophyly and Phylogenetic Relationships of *Prunum* + *Volvarina*

Within the Prunini, shell morphology, radular morphology, internal anatomy, molecular (lbs), and stratophenetic data provide strong support for the monophyly of *Prunum* + *Volvarina* (Coovert and Coovert 1995;

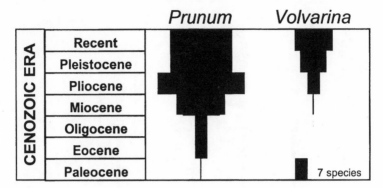

Fig. 6.2. Diversity of *Prunum* and *"Volvarina"* in the Cenozoic.

Nehm 1996; Nehm and Tran 1997). *Prunum* and *Volvarina* are differentiated by shell morphology, radular morphology, radular cartilage morphology, mantle morphology, and internal anatomy (Coovert and Coovert 1995; Coovert 1989; Strong and Nehm, unpublished data). The lineages of *Prunum* and *Volvarina* studied here include extinct species, so conchological features considered to be of evolutionary and systematic importance are the primary focus of comparisons between the two genera. In general, *Prunum* shells are large, thick, obovate, heavily callused, and denticulate, with thick and externally varixed lips. *Volvarina* shells are usually small, thin, and cylindrical and often lack many features present in *Prunum,* including callusing, denticulations, lip thickening, and the external varix. However, the conchological features of a few species of *Prunum* are "intermediate" between these two morphological extremes (Coovert 1988).

The fossil record of *Prunum* and *Volvarina* is remarkably rich throughout the Upper Cenozoic. Extant *Prunum* and *Volvarina* are most diverse in the eastern and western Atlantic (Lipe 1991), and Neogene sediments from the same region preserve a rich record of marginellids. The fossil record demonstrates different diversity patterns in the two groups (fig. 6.2). From the Eocene to the Pliocene, the diversity of *Prunum* increased, but it remained relatively constant from the Pleistocene to the Recent. This pattern is generally concordant with that of the molluscan fauna: the late Neogene was a time of generally steady diversity but considerable evolutionary turnover (Jackson et al. 1993; Allmon et al. 1993; Vermeij 1993; Nehm 1994). In contrast, species richness in *Volvarina* increased throughout the Neogene. I focus on the evolutionary and developmental origins of three lineages of *Volvarina* during this radiation.

Phylogeny of Three Western Atlantic *Prunum* + *Volvarina* Clades

Shell morphology, radular morphology, internal anatomy, and stratophenetic data provide strong support for the monophyly of *Prunum* + *Volvarina*. Within this clade three well-supported monophyletic groups (referred to as clades A, B, and C) contain both *Volvarina* and *Prunum* species (figs. 6.3 and 6.4). The discovery of stratophenetic patterns clearly demonstrating the evolution of *Volvarina* species from *Prunum* species was surprising and remarkable, considering the purported monophyly of the two genera; their morphological cohesion throughout different environments, geographies, and times; and their distinct and generally nonoverlapping occupation of morphospace (Nehm and Geary 1994; Nehm 1998). A systematic revision of the genus *"Volvarina"* is necessary and will be addressed elsewhere. In the present study, *"Volvarina"* will be retained as a convenient descriptor of a phenetic entity possessing a suite of morphological features. I focus on three clades that contain both *Prunum* and *"Volvarina"* species.

Clade A comprises the sister taxa *"V."* *christineladdae* (Maury 1917a, 1917b) and *P. coniforme* (Sowerby 1850), and the outgroup *P.* cf. *coniforme* (Maury 1917a, 1917b). The systematics, stratigraphic distributions, and phylogenetic relationships of this clade are discussed in detail by Nehm (1998, 2001), Maury (1917a, 1917b), and Woodring (1928). In clade A, an exceptionally complete morphological transition from *Prunum* to *"Volvarina"* is preserved in the Neogene stratigraphic sections of the Dominican Republic (Nehm and Geary 1994). Clade A is extinct, but ranged temporally from the Lower Middle Miocene Baitoa Formation to the Upper Pliocene Mao Formation and was endemic to the Neogene Gatunian faunal province of the western Atlantic. A summary of the stratigraphic distributions and phylogenetic relationships of clade A is shown in figure 6.4 (1 and 2).

Clade B comprises the sister taxa *P. marginatum* (Born 1778) and *"V."* species B, and the outgroup *P. circumvittatum* (Weisbord 1962). The systematics, biostratigraphy, and phylogenetic relationships of this clade are discussed by Gibson-Smith and Gibson-Smith (1979), Weisbord (1962), and Petuch (1981, 1982). L. Collins (pers. comm. 1995) provided additional geographic and stratigraphic data from the Panama Paleontological Project. *P. circumvittatum* is extinct, and the other two species in clade B are extant. This clade is endemic to the western Atlantic and first originated in the Miocene Cantaure Formation of Venezuela. The synapomorphies that unite clade B are two broad dorsal body-whorl color bands and a unique club-shaped apical whorl. Both of these features are absent from all other living and extinct species of *Prunum* and

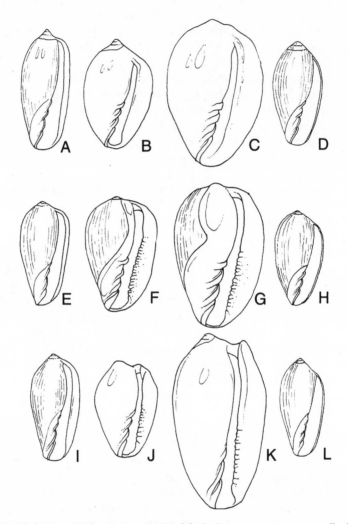

Fig. 6.3. Clade B: *A, "Volvarina"* species B adult; *B, Prunum marginatum* small adult; *C, P. marginatum* large adult; *D, P. marginatum* juvenile. Clade A: *E, "V." christineladdae* (Maury 1917a) adult; *F, P. coniforme* (Sowerby 1850) small adult; *G, P. coniforme* large adult; *H, P. coniforme* juvenile. Clade C: *I, "V."* species C adult; *J, P. macdonaldi* (Dall 1912) small adult; *K, P. macdonaldi* large adult; *L, P. macdonaldi* juvenile.

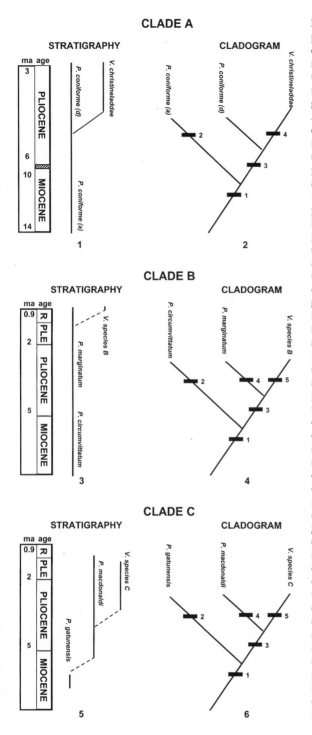

Fig. 6.4. Stratigraphic ranges and phylogenetic hypotheses for *Prunum* clades A, B, and C. *Black boxes* represent synapomorphies. *a* = ancestral; *d* = derived. *1*, Stratigraphic ranges of species in clade A. *2*, Phylogenetic hypothesis for clade A. *3*, Stratigraphic ranges of species in clade B. *4*, Phylogenetic hypothesis for clade B. *5*, Stratigraphic ranges of species in clade C. *6*, Phylogenetic hypothesis for clade C. *Clade A synapomorphies and autapomorphies: 1*, Four shell whorls; denticulations present; lip crenulations present; bilobed aperture margin callus present; posterior inner lip sulcus present; stout shell shape. *2*, Denticulations absent; lip crenulations absent; tall thin shell shape; dorsal lip callus reduced to absent; posterior callus margin reduced; external varix reduced; internal lip reduced; anterior callus margin reduced and thin. *Clade B synapomorphies and autapomorphies: 1*, Dorsal callus ring, two broad dorsal body-whorl stripes, club-shaped apical whorl, and oval shell shape present. *2*, Bilayered dorsal callus ring and apical callus present. *3*, Apical callus absent, single-layer dorsal callus ring present; stout/obovate shell shape. *4*, Dorsal callus ring absent; tall thin shell shape; posterior callus margin reduced to absent; external varix reduced; internal lip reduced; anterior callus margin reduced and thin. *Clade C synapomorphies and autapomorphies: 1*, Dorsal callus ridge, flanging of the ventral shell surface; upturn of the siphonal canal present. *2*, Apical callus present. *3*, Lip attachment (to body whorl) sulcus, posterior and medial external lip angle, flat and expanded dorsal surface present. *4*, Tall thin shell shape, small size, apical callus absent, ventral callus reduced to absent; dorsal lip callus reduced; posterior callus margin reduced; lip crenulations absent; external varix reduced; internal lip reduced. See text for discussion of these patterns.

"Volvarina." A summary of the stratigraphic distributions and phylogenetic relationships of clade B is shown in figure 6.4 (3 and 4).

Clade C comprises the sister taxa *P. macdonaldi* (Dall 1912) and *"V."* species C, and the outgroup *P. gatunensis* (Brown and Pilsbry 1911). The systematics, biostratigraphy, and phylogenetic relationships of this clade are discussed by Olsson (1922) and Woodring (1970). L. Collins (pers. comm. 1995) and Robinson (1991) provided additional geographic and stratigraphic data. In clade C, morphological intermediates connect *Prunum* and *"Volvarina"* (Nehm 1998). Clade C is extinct, and ranged from the Miocene Gatun Formation to the Pleistocene Armuelles Formation of Central America. The synapomorphies that unite clade C are an extensive flattening and flanging of the ventral shell surface (constructed from surficially deposited callus), and a dorsal upturn of the siphonal canal. A summary of the stratigraphic distributions and phylogenetic relationships of clade C is shown in figure 6.4 (5 and 6).

Stratophenetic, Molecular, and Geographic Evidence for the Monophyly of the Three Clades

All three clades containing *Prunum* and *"Volvarina"* species contain morphological features that do not collectively occur in any other living or fossil marginellid species (see above). Nevertheless, based on all of the morphological characters studied, it would be more parsimonious to conclude that the three species of *"Volvarina"* are members of one clade and that the unique morphological features uniting the three *Prunum* clades are homoplastic. However, the hypothesis that *"Volvarina"* species are members of a single clade is inconsistent with four independent sources of data (stratigraphic, stratophenetic, geographic, and molecular).

Stratigraphic and stratophenetic data are considered to be independent of maximum-parsimony phylogenetic analyses because (1) different features are used in the morphometric analysis and the maximum-parsimony phylogenetic analysis; and (2) the temporal distributions of fossils are independent of morphology. The exceptionally studied stratigraphic contexts of clades A, B, and C were established by Dominican Republic Project and Panama Paleontology Project researchers (Saunders et al. 1986; Nehm and Geary 1994; Nehm 2001; Collins 1993; L. Collins, pers. comm. 1995). Detailed univariate, multivariate, and geometric morphometric studies of the evolution of clade A and clade C, and the stratigraphic distributions of more than 4,000 specimens from these clades, provide strong evidence of the evolution of *"Volvarina"* from *Prunum* (fig. 6.5). Rejection of these biostratigraphic and stratophenetic patterns is equivalent to rejecting paleontological evidence in evolutionary research.

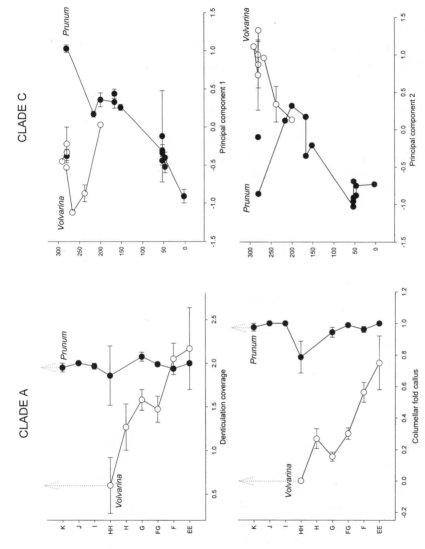

Fig. 6.5. Stratophenetic patterns illustrating the derivation of "*Volvarina*" from *Prunum* in clade A and clade C. See Nehm and Geary 1994 for illustrations of multivariate patterns in clade A and for precise definitions of stratigraphic intervals A–I. See Nehm 1998 for intervals J and K. Stratophenetic patterns in clade C are from the Bocas del Toro Basin, and the temporal axis represents meters in the stratigraphic section (Panama Paleontology Project data).

Studies of variation in 16s gene sequences in extant species of *Prunum* and *"Volvarina"* indicate that *"Volvarina"* is derived from *Prunum* (Nehm and Tran 1997; Nehm and Simison, unpublished data). Although this evidence does not directly address whether the three *"Volvarina"* species studied here are members of a single clade, it does indicate that *"Volvarina"* is not a basal taxon and that the stratigraphic distributions of the two genera in the fossil record match patterns indicated by molecular data.

Finally, the geographic distributions of the species and their developmental modes corroborate hypotheses of the monophyly of clades A, B, and C. *Prunum* and *"Volvarina"* are direct developers and lack a planktonic stage (Boss 1982; Coovert 1986). Consequently, marginellid species have very restricted geographic distributions and appear to have undergone radiations in local geographic regions corresponding to marine provincial and subprovincial boundaries (Petuch 1981, 1982). All three *"Volvarina"* species occur in separate basins and marine subprovinces, and therefore it is less parsimonious to conclude that the three *"Volvarina"* species are members of a single clade. The *Prunum* and *"Volvarina"* species in clades A, B, and C occur respectively in isolated basins and different marine subprovinces.

In summary, morphological, biostratigraphic, stratophenetic, molecular, and geographic data support the independent origins of *"Volvarina"* from *Prunum*. The species in each clade share (1) morphological features that do not occur in any other marginellid species, (2) restricted dispersal capabilities and distributions in the same basins and marine subprovinces, (3) stratophenetic patterns of *"Volvarina"* evolving from *Prunum,* and (4) molecular evidence that *"Volvarina"* is derived from *Prunum*. The developmental processes that generated these macroevolutionary patterns are the focus of this paper.

Methods

Criteria for Recognition of Determinate Growth

Specimens from clades A, B, and C are differentiated as juvenile or adult, using morphological features representative of sexual maturity and the accompanying cessation of growth. The termination of growth in living and fossil marginellids, and many other gastropod clades, is accompanied by several morphological changes in the shell (Vermeij and Signor 1992; Vermeij 1993). These changes occur in varix morphology, the internal lip, suture morphology, and callusing patterns. In general, the presence of a lip varix, a thickened inner lip, an ascending body-whorl suture, and inner lip callusing are indicative of sexual maturity and the cessation of surficial and marginal shell growth. Shell remodeling occurs before and after sexual maturity in several Neogastropod groups (Columbellidae,

Cystiscidae, and Conidae, for example) but does not occur in marginellids.

Morphometric Measurement and Statistical Analyses of Adult and Juvenile Shells

Seven distance measurements and four categorical measurements were made on approximately 2,500 juvenile and adult specimens from the three marginellid clades (fig. 6.6). Distance measurements include (1) *shell height:* distance from the apex to the anterior columellar fold; (2) *shell width:* distance between the point of greatest body-whorl convexity to the point of greatest lateral lip convexity; (3) *columellar fold height:* distance from the posterior plication to anterior plication; (4) *lip width:* greatest width of the lip; (5) *aperture height:* distance from the anterior plication to the most posterior point in the aperture; (6) *spire height:* distance from the apex to the body-whorl suture; and (7) *shell girth:* distance from the greatest convexity in the dorsal body whorl to the plane of the ventral shell surface. The four categorical measurements all pertain to the callus and are assigned to one of three categories (0, 1, or 2) based on their thickness and area: (1) *callus ridge area,* (2) *anterior callus margin area,* (3) *posterior callus margin area,* and (4) *dorsal lip callus.* Outer lip denticulations are recorded as present or absent (0 or 1).

Means and standard errors were calculated for all variables and compared among species and ontogenies in all three clades. Canonical variates analysis (CVA) was used to examine the concordance between a priori species and ontogenetic (juvenile or adult) designations and morphometric categorization of species and ontogenies. In cases of differences between a priori and predicted group membership the CVA classification was accepted. Principal component analyses (PCAs) of character correlation matrices were then performed in order to assess morphological variability among adults and juveniles from all three clades.

In order to examine patterns of morphological diversity in living and fossil marginellids of the monophyletic group *Prunum* + *"Volvarina,"* morphometric data were used to construct a simple morphospace. The primary morphological differences in this clade are shell size (shell height), shell shape (shell height/shell width), and the extent of ventral shell callusing (anterior callus margin area). Large-scale patterns of morphological diversity in the clade are examined by mapping all available fossil and living *Prunum* and *"Volvarina"* species within this simple morphospace (*n* = 120 species).

Characterization of Shell Composition

SEM is used to examine modes of deposition and microstructural composition of shell layers and features through ontogeny in the three clades.

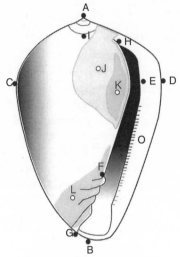

Adult *Prunum* shell

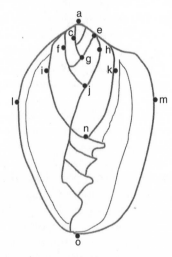

X-ray of adult shell

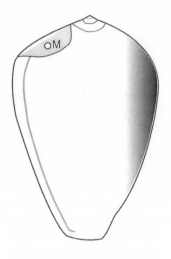

ADULT SHELL MEASUREMENTS

Shell height (distance A-B)
Shell width (distance C-D)
Fold height (distance F-G)
Lip width (distance E-D)
Aperture height (distance H-G)
Spire height (distance A-I)
Callus ridge area (K)
Anterior callus margin area (L)
Posterior callus margin area (J)
Dorsal lip callus (M)
Denticulations (O)

X-RADIOGRAPH MEASUREMENTS

Whorl 1
Shell height (distance a-g)
Shell width (distance c-e)
Whorl 2
Shell height (distance a-j)
Shell width (distance f-h)
Whorl 3
Shell height (distance a-n)
Shell width (distance i-k)
Whorl 4
Shell height (distance a-o)
Shell width (distance l-m)

Fig. 6.6. Morphometric template used in this study.

Microstructural analyses focus on shell layering, the external varix, the inner lip, dorsal lip callus, the anterior aperture margin callus, and the posterior aperture margin callus. Hard-tissue histology studies focus on the differential deposition of shell material through ontogeny and the quantitative differences in shell material volume.

Descriptions of shell microstructures follow Carter and Clark (1985). Simple prismatic structures have "mutually parallel, adjacent structural units (first order prisms) that do not strongly interdigitate along their mutual boundaries" (Carter and Clark 1985, 52). This structure resembles a layer of closely packed rods, with the long axes of the rods perpendicular to the layer. Crossed lamellar structures have first-order lamellae composed of thin, mutually parallel laths or rods, which show two non-horizontal dip directions of their elongate subunits, alternating in adjacent lamellae (Carter and Clark 1985). The lamellae are parallel or perpendicular to the shell margin in comarginal and radial crossed lamellar structures, respectively.

Determination of the Rate and Environmental Context of Marginellid Evolution

Sample dates and stratigraphic relationships were established by Dominican Republic Project and Panama Paleontology Project researchers (Saunders et al. 1986; Collins 1993; L. Collins, pers. comm. 1995; Nehm and Geary 1994; appendix). Each fossil or recent sample was categorized lithologically (sand, pebbly silt, or silt) and environmentally (brackish, shallow marine, marine, deep marine) in order to examine the relationship of morphological evolution with lithology and paleoenvironment. These seven categorical designations (sand, pebbly silt, silt, brackish, shallow marine, marine, deep marine) were made from observations in the field and/or paleoecological data from Saunders et al. (1982, 1986), van den Bold (1988), Vokes (1979, 1989), Nehm and Geary (1994), and Anderson (1994). Environmental, geographic, and bathymetric data for living species in clade B were extracted from the R. V. Pilsbury expedition ship logs from the University of Miami Rosenstiel School of Marine and Atmospheric Science.

Documentation of Ontogenetic Evolution

Growth series (*sensu* Raup and Stanley 1981) are constructed for *Prunum* and *"Volvarina"* by comparing individuals from the same population at different sizes and whorl volutions (developmental stages). This method is less accurate and informative than X-ray studies of individual ontogenies, but it is necessary because callus development through ontogeny is poorly visible in X-rays.

X-ray studies of individual ontogenies complement data obtained in growth series (fig. 6.6). The ventral surface of *Prunum* and *"Volvarina"* shells is generally flat and smooth. Shells were X-rayed in this life position (ventral surface down) in order to provide a consistent orientation for comparing the size and shape of individuals through ontogeny both within and among clades. *Prunum* and *"Volvarina"* shells were X-rayed at 30 kW for 60 seconds and 30 kW for 30 seconds, respectively, on Kodak Industrial M paper and developed using a Kodak Medical Autodeveloper.

Patterns of Morphological Evolution

Patterns of Morphospace Occupation in *Prunum* + *"Volvarina"*

Construction of a simple three-parameter morphospace, and mapping all known living and fossil *Prunum* and *"Volvarina"* species within it, produces a summary of general patterns of morphological diversity (fig. 6.7). In general, species in the *Prunum* + *"Volvarina"* clade represent two morphological "themes" that have existed throughout the Neogene. One theme is characterized by large, oval, and extensively callused morphologies, and the other theme is characterized by small, cylindrical, and poorly callused morphologies. Each theme corresponds to the systematic designations *Prunum* and *"Volvarina,"* respectively.

Morphological Evolution in the Three *Prunum* + *"Volvarina"* Clades

Univariate and multivariate analyses of distance and categorical variables indicate that patterns of morphological evolution are remarkably similar in all three clades (fig. 6.8). Univariate comparisons of distance variables indicate a reduction in magnitude of shell height, width, aperture height, columellar fold height, lip width, girth, and spire height in all three clades. Alternatively, the ratio of shell height/width increases in all three clades. Overall, large and oval shells evolve into small and cylindrical shells in all three clades. In separate PCAs for each clade, all distance variables score high (>0.80) on the first axis, supporting the conclusion that size and size-correlated shape changes best summarize the variation among species in each clade. Univariate comparisons of categorical variables also indicate that patterns of morphological evolution are similar in all three clades. Considerable reduction occurs in the area and thickness of the callus ridge, callus spoon, and posterior lip callus.

Although the direction of morphological evolution is similar among all three clades, the magnitude of change is slightly different (fig. 6.8). For example, callus spoon reduction is large in clade A, but small in clade B. Likewise, shape changes are minor in clade B, but major in clade A. Overall, the magnitude of morphological evolution is most different be-

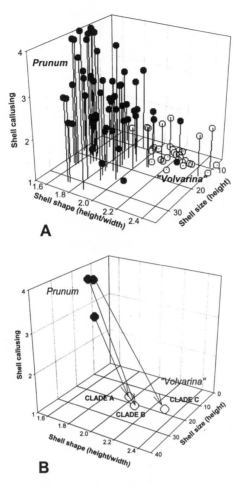

Fig. 6.7. *Prunum* + *"Volvarina"* morphospaces. Axes represent shell shape (height/width), shell size (total ventral area), and callus coverage on the body whorl. Each dot represents one species. *n* = 120 species. *A,* In general, Neogene-Recent *Prunum* and *"Volvarina"* are distinct morphological "themes." *B,* Speciation events in clades A, B, and C are mapped within this morphospace.

tween clades A and B; clade C is intermediate in terms of the magnitude of morphological change from large and oval to small and cylindrical shells. In summary, all three clades traverse the maximum morphological distance observed in the *Prunum* + *"Volvarina"* morphospace (fig. 6.7). In other words, each speciation event exhibits as much morphological evolution as has been achieved during the entire history of the *Prunum* + *"Volvarina"* clade.

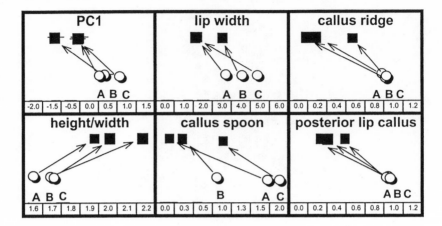

"Volvarina"■
Prunum ○

Fig. 6.8. Comparison of patterns of adult morphological evolution in six variables from *Prunum* to *"Volvarina."* *Circles* represent *Prunum* species, and *squares* represent descendant *"Volvarina"* species. All three clades exhibit similar patterns of morphological evolution. See text for morphometric variable descriptions. *PC1* = principal component analysis axis 1 (see text). Overall, remarkably similar evolutionary patterns occur in the three clades.

Intraspecific versus Interspecific Morphological Differences

Patterns of morphological variation within ancestral *Prunum* species are different from patterns of variation between ancestral *Prunum* and derived *"Volvarina."* In all three clades, the evolution of *"Volvarina"* from *Prunum* is characterized by a decrease in size and the production of a cylindrical shell form. Intraspecific variation in ancestral *Prunum* is characterized by small and spherical shells, or large and oval shells (fig. 6.3C and D, for example), whereas intraspecific variation in *"Volvarina"* is characterized by cylindrical shells with little adult size variation. Although adult *Prunum* shells vary considerably in size, small adult *Prunum* shells are spherical and never approach the cylindrical shape of small adult *"Volvarina."* In summary, patterns of intraspecific morphological variation are different from patterns of interspecific morphological variation in the three clades.

Microstructural Evolution from *Prunum* to *"Volvarina"*

SEM analyses of the microstructural composition of the shell whorls, external varix, inner lip, posterior and dorsal lip callus, anterior aperture margin callus, and posterior aperture margin callus reveal, as in morphometric comparisons, similar patterns among clades A, B, and C. Table

Table 6.1. Distribution of shell microstructures among species in *Prunum* clades A, B, and C

Clade	Phylogeny	Species	Morphology	SPR	CCL	RCL	SCCL
A	Outgroup	*P.* cf. *coniforme*	Adult	X	X	X	X
	Ingroup	*P. coniforme*	Adult	X	X	X	X
	Ingroup	*P. coniforme*	Juvenile	X	X	X	
	Ingroup	*"V." christineladdae*	Adult	X	X	X	
B	Outgroup	*P. circumvittatum*	Adult	X	X	X	X
	Ingroup	*P. marginatum*	Adult	X	X	X	X
	Ingroup	*P. marginatum*	Juvenile	X	X	X	
	Ingroup	*"V."* species B	Adult	X	X	X	
C	Outgroup	*P. gatunensis*	Adult	X	X	X	X
	Ingroup	*P. macdonaldi*	Adult	X	X	X	X
	Ingroup	*P. macdonaldi*	Juvenile	X	X	X	
	Ingroup	*"V."* species C	Adult	X	X	X	

Notes: SPR = simple prismatic layer; CCL = comarginal crossed lamellar layer deposited at the shell margin; RCL = radial crossed lamellar layer; SCCL = surficially deposited inner comarginal crossed lamellar layer; X = presence of a layer.

6.1 lists the microstructural layers of the primary shell in adult *Prunum* and adult *"Volvarina." Prunum* and *"Volvarina"* in all three clades share (from the exterior to the interior of the shell) simple prismatic, comarginal crossed lamellar, and radial crossed lamellar layers. In addition, *Prunum* possesses a thick inner comarginal crossed lamellar layer which is absent altogether in *"Volvarina."*

Shell shape is correlated with the number of shell microstructure layers. As indicated above, *"Volvarina"* shells are small and cylindrical, and lack an inner comarginal crossed lamellar layer. *Prunum* shells are large and robust, and have a thick inner comarginal crossed lamellar layer. The inner layer cannot account for shell shape, however, because it is surficially deposited after shell shape has already been established. It is the incremental deposition of simple prismatic, comarginal crossed lamellar, and radial crossed lamellar layers at the lip margin that generates shell shape in all clades.

"Volvarina" species in all three clades lack the dorsal lip callus and the posterior aperture margin callus, but other microstructural features of the shell are similar. The external varix and inner lip thickening of *Prunum* and *"Volvarina"* are formed by an expansion of the comarginal crossed lamellar layers in all three clades. *"Volvarina"* species deposit less material than *Prunum* species, but the lips of each are positionally and compositionally similar.

Shell callus is present in *Prunum* but reduced in volume or absent in *"Volvarina."* When present, the callus of *"Volvarina"* is compositionally the same as *Prunum*. The anterior and posterior aperture margin callus is composed of simple prismatic and comarginal crossed lamellar layers, whereas the dorsal lip callus is .composed exclusively of comarginal crossed lamellar layers. The dorsal lip callus, the anterior aperture margin callus, and the posterior aperture margin callus are surficially deposited external to the simple prismatic shell layer of the primary shell. In summary, as in morphometric studies, similar microstructural evolution occurs in all three clades.

The Rate and Environmental Context of Morphological Evolution

The tempo of morphological and microstructural evolution from *Prunum* to *"Volvarina"* is well-constrained in clade A and clade B, but poorly constrained in clade C. In clade A the morphological changes between *Prunum* and *"Volvarina"* took place in 73,000 to 275,000 years (Nehm and Geary 1994). This transitional interval is marked by a series of morphological intermediates. In clade B the evolution of *"Volvarina"* takes place in less than 1 million years. These changes probably occurred in the late Pleistocene or Recent, inasmuch as no fossils of *"Volvarina"* have been found despite a considerable amount of sampling in appropriate geographic regions and paleoenvironmental settings (such as shallow-marine sediments of Venezuela, Colombia, and Panama). In clade C *"Volvarina"* originated in the Upper Miocene, but no information is available on the duration of the transition between forms. Because ancestral *Prunum* species in all three clades lived for approximately 6 to 12 million years, the morphological evolution of *"Volvarina"* from *Prunum* took place rapidly relative to the duration of the ancestral species.

Are the similar patterns of morphological and microstructural evolution observed in the three western Atlantic clades a product of similar environmental or ecological conditions? We can address this question comprehensively by constructing a marginellid design space (Hickman 1993). Theoretical design space subsumes morphospace through the addition of function, ecology, environment, or behavior. The design space used here incorporates lithology (biogenic or siliciclastic), paleobathymetry (shallow or deep marine), and morphology (height/width). Placing the western Atlantic clades in a design space creates a framework for tracking and comparing actual evolutionary pathways and their extrinsic contexts relative to possible evolutionary pathways (fig. 6.9).

In clade A, ancestral *Prunum* are large and robust in shallow-marine siliciclastic paleoenvironments (<50 m paleodepth) and evolve into small and cylindrical *"Volvarina"* in deep-marine siliciclastic paleoenvironments (>200 m paleodepth). In clade B, *Prunum* are large and robust in deep-marine muddy and siliciclastic environments (>50 m depth) and evolve into small and cylindrical *"Volvarina"* in shallow-marine (<30 m

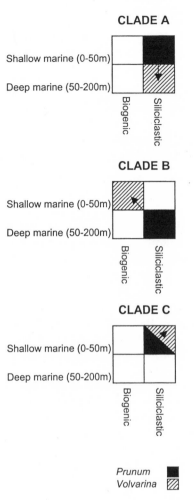

Fig. 6.9. Design space for clades A, B, and C, illustrating the bathymetric and lithological context of evolutionary change from ancestral *Prunum* to derived *"Volvarina."*

depth) sponge-dominated substrates. In clade C, *Prunum* are large and robust in shallow-marine siliciclastic paleoenvironments and evolve into small and cylindrical *"Volvarina"* in the same conditions. Design space provides an interesting picture of the extrinsic context of morphological evolution in the western Atlantic clades, and demonstrates that, although all clades follow similar directions of morphological evolution, they exhibit different trajectories in design space (fig. 6.9).

Thus far it has been demonstrated that (1) *"Volvarina"* is a polyphyletic group derived from several ancestral *Prunum* lineages, (2) nearly identical patterns of morphological evolution occurred in each clade, (3) the morphological transitions between species were rapid relative to the duration of the ancestral species, (4) intraspecific variation follows a different morphological "direction" than interspecific evolution, and (5) the physical environmental and sedimentological context of morphological evolution were different in each clade.

Patterns of Developmental Evolution

Morphometric Comparisons of Ontogenetic Evolution

Are all clades characterized by similar patterns of developmental evolution? Specifically, how do the ontogenies of *Prunum* and *"Volvarina"* differ in each clade? Ontogenetic comparisons of the morphological features of *Prunum* and *"Volvarina"* within each clade produce a striking pattern: nearly all features of ancestral juvenile *Prunum* are shared with derived adult *"Volvarina"* in all three clades (table 6.2). Adult *"Volvarina"* and juvenile *Prunum* share small, thin, cylindrical shells lacking extensive callusing, inner lip thickening, and lip denticulations. Yet adult *"Volvarina"* are differentiated from juvenile *Prunum* by the presence of a weak external varix and an ascending body-whorl suture. These features indicate sexual maturity and the cessation of growth have occurred in *"Volvarina."* Overall, however, juvenile features of ancestral *Prunum* are retained in the adult of *"Volvarina."*

Morphometric comparisons of adult *Prunum* and *"Volvarina"* produce good separation in bivariate morphospace for all three clades (fig. 6.10). Clade A demonstrates overlap between adult *Prunum* and *"Volvarina."* This pattern is due to a temporally brief (73,000 to 275,000 year) series of abundant transitional forms. Clades B and C separate well, with minimal overlap. Unlike the comparison of adult forms, juvenile *Prunum* and adult *"Volvarina"* from all three clades overlap in bivariate

Table 6.2. The distribution of conchological characters among juvenile *Prunum*, adult *"Volvarina,"* and adult *Prunum*

Character	Juvenile *Prunum*	Adult *"Volvarina"*	Adult *Prunum*
Shell size	Small	Small	Large
Shell thickness	Thin	Thin	Thick
External varix	Absent	Reduced or absent	Present
Lip thickening	Absent	Present thin	Present thick
Shell shape	Cylindrical	Cylindrical	Obovate
Spire	Low	Low	Low
Denticulations	Absent	Absent	Present or absent
Ventral callusing	Absent	Absent	Present

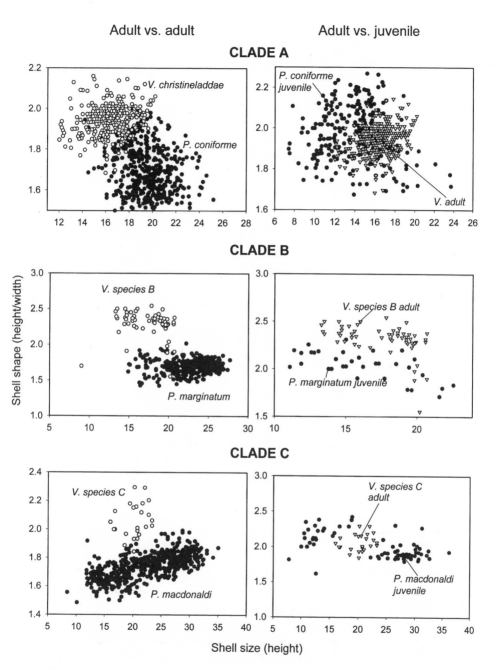

Fig. 6.10. Morphometric comparisons among adult *Prunum* and adult *"Volvarina"* and juvenile *"Volvarina"* and adult *Prunum* from clades A, B, and C. Adult *Prunum* and adult *"Volvarina"* are well differentiated, whereas adult *"Volvarina"* and juvenile *Prunum* are similar sizes and shapes.

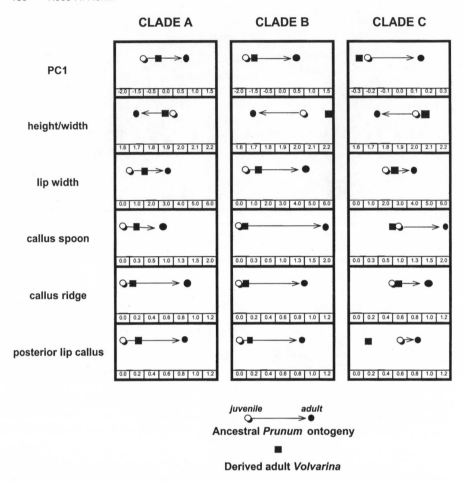

Fig. 6.11. Ontogeny and phylogeny in clades A, B, and C. *Circles* represent the ontogeny of ancestral *Prunum*: *open circles* are mean values for juvenile specimens, and *filled circles* are mean values for adult *Prunum*. An *arrow* connects the ontogenetic sequence. Mean values for adult *"Volvarina"* are represented by *filled squares*. In all features of all clades, derived adult *"Volvarina"* are most similar to ancestral juvenile *Prunum*.

plots using the diagnostic features of shell size and shape, supporting qualitative assessments of similarity (table 6.2).

Figure 6.11 summarizes the relationship between ontogeny and phylogeny in the three clades using bivariate and multivariate (PCA) morphometric analyses. Circles are used to represent the ontogeny of ancestral *Prunum*: open circles are mean values for juvenile specimens, and filled circles are mean values for adult *Prunum*. An arrow connects the ontogenetic sequence. Mean values for adult *"Volvarina"* are represented by

Table 6.3. Discriminant function analysis classification table for adult *Prunum*, juvenile *Prunum*, and adult *"Volvarina"* in clades A, B, and C

Morphology	Total	A (%)	n	J (%)	n	V (%)	n
				Predicted Placement			
Clade A (*n* = 1007)							
Adult *Prunum*	455	94.51	430	2.42	11	3.08	14
Juvenile *Prunum*	242	1.24	3	96.28	233	2.48	6
Adult *"Volvarina"*	310	8.71	27	16.77	52	74.52	231
Clade B (*n* = 501)							
Adult *Prunum*	406	97.54	396	2.46	10	0	0
Juvenile *Prunum*	33	0	0	100.00	33	0	0
Adult *"Volvarina"*	62	1.61	1	3.23	2	95.16	59
Clade C (*n* = 694)							
Adult *Prunum*	596	100.00	596	0	0	0	0
Juvenile *Prunum*	74	6.76	5	93.24	69	0	0
Adult *"Volvarina"*	24	41.67	10	4.17	1	54.17	13

Notes: Boxes contain the percentage of specimens correctly classified. Other numbers and percentages are for specimens that were predicted to be in other categories. A = adult *Prunum*; J = juvenile *Prunum*; V = adult *"Volvarina."*

filled squares. In nearly all features in all clades derived adult *"Volvarina"* are most similar to ancestral juvenile *Prunum*.

Are *"Volvarina"* well-defined morphologically, or are they predicted to be juvenile or adult *Prunum*? CVA is used to compare actual versus predicted group membership in adult and juvenile categories for each clade (table 6.3). Overall, adult and juvenile specimens of *Prunum* in clades A, B, and C, and *"Volvarina"* in clade B, are correctly placed (actual = predicted) in over 93% of all cases. However, there are only moderate levels of similarity between actual and predicted group membership for adult *"Volvarina"* in clades A and C. Many of the specimens identified as adult *"Volvarina"* are predicted to be juvenile *Prunum* in clade A but are predicted to be adult *Prunum* in clade C. This is not surprising considering that adult *"Volvarina"* have a mixture of the features from juvenile and adult *Prunum*. Overall, the CVA corroborated the initial classifications of 94% of the 2,500 specimens.

Species of *Prunum* and *"Volvarina"* in all three western Atlantic clades possess four whorls. It is therefore possible to study the ontogenetic evolution from *Prunum* to *"Volvarina"* by comparing the size and shape of whorls 1 to 4 through ontogeny and phylogeny. Are similar patterns of morphological evolution from adult *Prunum* to adult *"Volvarina"* across clades the result of similar ontogenetic changes? When do differences between *Prunum* and *"Volvarina"* first occur in ontogeny? Morphometric study of X-rays provides the answers to these questions.

Adult *"Volvarina"* are smaller than adult *Prunum.* These size differences (measured by shell height) originate at different stages of ontogeny in the three clades (fig. 6.12). In clade A the size of the first whorl (the protoconch) is the same for both *Prunum* and *"Volvarina."* Divergence in shell size between *Prunum* and *"Volvarina"* first occurs between whorl 1 and 2, and continues through whorl 4, in which the greatest size difference develops. In clade B whorl 1 of *Prunum* is slightly larger than *"Volvarina,"* but whorl 2 is the same size in the two species. Whorl 3, however, is of considerably different size among taxa, and, once again, size differences are greatest at whorl 4. Clade C is unique in that *"Volvarina"* and *Prunum* are nearly the same height through ontogeny and phylogeny. However, shell height is not an accurate estimate of size in this case; *"Volvarina"* is remarkably narrow compared to *Prunum* of similar height.

Comparisons of shell shape (measured as shell height/shell width) between whorls and species reveal similar ontogenetic patterns across all three clades. Protoconch shape (whorl 1) is stout in *"Volvarina"* and *Prunum* from clades A, B, and C. There is considerable shape variation in whorl 1, primarily as a result of measurement difficulty. In all three clades, shells become more cylindrical in whorls 2 and 3 because height is increasing more rapidly than width. In general, the shape of whorl 4 returns to the more oval shape of whorl 1. This pattern is best exemplified in clade C. The increase in width relative to height at adulthood is primarily a result of a wider aperture and the formation of the lip varix (which widens the body whorl).

X-ray studies of ontogenetic and phylogenetic changes in all three western Atlantic clades indicate that adult *"Volvarina"* retain the shape of the early whorls of ancestral *Prunum,* and that differences in size occur at different whorl numbers. Similar patterns of shape change also occur through ontogeny and phylogeny in all three clades. These observations corroborate the previous conclusions based on qualitative and quantitative (bivariate and multivariate) comparisons. In summary, the evolutionary retention of ancestral juvenile features occurs in descendant adults: global paedomorphosis occurs in all three clades, and is correlated with the repeated generation of the *"Volvarina"* morphology.

Microstructural Evolution through Ontogeny

How are different morphological features of the shell constructed by the mantle through ontogeny? Are similar adult morphologies a product of similar structural and compositional changes during ontogeny? As in adult morphological and ontogenetic evolution, patterns of microstructural deposition through ontogeny are remarkably similar for all the characters studied in all three *Prunum* clades. Figure 6.13 summarizes the

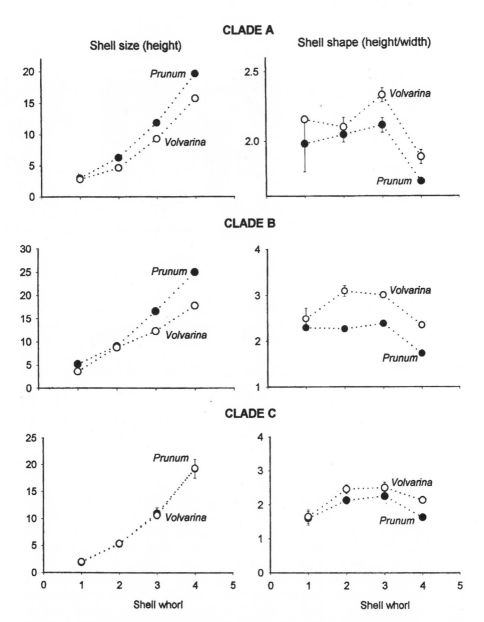

Fig. 6.12. Ontogenetic comparisons between ancestral *Prunum* and derived *"Volvarina"* from clades A, B, and C. Mean values and two standard errors are indicated for whorl height and whorl height/width. Both *Prunum* and *"Volvarina"* in the three clades have four shell whorls.

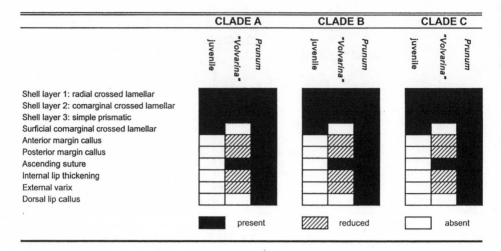

Fig. 6.13. The distribution of shell microstructure features in juvenile *Prunum*, adult *"Volvarina,"* and adult *Prunum* in clades A, B, and C.

presence/absence, microstructural composition, and general order of appearance of shell features through ontogeny in juveniles and adults of *Prunum* and *"Volvarina"* in each clade. Juvenile shells in all three clades possess simple prismatic, comarginal crossed lamellar, and radial crossed lamellar layers. These layers are deposited by incremental growth in a zone close to the margin during early ontogeny. The absolute thickness of these layers increases through ontogeny, but the relative thickness remains largely constant. Juveniles possess a very faint surficially deposited inner crossed lamellar layer. Adult *Prunum* possess a thick surficially deposited inner crossed lamellar layer in addition to simple prismatic, comarginal crossed lamellar, and radial crossed lamellar layers.

Morphology, X-ray, and microstructure studies all indicate that *"Volvarina"* species retain the form of the ancestral juvenile shell, but differ from ancestral adults in both the presence of shell features and the magnitude of microstructure deposition (but not composition). *"Volvarina"* shells, like juveniles and normal adults, are a product of the deposition of simple prismatic, comarginal crossed lamellar, and radial crossed lamellar layers at the shell margin. All subsequent shell deposition differs from ancestral adults: the dorsal lip callus, posterior margin callus, and internal crossed lamellar shell layer do not develop, and the external varix, internal lip, and anterior margin callus do develop but are considerably reduced. Those features that are present in both adult *Prunum* and *"Volvarina"* are similar in position and microstructural composition. Overall, derived adult *"Volvarina"* possess shapes and sizes similar to ancestral juvenile *Prunum,* but exhibit characteristics indicative of maturity in adult *Prunum* (e.g., a thickened lip, an ascending suture, and an external varix).

Discussion: Linking Pattern and Process in Marginellid Evolution

The foregoing analysis of morphological evolution in adult marginellid gastropods has demonstrated that (1) *"Volvarina"* is a polyphyletic group derived from several *Prunum* lineages; (2) morphological evolution from *Prunum* to *"Volvarina"* is similar in direction and magnitude in all clades; (3) patterns of morphological variation within species are different from patterns of evolution between species; (4) interspecific morphological differences in all three clades are a result of similar ontogenetic and microstructural evolution; (5) the ecological, environmental, and geographic contexts of developmental and morphological evolution are different in all three clades; and (6) morphological evolution of *"Volvarina"* morphologies is rapid relative to the duration of ancestral *Prunum* species. Several developmental patterns may now be added to these conclusions: (1) juvenile features of ancestral *Prunum* are retained in adults of descendant *"Volvarina"*: paedomorphosis occurred in all three clades; (2) paedomorphosis produced nearly identical morphogenetic (microstructural) changes through ontogeny in all clades; and (3) similar paedomorphic patterns occur regardless of phylogenetic, ecological, or environmental differences.

Paedomorphosis clearly explains many aspects of morphological evolution in *Prunum* and *"Volvarina,"* but what extrinsic and intrinsic processes initially *generated* and ultimately *maintained* paedomorphs? Phylogenetic, ontogenetic, and morphogenetic patterns are used to address two questions: First, what genetic, developmental, and epigenetic processes initially *generated* paedomorphs? Second, what processes *maintained* paedomorphs after they appeared?

What Genetic, Developmental, and Epigenetic Processes Initially Generated Paedomorphs?

It is difficult, if not impossible, to determine the processes that *initially* generated paedomorphosis using paleontological data. Neontological research has not yet established how genetic, epigenetic, and developmental processes collectively interact to produce a paedomorphic *result,* although substantial progress has been made toward answering this question (Gerhart and Kirschner 1997; Raff 1996). Despite these difficulties, paleontological patterns can contribute to an understanding of generating processes by constraining the realm of likely hypotheses, providing data for the testing of these hypotheses, and establishing direction for neonto-logical testing of such processes (Cheetham et al. 1993, 1994; Cheetham and Jackson 1995).

In all three clades, remarkably similar patterns of paedomorphosis oc-cur in very different ecological and environmental settings, suggesting that intrinsic processes are an important component of these patterns. I focus on two alternative hypotheses that may explain the nearly identical patterns of morphological, morphogenetic (microstructural), and ontoge-netic evolution in the three marginellid clades. First, did the same genetic changes occur in each clade to produce one phenotypic result? Paedomor-phosis in some salamander clades, for example, has been suggested to be a result of changes in a single gene (Tompkins 1978; however, see Voss 1995 for an opposing view). Second, did developmental channeling of a host of different genetic changes produce one phenotypic result? For example, similar paedomorphic morphologies in other salamander clades have been demonstrated to be the result of several interacting genetic and developmental factors (Voss 1995; Whiteman 1994; references therein).

The developmental channeling hypothesis is supported by phyloge-netic and morphological data. The western Atlantic lineages are not sister clades and have been separated since the Miocene. It seems unlikely that the same genetic changes occurred in all three clades to produce nearly identical morphological results. A more likely explanation is that differ-ent genetic changes occurred and were channeled by the developmental program to produce one morphological result. This hypothesis may be tested by comparing genetic differences among living species of "Volva-rina" and Prunum and their morphological and developmental effects. In addition to exploring the generation of paedomorphs, paleontological data can more extensively contribute to our understanding of the pro-cesses that selectively maintained paedomorphs after they arose.

What Processes Maintained Paedomorphs after They Evolved?

The adaptive significance of paedomorphosis may include aspects of mor-phology, life history, or both (Gould 1977). Morphological changes from Prunum to "Volvarina" may be a result of selection for reduced energy expenditure. The shell is a costly structure in terms of the metabolic en-ergy required for its construction (Palmer 1983). In all three clades, the truncation of the ancestral Prunum ontogeny occurs at a stage just prior to the deposition of the posterior callus, the dorsal lip callus, and the inner surficially deposited comarginal crossed lamellar layer. Three other morphological features, the external varix, internal lip thickening, and anterior margin callus, are also greatly reduced in area and volume. Trun-cation and reorganization of the morphogenetic program just prior to the construction of these shell features may have been maintained by selection for energy conservation through reduced shell deposition and smaller shell size.

Size reduction, and the loss or reduction of callus features in all three clades, may also be tied to duraphagous predation. The ecological and evolutionary dynamics of gastropod mollusks and duraphagous predators have figured prominently in contemporary models of molluscan morphological evolution. Vermeij (1978, 1987) hypothesized that many gastropod shell features, including thick shell whorls, well-developed lips, and narrow apertures, bestow resistance to duraphagous predators. In marginellids (including *Prunum*) the smooth, polished, and globose shape of the shell (low height/width values) has been hypothesized to prevent crushing and peeling (Vermeij, pers. comm. 1998). *Prunum* shells in all three clades exhibit high percentages of repair scars (regardless of bathymetry or substrate), which represent unsuccessful attacks by duraphagous predators (Vermeij 1978). Predation may therefore have played a role in the morphological evolution of callus features and shell size. It is possible that the extensive callusing, lip thickening, and oval form of *Prunum* were maintained by strong selection as a result of attacks by durophagous predators. In the absence of predators over long time intervals, reducing the suite of metabolically expensive shell features (as occurs in "*Volvarina*") may be most easily accomplished by maintaining thinner and less callused juvenile morphology into adulthood. This hypothesis also explains why bathymetric and sedimentologic variables are not correlated with patterns of paedomorphosis in the three clades: the selective advantage of paedomorphosis may be associated with the reduction of predation intensity rather than other environmental parameters. Further research is necessary to test this hypothesis.

The morphological changes associated with paedomorphosis may also be by-products of selection for life-history strategies (Gould 1977; McKinney and McNamara 1991, 271). Two different developmental paths, each with different life-history consequences, may have been taken by the three clades to produce a paedomorphic result. Progenesis is thought to produce small individuals and favor early maturation in unpredictable and fluctuating environments (shallow water) with abundant resources. Neoteny, on the other hand, is thought to produce large individuals and favor "K selected" life histories in environmentally stable (deep water), crowded, and high-diversity habitats. It is not possible to determine the changes in the timing of reproduction in fossil marginellids and thereby distinguish neoteny from progenesis. Using environmental data, however, it is possible to predict which mode would be advantageous for each clade. In clade A, paedomorphosis occurred in very deep (>200 m), stable, and low-diversity environments. In clade B, paedomorphosis occurred in shallow (<30 m) and fluctuating high-diversity environments.

In clade C, paedomorphosis occurred in shallow-marine sponge-dominated substrates with generally low diversity. Based on environmental stability alone, neoteny may have been advantageous in clade A, whereas progenesis may have been advantageous in clades B and C. However, other environmental parameters, such as community diversity, provide different predictions. Overall, different selective regimes for life history may have been acting on the three clades, resulting in the maintenance of different paedomorphic results (neoteny or progenesis). Nevertheless, the morphological results were the same across all three clades despite potentially different selective scenarios.

Although developmental canalization is used as an explanation for the *generation* of similar paedomorph morphologies, it may also be invoked as a process that *maintains* paedomorph morphologies (Riedl 1978). The nonrandom apportionment of intraspecific and interspecific variation in morphospace, and the stability of such patterns across different environments and ecologies, is considered to be a result of the structure and mechanistic rules of development (Oster and Alberch 1982). Bounded domains of developmental stability are thought to maintain morphological patterns through time and space. Mutation and selection cause dispersion within these developmental domains, but variation is filtered and buffered to maintain morphological "themes" (such as the nearly identical paedomorph construction in all three clades). Gould (1984) provided an example of such developmental rules in *Cerion* by demonstrating that a variety of different perturbations resulted in the same morphological outcome. This view of development must also be considered as an explanation for the maintenance of paedomorphosis in all three clades.

Summary

Paleontological investigations of morphological evolution in three *Prunum* + "*Volvarina*" clades from the Neogene to Recent of the western Atlantic demonstrate several striking patterns that contribute substantially to our understanding of the developmental processes at work in large-scale morphological evolution: (1) "*Volvarina*" is a polyphyletic group derived from several *Prunum* lineages. (2) Morphological evolution from *Prunum* to "*Volvarina*" was similar in direction and magnitude in three western Atlantic clades. (3) Patterns of morphological and developmental variation within species are different from patterns of change between species. (4) Interspecific morphological differences in all clades were achieved by similar ontogenetic and microstructural changes. (5) The ecological, environmental, and geographic context of developmental and morphological evolution was different in all three clades. (6) The evolution of *"Volvarina"* from *Prunum* was rapid relative to the temporal duration of ancestral *Prunum*. (7) Juvenile features of ancestral *Prunum* are retained in adults of descendant *"Volvarina"*: paedomorphosis occurred in all three clades. (8) Paedomorphosis produced nearly identical morphogenetic (microstructural) changes through ontogeny in all clades. (9) Similar paedomorphic patterns occurred regardless of phylogenetic, ecological, or environmental differences. (10) Duraphagous predation is documented in all three clades, and may have played a role in the selective maintenance of paedomorphic morphologies.

Studies of morphological disparity have triggered considerable debate regarding the role of developmental processes in macroevolutionary change. This study uses marginellid gastropods to explore how paleontologists may analyze the role of developmental processes in patterns of morphological disparity. Morphological disparity in the *Prunum* + *"Volvarina"* clade is clustered into two generally nonoverlapping groups that have existed since the early Neogene. Nearly identical developmental changes occurred in three independent origins of *"Volvarina"* species from *Prunum* species. Whether all species in the *"Volvarina"* cluster are a product of similar developmental evolution requires further phylogenetic, morphometric, and morphogenetic study. Nevertheless, developmental processes appear to play a fundamental role in patterns of morphospace occupation in the *Prunum* + *"Volvarina"* clade.

Acknowledgments

I am grateful to J. B. C. Jackson, F. K. McKinney, and S. Lidgard for the opportunity to include my work in this compendium, and to A. Cheetham and D. Geary for sparking my interest in both speciation and the Neogene of the Dominican Republic. The thousands of marginellid samples used for this study were provided by Peter Jung and René Panchaud of the Naturhistorisches Museum, Basel, Switzerland; Jack and Winifred Gibson-Smith of Surrey, England; Emily Vokes of Tulane University; and Tom Waller and Warren Blow of the Smithsonian Institution. I am grateful for the access to these collections and the generous assistance provided by these individuals. Mark Groves, Carole Hickman, Ken McKinney, Kevin Padian, Jim Valentine, and two anonymous reviewers made helpful comments on the manuscript. Karen Wetmore provided technical assistance with the University of California Museum of Paleontology's Electroscan environmental scanning electron microscope (ESEM), and Claus Hedegaard helped interpret SEM images. Financial support by the Geological Society of America, Sigma Xi, and the National Science Foundation (Dissertation Improvement Grant DEB 9520457) is gratefully acknowledged.

Appendix

Table 6A.1. Specimen data for samples examined microstructurally from clades A, B, and C

Clade	Species	Ontogeny	Sample	Age	Formation	Locality
A	*P.* cf. *coniforme*	A	NMB 15865	Miocene	Baitoa	Dominican Republic
	P. coniforme	J	TU 1227	Pliocene	Gurabo	Dominican Republic
	P. coniforme	A	TU 1206	Pliocene	Gurabo	Dominican Republic
	"*V.*" *christineladdae*	A	TU 1211	Pliocene	Gurabo	Dominican Republic
B	*P. circumvittatum*	A	NMB 17512	Pliocene	Mare	Venezuela
	P. marginatum	J	UMML 30-6890	Recent	Recent	Southern Caribbean
	P. marginatum	A	UMML 30-7224	Recent	Recent	Southern Caribbean
	"*V.*" species B	A	UMML 30-3988	Recent	Recent	Southern Caribbean
C	*P. gatunensis*	A	NMB 17808	Miocene	Rio Banano	Costa Rica
	P. macdonaldi	J	TU 589	Pliocene	Rio Banano	Costa Rica
	P. macdonaldi	A	TU 589	Pliocene	Rio Banano	Costa Rica
	"*V.*" species C	A	NMB 17810	Pliocene	Rio Banano	Costa Rica

Notes: A = adult *Prunum*; J = juvenile *Prunum*; NMB = Naturhistorisches Museum Basel; UMML = University of Miami Marine Labs (RSMAS) Invertebrate Museum; TU = Tulane University.

References

Allmon, W. D. 1994. Patterns and processes of heterochrony in lower Tertiary Turitelline gastropods, U.S. Gulf and Atlantic coastal plains. *Journal of Paleontology* 68 (1): 80–95.

Allmon, W. D., G. Rosenberg, R. Portell, and K. Schindler. 1993. Diversity of Atlantic coastal plain mollusks since the Pliocene. *Science* 260:1626–29.

Anderson, L. C. 1994. Paleoenvironmental control of species distributions and intraspecific variability in Neogene Corbulidae (Bivalvia: Myacea) of the Dominican Republic. *Journal of Paleontology* 68 (3): 460–73.

Born, I. 1778. *Index Rerum Naturalium Musei Caesarei Vindobonensis Pars: I, Testacea.*

Boss, K. J. 1982. Mollusca. In *Synopsis and classification of living organisms,* ed. S. P. Parker, 1:430–73. New York: McGraw-Hill.

Briggs, D. E. G., R. A. Fortey, and M. A. Willis. 1992. Morphological disparity in the Cambrian. *Science* 256:1670–73.

Brown, A. P., and H. A. Pilsbry. 1911. Fauna of the Gatun Formation, Isthmus of Panama. *Proceedings of the Academy of Natural Sciences, Philadelphia* 63: 336–73, pl. 22–29.

Carter, J. G., and G. R. Clark II. 1985. Classification and phylogenetic significance of molluscan shell microstructure. In *Mollusks: Notes for a short course,* organized by D. J. Bottjer, C. S. Hickman, and P. D. Ward, ed. T. W. Broadhead, 50–71. University of Tennessee Department of Geological Sciences Studies in Geology 13.

Cheetham, A. H., and J. B. C. Jackson. 1995. Process from pattern: Tests for selection versus random change in punctuated bryozoan speciation. In *New approaches to speciation in the fossil record,* ed. D. H. Erwin and R. L. Anstey, 184–207. New York: Columbia University Press.

Cheetham, A. H., J. B. C. Jackson, and L. C. Hayek. 1993. Quantitative genetics of bryozoan phenotypic evolution: 1, Rate tests for random change versus selection in differentiation of living species. *Evolution* 47:1526–38.

———. 1994. Quantitative genetics of bryozoan phenotypic evolution: 2, Analysis of selection and random change in fossil species using reconstructed genetic parameters. *Evolution* 48:360–75.

Collins, L. S. 1993. Neogene paleoenvironments of the Bocas del Toro Basin, Panama. *Journal of Paleontology* 67 (5): 699–710.

Coovert, G. A. 1986. A review of marginellid egg capsules. *Marginella Marginalia* (Dayton Museum of Natural History) 1 (2): 5–7.

———. 1987. A literature review and summary of marginellid external anatomy. *Marginella Marginalia* (Dayton Museum of Natural History) 3 (2–3): 8–25.

———. 1988. Marginellidae of Florida: 2, *Prunum succinea* with a discussion of *Prunum* and *Volvarina. Marginella Marginalia* (Dayton Museum of Natural History) 4 (5): 35–42, pl. 1.

———. 1989. A literature review and summary of published marginellid radulae. *Marginella Marginalia* (Dayton Museum of Natural History) 7 (1–6): 1–37.

Coovert, G. A., and H. K. Coovert. 1990. A study of marginellid radulae: 1, Type 6 radula, *"Prunum/Volvarina"* type. *Marginella Marginalia* (Dayton Museum of Natural History) 8–9:1–68.

———. 1995. Revision of the suprageneric classification of marginelliform gastropods. *Nautilus* 109 (2–3): 43–110.

Dall, W. H. 1912. New species of fossil shells from Panama and Costa Rica. *Smithsonian Miscellaneous Collections* 29 (2): 7.

Davidson, E. H., K. J. Peterson, and R. A. Cameron. 1995. Origin of bilateral body plans: Evolution of developmental regulatory mechanisms. *Science* 270: 1319–25.

Foote, M. 1991. Morphological and taxonomic diversity in a clade's history: The blastoid record and stochastic simulations. *Contributions from the Museum of Paleontology, University of Michigan* 28:101–40.

———. 1993a. Contributions of individual taxa to overall morphological disparity. *Paleobiology* 19 (4): 403–19.

———. 1993b. Discordance and concordance between morphological and taxonomic diversity. *Paleobiology* 19:185–204.

———. 1994. Morphological disparity in Ordovician-Devonian crinoids and the early saturation of morphological space. *Paleobiology* 20:320–44.

Foote, M., and S. J. Gould. 1992. Cambrian and Recent morphological disparity. *Science* 258:1816.

Geary, D. H. 1988. Heterochrony in gastropods: A paleontological view. In *Heterochrony in evolution: A multidisciplinary approach*, ed. M. L. McKinney, 183–96. New York: Plenum.

———. 1990a. Exploring the roles of intrinsic and extrinsic factors in the evolutionary radiation of *Melanopsis*. In *Causes of evolution: A paleontological perspective*, ed. R. M. Ross and W. A. Allmon, 305–21. Chicago: University of Chicago Press.

———. 1990b. Patterns of evolutionary tempo and mode in the radiation of *Melanopsis* (Gastropoda: Melanopsidae). *Paleobiology* 16:492–511.

Gerhart, J., and M. Kirschner. 1997. *Cells, embryos, and evolution.* Malden: Blackwell Science.

Gibson-Smith, J., and W. Gibson-Smith. 1979. The genus *Arcinella* (Mollusca: Bivalvia) in Venezuela and some associated faunas. *Geos* 24:11–32.

Gould, S. J. 1969. An evolutionary microcosm: Pleistocene and Recent history of the land snail P. (*Poecilozonites*) in Bermuda. *Bulletin of the Museum of Comparative Zoology* 138:407–532.

———. 1977. *Ontogeny and phylogeny.* Cambridge: Harvard University Press.

———. 1984. Morphological channeling by structural constraint: Convergence in styles of dwarfing and gigantism in *Cerion*, with a description of two new species and a report on the discovery of the largest *Cerion*. *Paleobiology* 10: 172–94.

Hall, B. K. 1999. *Evolutionary developmental biology.* Second edition. London: Chapman and Hall.

Harvey, P. H., and M. D. Pagel. 1991. *The comparative method.* Oxford: Oxford University Press.

Hickman, C. S. 1993. Theoretical design space: A new program for the analysis of structural diversity. *Neues Jahrbuch für Geologie und Palaontologie, Abhandlungen* 190 (2/3): 169–82.

Jackson, J. B. C., P. Jung, A. G. Coates, and L. S. Collins. 1993. Diversity and extinction of tropical American mollusks and emergence of the Isthmus of Panama. *Science* 260:1624–26.

Lee, M. S. Y. 1992. Cambrian and Recent morphological disparity. *Science* 258: 1816–17.

Lindberg, D. R. 1988. Heterochrony in gastropods: A neontological view. In *Heterochrony in evolution: A multidisciplinary approach,* ed. M. L. McKinney, 197–216. New York: Plenum.

Lipe, R. 1991. *Marginellas.* St. Petersburg: Sanibel Press.

Maury, C. J. 1917a. Santo Domingo type sections and fossils: Part 1. *Bulletins of American Paleontology* 5 (29): 1–251.

———. 1917b. Santo Domingo type sections and fossils: Part 2. *Bulletins of American Paleontology* 5 (30): 1–43.

McKinney, M. L., ed. 1988. *Heterochrony in evolution: A multidisciplinary approach.* New York: Plenum.

McKinney, M. L., and K. J. McNamara. 1991. *Heterochrony: The evolution of ontogeny.* New York: Plenum.

McNamara, K. J., ed. 1995. *Evolutionary change and heterochrony.* New York: John Wiley and Sons.

Nehm, R. H. 1994. A biogeographic comparison of evolutionary turnover in *Prunum* (Marginellidae: Neogastropoda) from the Neogene American tropics. *Geological Society of America Abstracts with Programs* 26 (7): A375.

———. 1996. Estimating phylogenetic relationships in marginelliform gastropods: Revisiting replicability and testability in molluscan systematics. *American Malacological Union Programs and Abstracts.*

———. 1998. Macroevolutionary pattern and developmental process in marginellid gastropods from the Neogene of the Caribbean Basin. Ph.D. diss., University of California, Berkeley.

———. 2001. Neogene paleontology of the northern Dominican Republic: 21, The genus *Prunum* (Gastropoda: Marginellidae). *Bulletins of American Paleontology* 359:1–71.

Nehm, R. H., and D. H. Geary. 1994. A gradual morphologic transition during a rapid speciation event in marginellid gastropods (Neogene: Dominican Republic). *Journal of Paleontology* 68 (4): 787–95.

Nehm, R. H., and C. S. Hickman. 1995. Size change in Neogene gastropods from the Dominican Republic: Taphonomic, ecologic, or evolutionary controls? *Geological Society of America Abstracts with Programs* 27 (6): A446.

Nehm, R. H., and W. B. Simison. 2000. Molecular phylogeny of marginelliform gastropods. Unpublished data.

Nehm, R. H., and C. Tran. 1997. Molecular phylogeny of marginelliform gastropods: A progress report. *American Malacological Union Program with Abstracts.*

Olsson, A. A. 1922. The Miocene of Costa Rica with notes on its general stratigraphic relations. *Bulletins of American Paleontology* 9:1–168.

Oster, G., and P. Alberch. 1982. Evolution and bifurcation of developmental programs. *Evolution* 36:444–59.

Palmer, A. R. 1983. Relative cost of producing skeletal organic matrix versus calcification: Evidence from marine gastropods. *Marine Biology* 75:287–92.

Petuch, E. J. 1981. A relict Neogene caenogastropod fauna from northern South America. *Malacologia* 20 (2): 307–47.

———. 1982. Geographical heterochrony: Contemporaneous coexistence of Neogene and Recent molluscan faunas in the Americas. *Palaeogeography, Palaeoclimatology, Palaeoecology* 37:277–312.

Ponder, W. F., and D. R. Lindberg. 1997. Towards a phylogeny of gastropod molluscs: An analysis using morphological characters. *Zoological Journal of the Linnaean Society* 119:83–265.

Rachootin, S. P., and K. S. Thomson. 1981. Epigenetics, paleontology, and evolution. In *Evolution today,* ed. G. G. E. Scudder and C. L. Reveal. Pittsburgh: Hunt Institute.

Raff, R. A. 1996. *The shape of life: Genes, development, and the evolution of animal form.* Chicago: University of Chicago Press.

Raup, D. M., and S. M. Stanley. 1981. *Principles of paleontology.* New York: W. H. Freeman and Company.

Riedl, R. 1978. *Order in living systems.* New York: Wiley.

Robinson, D. 1991. Systematics of gastropods from the Moin Formation. Ph.D. diss., Tulane University, New Orleans.

Saunders, J. B., P. Jung, and B. Biju-Duval. 1986. Neogene paleontology of the northern Dominican Republic: 1, Field surveys, lithology, environment, and age. *Bulletins of American Paleontology* 89 (323): 1–79.

Saunders, J. B., P. Jung, J. Geister, and B. Biju-Duval. 1982. The Neogene of the south bank of the Cibao Valley, Dominican Republic: A stratigraphic study. *Ninth Caribbean Geological Conference Transactions,* Santo Domingo, 1980, 1:151–60.

Smith, L. H., and P. M. Bunje. 1999. Morphologic diversity of inarticulate brachiopods through the Phanerozoic. *Paleobiology* 25 (3): 396–408.

Thomson, K. S. 1988. *Morphogenesis and evolution.* Oxford: Oxford University Press.

Tissot, 1988a. Geographic variation and heterochrony in two species of cowries (genus *Cypraea*). *Evolution* 42 (1): 103–7.

———. 1988b. Multivariate analysis. In *Heterochrony in evolution: A multidisciplinary approach,* ed. M. L. McKinney, 35–52. New York: Plenum.

Tompkins, R. 1978. Genic control of axolotl metamorphosis. *American Zoologist* 18:313–19.

Valentine, J. W., and D. H. Erwin. 1991. Interpreting great developmental experiments: The fossil record. In *Development as an evolutionary process,* ed. R. A. Raff and E. C. Raff, 71–107. New York: Alan A. Liss.

van den Bold, W. A. 1988. Neogene paleontology of the northern Dominican Republic: 7, The subclass Ostracoda (Arthropoda: Crustacea). *Bulletins of American Paleontology* 94 (329): 1–105.

Vermeij, G. J. 1978. *Biogeography and adaptation.* Cambridge, MA: Belknap Press.

————. 1987. *Evolution and escalation: An ecological history of life.* Princeton: Princeton University Press.

————. 1993. *A natural history of shells.* Princeton: Princeton University Press.

Vermeij, G. J., and P. W. Signor. 1992. The geographic, taxonomic, and temporal distribution of determinate growth in marine gastropods. *Biological Journal of the Linnaean Society* 47:233–47.

Vokes, E. H. 1979. The age of the Baitoa Formation, Dominican Republic, using Mollusca for correlation. *Tulane Studies in Geology and Paleontology* 15: 105–16.

————. 1989. Neogene paleontology of the northern Dominican Republic: 8, The family Muricidae (Mollusca: Gastropoda). *Bulletins of American Paleontology* 97:5–94.

Voss, S. R. 1995. Genetic basis of paedomorphosis in the axolotl, *Ambystoma mexicanum:* A test of the single-gene hypothesis. *Heredity* 86:441–47.

Wagner, P. J. 1995. Testing evolutionary constraint hypotheses with early Paleozoic gastropods. *Paleobiology* 21:248–72.

Weisbord, N. E. 1962. Late Cenozoic gastropods from northern Venezuela. *Bulletins of American Paleontology* 42 (193): 1–672.

Whiteman, H. H. 1994. Evolution of facultative paedomorphosis in salamanders. *Quarterly Review of Biology* 69 (2): 205–21.

Williamson, P. G. 1987. Selection or constraint? A proposal on the mechanism for stasis. In *Rates of evolution,* ed. K. S. W. Campbell and M. F. Day, 129–42. London: Allen and Unwin.

Willis, M. A. 1998. Cambrian and recent disparity: The picture from priapulids. *Paleobiology* 24 (2): 177–99.

Willis, M. A., D. E. G. Briggs, and R. A. Fortey. 1994. A comparison of Cambrian and Recent arthropods. *Paleobiology* 20:93–130.

Woodring, W. P. 1928. Miocene mollusks from Bowden, Jamaica: 2, Gastropods and discussion of results. *Carnegie Institution of Washington* 385:1–564, pls. 1–40.

————. 1970. Geology and paleontology of Canal Zone and adjoining parts of Panama: Description of Tertiary mollusks (Gastropods: Eulimidae, Marginellidae to Helminthoglyptidae). *U.S. Geological Survey Professional Paper* 306-D:299–452.

Wray, G. A., and A. E. Bely. 1994. The evolution of echinoderm development is driven by several distinct factors. *Development* 1994, suppl.: 97–106.

Wray, G. A., J. S. Levinton, and L. H. Shapiro. 1996. Molecular evidence for deep Precambrian divergences among metazoan phyla. *Science* 274:568–73.

7 The Interrelationship of Speciation and Punctuated Equilibrium

STEPHEN JAY GOULD

Critiques Based on the Definability of Paleontological Species

Empirical Affirmation

The issue of whether true biospecies (or entities operationally close enough to biospecies) can be recognized in fossils has prompted long and intense debate in paleontology (see Sylvester-Bradley 1956), and does not represent a new or special difficulty raised by punctuated equilibrium. But given the reliance of punctuated equilibrium on speciation as the mechanism behind the pattern, this old problem does legitimately assume a central place in debates about our theory (as emphasized in all negative commentary, particularly clearly by Turner 1986, and in the book-length critiques of Levinton 1988 and Hoffman 1989).

At least we may begin by exposing the most widely treated issue of the older literature as a *Scheinproblem* (literally an "appearance problem" with no real content): the logical impossibility of defining a species boundary within a gradualistic continuum. I think we may now accept that the punctuational pattern exists at high relative frequency, and that few gradualistic and anagenetic continua have been documented between fossil species (see symposia of case studies in Cope and Skelton 1985; and Erwin and Anstey 1995). Turner's sharp critique (1986), for example (and I do accept his formulation, though not his resolution), depicts the chief claims of punctuated equilibrium as a three-pronged fork. He accepts the first tine—the existence of the punctuational pattern itself—as sufficiently demonstrated by enough empirical cases in the fossil record. He regards the third tine—macroevolutionary invocation of the theory to explain trends by species sorting—as "an important extension of evolutionary theory into a hitherto little explored territory" (206). But he then rejects the second tine as both unlikely and too difficult to test in any case—explanation of the punctuational pattern as a result of speciation scaled into geologic time.

If we accept that fossils generally don't appear in the geologic record as unbreakable continua, but usually as morphological "packages" with reasonably defined temporal boundaries and sufficient stability within an extended duration, how can we assert that these packages are biospecies, or at least close enough to these neontologically defined units to bear comparison? After all, we cannot apply conventional tests of observed ecological interaction or interbreeding to fossils—and, whereas biospecies may be *recognized* by morphological differentia in everyday practice, they are not supposed to be so *defined*. Can the temporally extended "morphospecies" of paleontology really be equated with the "nondimensional species concept" (Mayr's words) of neontology?

I certainly accept the centrality and difficulty of these issues, but I do not regard them as insuperable, and I do not view the species concept as untestable with fossils. After all, the overwhelming majority of modern species in our literature and museum drawers have also been phenotypically, not ecologically, defined. Once we accept that no special paleontological riddles arise from the *Scheinproblem* of temporal continua, then most paleospecies have been no worse characterized than the majority of neospecies. Still, I will not advance this excuse as exculpatory for the fossil record, for a neontologist could reply, with impeccable logic, that such neospecies must also be regarded as uncertain, if not vacuous, and that no paleontological defense can be mounted by arguing that ordinary practice with fossils follows the worst habits (majoritarian though they may be) of neontological taxonomy.

But a best defense of phenotypically defined neospecies would follow from demonstrations that taxa so established usually do match true biospecies upon proper behavioral and ecological study—as has, indeed, often been accomplished (see references in Jackson and Cheetham 1994). Similarly, my main source for confidence about paleospecies arises from proven correspondences with true biospecies in favorable cases providing sufficient information for such a test (particularly when species with lengthy fossil records are still extant). I do not, of course, argue that all paleospecies are true biospecies, or that we can even estimate the percentage so defined (anymore than we know the relative frequency of modern taxa that represent true biospecies). But I do not see why the probability of equating well-defined paleospecies with proper biospecies should be any lower than the corresponding figure for equally well documented modern taxa, similarly recognized by morphological criteria. (In fact, one might even argue that well-documented paleospecies may represent biospecies with higher probability because we know their phenotypes, and have measured their stability, across long periods of time and wide ranges of environment—whereas modern morphospecies may be ecophenotypic expressions of a single time and place, therefore ranking only as local populations, and providing only a limited sample of the species' potential range of variability.)

When well-defined paleospecies have been tested for their correspondence with modern biospecies, such status has often been persuasively affirmed. Two recent studies seem particularly convincing. Michaux (1989) examined four living species of the marine gastropod genus *Amalda* from New Zealand. Fossils of this genus date to the Upper Eocene of this region, while all four species extend at least to the Miocene-Pliocene boundary. The four taxa represent good biospecies, based on absence of hybrids in sympatry, and on extensive electrophoretic study (Michaux 1987) showing distinct separation among species and "no detectable cryptic groupings" (Michaux 1989, 241) within any species. Michaux then used canonical discriminant analysis to achieve clear morphometric distinction among the species based on ten shell measurements for each of 671 live specimens.

He then made the same measurements on 662 fossil specimens from three of the species (the fourth did not yield enough shells for adequate characterization). Mean values, in multivariate expression based on all ten variables, fluctuated mildly through time, but never departed from the range of variation within extant populations—an excellent demonstration of stasis as dynamic holding within well-defined biospecies through several million years. Michaux concluded (1989, 246–48): "Fossil members of three biologically distinct species fall within the range of variation that is exhibited by extant members of these species. The phenotypic trajectory of each species is shown to oscillate around the modern mean through the time period under consideration. This pattern demonstrates oscillatory change in phenotype within prescribed limits, that is, phenotypic stasis."

Jackson and Cheetham's extensive studies (1990, 1994) of cheilostome bryozoan species provide even more gratifying affirmation, especially since these "simple," sessile and colonial forms potentially express all the attributes of extensive ecophenotypic variation (especially in molding of colonies to substrates, and in effects of crowding) and morphological simplicity (lack of enough or sufficiently complex skeletal characters for good definition of taxa) generally regarded as rendering the identification of biospecies hazardous, if not effectively impossible, in fossils. Moreover, Cheetham had begun his paleontological studies (see Gould and Eldredge 1993) under the assumption that careful work would reveal predominant gradualism and refute the "new" hypothesis of punctuated equilibrium— so the conclusions eventually reached did not arise from any a priori preference!

In a first study—devoted to determining whether biospecies could be recognized from skeletal phenotypes (of the sort used to define fossil taxa) in several species within three genera of extant Caribbean cheilostomes— Jackson and Cheetham (1990) examined heritability for skeletal characters in seven species. In a "common garden" experiment (under effectively identical conditions at a single experimental site), they grew F_1 and F_2 generations from embryos derived from known maternal colonies collected in disparate environments and places. All but 9 of 507 offspring were classified by discriminant analysis into the same morphospecies as their maternal parent. The authors then used electrophoretic methods to study enzyme variation in 402 colonies of eight species in the three genera. They found clear and complete correspondence between genetic and morphometric clusterings, and also determined that "genetic distances between morphospecies are consistently much higher than between populations of the same morphospecies" (581); moreover, they found no evidence for any cryptic division (potential "sibling species") within skeletally defined morphospecies.

In a concluding and gratifying observation—indicating that paleontologists need not always humble themselves before the power of neontological genetic analysis of biospecies—Jackson and Cheetham (1990, 582) made an empirical observation about the capacity of morphometric data (of the sort generated from fossils):

> The identity of quantitatively defined morphospecies of cheilostome bryozoans is both heritable and unambiguously distinct genetically. The importance of rigorous quantitative analysis was underlined by our discovery of three species of *Stylopoma* previously classified as one, a separation subsequently confirmed genetically. The widely supposed lack of correspondence between morphospecies and biospecies may result as much out of uncritical acceptance of outdated, subjectively defined taxa as from any fundamental biologic differences between the two kinds of species.

Jackson and Cheetham (1994) then followed this study with more extensive documentation, this time using large numbers of fossil species as well as living forms, of phylogenetic patterns in two Caribbean cheilostome genera, *Stylopoma* (included in the first study as well) and *Metrarabdotos* (the subject of Cheetham's earlier and elegant affirmations of punctuated equilibrium from morphometric data alone— Cheetham 1986 and 1987). Again they found strict correspondence between genetically defined clusters and taxa established by skeletal characters accessible from fossils.

With increased confidence that the taxa of his classical studies on punctuated equilibrium in *Metrarabdotos* represent true biospecies, Cheetham (now with Jackson) could affirm (1994, 420): "Morphological stasis over millions of years punctuated by relatively sudden appearances of new morphospecies was demonstrated previously for *Metrarabdotos.* Our updated results strengthen confidence in that pattern, with 11 morphospecies persisting unchanged for 2–6 m.y., all at $p > 0.99$, and no evidence that intraspecific rates of morphological change can account for differences between species."

For *Stylopoma,* where fossil evidence had not previously been analyzed morphometrically, results also affirmed punctuated equilibrium throughout (1994, 420): "The excellent agreement between morphologically and genetically defined species used in this taxonomy suggests that morphological stasis reflects genuine species survival over millions of years, rather than a series of morphologically cryptic species. Moreover, eleven of the 19 species originate fully formed at $p > 9$, with no evidence of morphologically intermediate forms, and all ancestral species but one survived unchanged all with their descendants."

In a concluding paragraph about both genera, Jackson and Cheetham wrote (407):

> Stratigraphically rooted trees suggest that most well-sampled *Metrarabdotos* and *Stylopoma* species originated fully differentiated morphologically and persisted unchanged for >1 to >16 m.y., typically alongside their putative ancestors. Moreover, the tight correlation between phenetic, cladistic, and genetic distances among living *Stylopoma* species suggests that changes in all three variables occurred together during speciation. All of these observations support the punctuated equilibrium model of speciation.

Despite the encouragement provided by these and other cases, problems certainly remain in the definition of paleontological species—a subject of central importance to punctuated equilibrium, given our invocation of speciation as the source of the primary pattern for life's history, and of the raw material for higher-level selection and sorting. These problems fall into three major categories (both by inherent logic of the case and by recorded debate in the literature): the first untroubling, the second potentially serious, and the third largely resolved in empirical terms. All three categories focus on the possibility that paleospecies systematically misrepresent the nature and number of actual biospecies. (If paleospecies don't correspond with biospecies in all cases—an undeniable proposition of course—but these discrepancies show no pattern and produce no systematic bias, then we need not be troubled unless the relative frequency of noncorrespondence be overwhelmingly high, an unlikely situation given the excellent alignments found in the few studies explicitly done to investigate this problem, as discussed just above.) I discuss these three issues sequentially in the rest of this section.

Reasons for a Potential Systematic Underestimation
of Biospecies by Paleospecies

Might we be missing a high percentage of actual speciation events because paleontologists can recognize only a cladogenetic branch with clear phenotypic consequences (for characters preserved as fossils), whereas many new species arise without substantial morphological divergence from their ancestors? In the strongest cases, paleontologists (obviously) cannot detect sibling species, a common phenomenon in evolution (see Mayr 1963 for the classic statement). Moreover, we may also miss subtle changes in phenotype, or substantial alterations (of color, for example) in features that are often important in species recognition, but not preserved in the fossil record.

Our harshest critics have urged this point as particularly telling against punctuated equilibrium. Levinton, for example (1988, 182), holds that "the vast majority of speciation events probably beget no significant change." He then views the consequences as effectively fatal for punctuated equilibrium: "The punctuated equilibrium model argues that morphological change is associated with speciation and that species are static during their history due to some internal stabilizing mechanism. There is no evidence coming from living species to support this. If anything, recent research has demonstrated that speciation occurs typically with little or no morphological change; hence the large-scale occurrence of sibling species" (211).

Hoffman (1989, 115) uses this argument to assert the untestability, hence the nonscientific status, of punctuated equilibrium:

> Long-term evolutionary stasis of species, however, simply cannot be tested in the fossil record. Paleontological data consist solely of a small sample of phenotypic traits—little more than morphology of the skeletal parts— which does not allow us to make any inference about changes in a species's genetic pool or even about changes of the frequency distribution of phenotypes in a phyletic lineage. The non-preserved portion of the phenotype of each fossil species is so extensive that it may always undergo considerable evolutionary changes that remain undetectable by the paleontologist. What appears then to the paleontologist as a species in complete evolutionary stasis may in fact represent a succession of fossil species or perhaps a whole cluster of species, a phylogenetic tree with a sizable number of branching points, or speciation events.

While fully allowing all these points about underrepresentation of true species in fossil data, I simply do not comprehend how punctuated equilibrium could be thus rendered untestable, or even seriously compromised (see further arguments in Gould 1982, 1989, and Gould and Eldredge

1993). I base this argument on two logical and methodological points, not on the probable empirical record (where I largely agree with our critics).

The proper study of macroevolution. As we all accept the principle of testability for granting scientific status to any inquiry, we do not include, within our compass, fascinating questions that cannot be answered (even if they treat potentially empirical subjects). For example, and for the moment at least, we know no way to ask a scientific question about what happened before the big bang, for compression of universal matter to a single point of origin wipes out all traces of any previous history. (Perhaps we will figure out a way to obtain such data some day, or perhaps the big bang theory will be discarded. The question might then become scientifically tractable.) Similarly, we know that many kinds of evolutionary events leave no empirical record—and that we therefore cannot formulate scientific questions about them. (For example, I doubt that we will be able to resolve the origins of human language, unless written expression occurred far earlier than current belief and evidence now indicate.)

Macroevolution is the study of phenotypic change (and any inferable correlates or sequelae) in lineages and clades throughout geologic time. Punctuated equilibrium proposes that such changes generally occur in discrete units or quanta in geologic time, and that these quanta represent events of branching speciation. Thus, we do argue for speciation as the source of raw material for macroevolutionary change in lineages. But we do not and cannot argue the quite different (and irrelevant) proposition that all speciation events produce measurable quanta of macroevolutionary change. The statement—our proposition—that nearly all macroevolutionary change occurs in increments of speciation carries no implications for the unrelated claim, often imputed to punctuated equilibrium by our critics (but largely irrelevant to our theory), that nearly all events of speciation produce an increment of macroevolutionary change. This conclusion flows from elementary logic, not from empirical science. The argument that all B comes from A does not imply that all A leads to B. All human births (at least before modern interventions of medical technology) derive from acts of sexual intercourse, but all acts of intercourse don't lead to births.

To make a more relevant analogy, in the strict version of Mayr's peripatric theory of speciation, nearly all new species arise from small populations isolated at the periphery of the parental range. But the vast majority of peripheral isolates never form new species, for such imperiled and unpromising units either die or reamalgamate with the parental population. Similarly, most new species may be unrecorded in the fossil record. Nonetheless, when changes do occur in lineages of fossils, speciation may still provide the source of input in a great majority of cases. Thus, most speciation could be cryptic (and unknowable from fossil evidence), while effectively all macroevolutionary change still arises from the minority of speciation events with phenotypic consequences. Just as peripheral isolates might represent "the only game in town" for forming new species (though few isolates ever speciate), cladogenetic speciation may be "the only game in town" for inputting phenotypic change into macroevolution (though few new species exhibit such change).

The treatment of ineluctable natural bias in science. In an ideal world—the one we try to construct in controlled laboratory experiments—no systematic bias distorts the relative frequency of potential results. But the real world of nature meets us on her own terms, and we must accept any distortions of actual frequencies that directional biases of recording or preservation inflict upon the archives of our evidence. At best, we may be able to correct such biases if we can make a quantitative estimate of their strength. (This procedure, for example, has been widely used to correct the systematic underestimate of geologic ranges imposed by the evident fact that observed first and last occurrences of a fossil species can only provide an estimate for actual origins and extinctions—and that the observed geologic range of a species must be shorter [and at least cannot be longer] than the actual duration. Studies of "waiting times" between sequential samples within the observed range, combined with mathematical models for drawing error bars around first and last occurrences, have been widely used to treat this important problem—see Sadler 1981; Schindel 1982; Marshall 1994, 1995.)

Often, however, we can specify the direction of a bias, but do not know how to make a quantitative correction. In such cases, the sciences of natural history must follow a cardinal rule: if the direction of bias coincides with the predicted effect of the theory under test, then serious, perhaps insurmountable, problems arise; but if a systematic bias works *against* a favored theory, then we encounter an acceptable impediment—for if a favored theory can be affirmed in the face of unmeasurable biases working *against* a preferred explanation, then the case becomes strengthened.

For proponents of punctuated equilibrium, speciation serves as the primary source of morphological changes that, by summation of increments, build trends in the history of lineages. If a systematic bias in the nature of paleontological evidence leads us to underestimate the number of speciation events, and if we can still explain trends by an observed number that must be less than the actual frequency, then the case for punctuated equilibrium becomes stronger by affirmation in the face of a bias working against full expression of the theory's effect. Thus, although we regret the existence of any bias that we cannot correct, the systematic underestimate of speciation events does not subvert punctuated equilibrium, because such a natural skewing of evidence makes our hypothesis even more difficult to affirm—and support for punctuated equilibrium therefore emerges in a context even more challenging than the unbiased world of controlled experimentation.

Moreover, one may even look on the bright side and recognize that such biases might occur for interesting reasons in themselves—reasons that might even enhance the importance of punctuated equilibrium and its implications. Levinton (1988, 379) probably did not intend the following passage in such a positive light, but I would suggest such a reading: "One cannot rule out the possibility that speciation is rampant, but morphological evolution only occurs occasionally when a population is forced into a marginal environment and subjected to rapid directional selection. What then becomes interesting is why the character complexes evolved in the daughter species remain constant. This is, again, the issue of stasis, which I believe to be the legitimate problem spawned by the punctuated equilibrium model."

Finally, I am not sure that fossil species do strongly underestimate the frequency of true biospeciation—though I do accept that a bias, if present at all, probably works in this direction. The most rigorous empirical studies on correspondence between well-defined paleospecies and true biospecies—the work of Michaux and of Jackson and Cheetham discussed above—affirm a one-to-one link between paleontological morphospecies and extant, genetically defined biospecies.

Reasons for a Potential Systematic Overestimation of Biospecies by Paleospecies

If a bias did exist in this opposite direction, the consequences for punctuated equilibrium would be serious (by the argument made just above on acceptable and unacceptable forms of unavoidable natural biasing). For if we systematically name too many species by paleontological criteria, then we might be affirming punctuated equilibrium by skewing data in the direction of our favored theory, rather than by genuine evidence from the fossil record. I doubt that such a problem exists for punctuated equilibrium, however, especially since all experts seem to agree, both strong advocates and fierce critics alike (as the preceding discussion documented), that if any systematic bias exists, the probable direction lies in the acceptable opposite claim for underestimation of biospecies by paleospecies.

I don't doubt, of course, that past taxonomic practice, so often favoring the erection of a species name for every recognizable morphological variant (even for odd individuals rather than populations), has greatly inflated the roster of legitimate names in many cases, particularly for fossil groups last monographed several generations ago. (Our literature even recognizes the half-facetious term "monographic burst" for peaks of diversity thus artificially created. But this problem of past oversplitting cannot be construed as either uniquely or even especially paleontological, for neontological systematics then followed the same practices as well.) This grossly uneven, and often greatly oversplit, construction of species-level taxonomy in paleontology has acted as a strong impediment for the entire research program of the prominent "taxon-counting" school (Raup 1978; Raup and Sepkoski 1986). For this reason, the genus has traditionally been regarded as the lowest unit of rough comparability in paleontological data (see Newell 1949). Sepkoski (1982) therefore compiled his two great compendia—the basis for so much research in the history of life's fluctuating diversity—at the family, and then at the genus level (but explicitly not at the species level in recognition of extreme imbalance in practice among groups, and frequent oversplitting).

Although this problem has proved far more serious for taxon counters than for proponents of punctuated equilibrium, a potential bias toward overrepresentation also poses a threat for our theory, as Levinton (1988, 364) properly recognizes: "The problem is not very new. Meyer (1878) claimed that the ability to recognize gradual evolutionary change in *Micraster* [a famous sequence of Cretaceous echinoids] was obscured by the rampant naming of separate species by previous taxonomists."

I would be seriously concerned about this issue—which does, after all, fall into the worrisome category of biases that favor a preferred hypothesis under test—but for two arguments and realities that obviate the danger. First, if supporters of punctuated equilibrium did try to affirm their hypothesis by using names recorded in the literature as primary data for judging the strength and effect of speciation upon evolutionary trends, then we would face a serious difficulty. But I know of no study employing this approach—for paleontologists recognize and avoid the dangers of such a directional bias. Punctuated equilibrium, to my knowledge, has never been asserted based on taxon counting at the species level. All confirmatory studies employ measured morphometric patterns, not the geologic ranges of names recorded in literature.

Second, as stated above, all students of this subject seem to agree that, if a systematic bias exists in correspondences between paleospecies and biospecies, fossil data are probably skewed in the opposite direction of recognizing *fewer* paleospecies than biospecies—an acceptable bias operating against the confirmation of punctuated equilibrium.

Reasons Why an Observed Punctuational Pattern Might Not Represent Speciation

Suppose that we have empirical evidence for a punctuational event separating two distinct morphological packages regarded as both different

enough to be designated as separate paleospecies by any standard criterion, and as genealogically close enough to be putative ancestor and descendant. What more do we need? Does this situation not affirm punctuated equilibrium ipso facto?

But critics charge (and I must agree) that such evidence cannot be per-
suasive by itself, because punctuated equilibrium explicitly links punctuational patterns to events of branching speciation. Therefore, recorded
punctuations produced for other reasons do not affirm punctuated equilibrium—and may even challenge the theory if their frequency be high
and, especially, if they cannot be distinguished in principle (or frequently
enough in practice) from events of cladogenetic branching.

I do believe that punctuational patterns often arise (at all scales in
evolutionary hierarchies of levels and times) for reasons other than geologically instantaneous speciation—and I welcome such evidence as an
affirmation of pervasive importance for a general style of nongradualistic
change, with punctuated equilibrium as its usual mode of expression at
the scale here under consideration. But I also affirm that testable, and
generally applicable, criteria have been formulated for distinguishing
punctuated equilibrium from other reasons for punctuational patterns—
and that available evidence amply confirms the importance and high relative frequency of punctuated equilibrium.

Of the two major reasons for punctuational patterns not due to speciation, Darwin's own classic argument of imperfection—geologic gradualism that appears punctuational because most steps of a continuum have
not been preserved in the fossil record—retains pride of place by venera-
ble ancestry. I do not, of course, deny that many (or most) breaks in
geologic sequences only reflect missing evidence. But proponents of punctuated equilibrium do not base their claims on such inadequate examples
that cannot be decided in either direction. The test cases of our best literature—whether their outcomes be punctuational or gradualistic—have
been applied to stratigraphic situations where temporal resolution and
density of sampling can make appropriate distinctions by recorded evidence, not conjecture about missing data.

The second reason has been highlighted by some critics, but unfairly
so I think—because punctuated equilibrium has always recognized the
argument and has, moreover, enunciated and explicitly tested proper criteria for making the necessary distinctions. To state the supposed problem: what can we conclude when we document a truly punctuational
sequence that cannot be attributed to imperfections of the fossil record?
How do we know that such a pattern records an event of branching speciation, as the theory of punctuated equilibrium requires? When ancestral
species A abruptly yields to descendant species B in a vertical sequence
of strata, we may only be witnessing an anagenetic transformation

through a population bottleneck, or perhaps an event of migration, where species B, having evolved gradualistically from species A in another region, invades the geographic range, and abruptly wipes out its ancestor. Hoffman (1989, 106) regards this point as fatal for punctuated equilibrium: "The problem is the same as with any series of points on a graph, which can represent a curve as well as a broken line. There is no way to resolve such a dilemma objectively until a criterion is given that determines the minimum duration of a period of stasis or the minimum gap to be interpreted as punctuation. Such criteria, however, can only be established by arbitrary conventions, not by inference from any biological data or theory."

But an appropriate and nonarbitrary criterion exists—and has been fully enunciated, featured as crucial, and subjected to frequent test, from the early days of punctuated equilibrium. We can distinguish the punctuations of rapid anagenesis from those of branching speciation by invoking the eminently testable criterion of ancestral survival following the origin of a descendant species. If the ancestor survives, then the new species has arisen by branching. If the ancestor does not survive, then we must count the case either as indecisive, or as good evidence for rapid anagenesis—but, in any instance, certainly not as evidence for punctuated equilibrium.

(By using these criteria, we obey the methodological requirement that existing biases must work against a theory under test. When ancestors do not survive following the first appearance of descendants, the pattern may still be recording an event of branching speciation—hence affirmation for punctuated equilibrium. But we may not count such cases in our favor, for the plausible alternative of rapid anagenesis cannot be disproven. By only counting cases where ancestors demonstrably survive, we accept only a subset of the events actually caused by speciation. Thus, we underestimate the frequency of punctuated equilibrium—as we must do in the face of an unresolvable bias affecting a hypothesis under test.)

In our first papers, we did not recognize or articulate the importance of tabulating cases with ancestral survival following punctuational origin of a descendant as a criterion for distinguishing punctuated equilibrium from other forms of punctuational change. (Both of our original examples in Eldredge and Gould 1972 did feature—and prominently discuss—ancestral survival as an important aspect of the total pattern. We did have a proper "gut feeling" about best cases, but we did not formalize the criterion.) But, beginning in 1983, and continuing thereafter, we have stressed the centrality of this criterion in claims for speciation as the mechanism of punctuated equilibrium. Contrasting the difference in paleontological expression between Wright's shifting balance and punctuated equilibrium by speciation, I wrote (Gould 1982, 100): "Since punctuational events can occur in the phyletic mode under shifting balance, but by branching speciation under punctuated equilibrium, the persistence of ancestors following the abrupt appearance of a descendant is the surest sign of punctuated equilibrium."

This criterion has been actively, and more and more routinely (as researchers recognize its importance), applied in the expanding literature on empirical study of evolutionary tempos and modes in well-documented fossil sequences. Cases of probable anagenetic transformation have been documented (no ancestral survival when good stratigraphic resolution should have recorded any genuine persistence), especially in planktic marine Foraminifera, where long oceanic cores often provide unusually complete evidence (Arnold 1983; Malmgren and Kennett 1981, who have coined the appropriate term "punctuated anagenesis" for this phenomenon).

However, abundant cases of ancestral survival, and consequent punctuational origin of descendant taxa by branching speciation, have also been affirmed as illustrations of punctuated equilibrium. These examples span the gamut of taxonomies and ecologies, ranging from marine microfossils (Cronin 1985 on ostracodes), to "standard" macroscopic marine invertebrates (with Cheetham's famous studies of bryozoans, 1986 and 1987, as classic and multiply documented examples), to freshwater invertebrates (Williamson's work [1981] on multiple events of speciation in African lake mollusks, where ancestral species reinvade upon coalescence of lakes following periods of isolation that provided conditions for speciation), to terrestrial vertebrates (Flynn 1986 on rodents; Prothero and Shubin 1989 on horses). I have also been particularly (if parochially) gratified by the increasing application of punctuated equilibrium to resolving hominid phylogeny. The criterion of ancestral survival has been prominently featured in this literature, as by McHenry (1994, 6785), who notes that "ancestral species overlap in time with descendants in most cases in hominid evolution, which is not what would be expected from gradual transformations by anagenesis."

In any case, punctuated equilibrium can be adequately and generally recognized by firm evidence linking observed punctuational patterns to branching speciation as a cause. The theory of punctuated equilibrium is eminently testable and has, indeed, passed such trials in cases now too numerous to deny a high relative frequency to this important evolutionary phenomenon (see Gould and Eldredge 1993).

Critiques Based on Denying Events of Speciation as the Primary Locus of Change

Once we overcome the problem of definability for species in the fossil record, punctuated equilibrium still faces a major issue rooted in the crucial subject of speciation. Punctuated equilibrium affirms, as a primary statement, that ordinary biological speciation, when properly scaled into geologic time, produces the characteristic punctuational pattern of our fossil record. We must therefore be able to defend the central implication that morphological change should be preferentially associated with events of branching speciation. Our critics have strongly argued that such a proposition cannot be justified by our best understanding of evolutionary processes and mechanisms.

I believe that our critics have been correct in this argument, and that Eldredge and I probably made a major error by advocating, in the original formulation of our theory, a direct acceleration of evolutionary rate by processes of speciation. This claim, I now think, represents one of the two most important fallacies that we have committed in advocating punctuated equilibrium during the past 25 years. (The other error lay first in our failure to recognize the phenomenon of species selection as distinct [by hierarchical reasoning] from classical Darwinian organismic selection, and then [see Gould and Eldredge 1977] in our decision to advocate an overly broad and purely descriptive definition rather than a properly limited meaning based on emergent fitnesses [as corrected in Lloyd and Gould 1993].)

We did not urge this correlation between speciation events and morphological change only because the pattern of punctuated equilibrium could be best defended thereby. We did, of course, recognize the logical link, as in the following statement from Gould 1982, 87 (see also Gould and Eldredge 1977, 137): "Reproductive isolation and the morphological gaps that define species for paleontologists are not equivalent. Punctuated equilibrium requires either that most morphological change arise in coincidence with speciation itself, or that the morphological adaptations made possible by reproductive isolation arise rapidly thereafter." But we based our defense of this proposition upon a large, and then quite standard, literature advocating a strong negative correlation between capacity for rapid evolutionary change and population size. Small populations, under these models, maintained maximal prospects for rapid transformation based on several factors, including potentially rapid fixation of favorable variants, and enhancement of differences from ancestral populations by interaction of rapid selection with stochastic reasons for change (particularly the founder effect) that can occur only with adequate speed and impact in small populations. Large and stable populations, by the converse of these arguments, should be sluggish and resistant to change.

This literature culminated in Mayr's spirited defense (first in a famous 1954 article, and then in the 1963 book that served as the closest analog to a "bible" for graduate students of my generation) for "genetic revolution" as a common component of speciation. Since Mayr (who coined the name "founder effect" in this context) also linked his concept of "genetic revolution" to the small, peripherally isolated populations that served as "incipient species" in his influential theory of peripatric speciation—and since we had invoked this theory in our original formulation of punctuated equilibrium (Eldredge and Gould 1972)—our defense of a link between speciation and concentrated episodes of genetic (and phenotypic) change flowed logically from the evolutionary views we had embraced. Thus, we correlated punctuations with the extensive changes that often occurred during events of speciation in small, peripherally isolated populations; and we linked stasis with the expected stability of large and successful populations following their more volatile and punctuational origins as small isolates.

I can claim no expertise in this aspect of neontological evolutionary theory, but I certainly acknowledge, and must therefore provisionally accept, the revised consensus of the past 20 years that has challenged this body of thought, and rejected any general rationale for equating the bulk of evolutionary change with events of speciation in small populations, or with small populations in any sense. As I read the current literature, most evolutionists now view large populations as equally prone to evolutionary transformation, and find no reason to equate times of speciation—the attainment of reproductive isolation—with acceleration in general rates of genetic or phenotypic change (see, for example, Ridley 1993 and Williams 1992). (I do, however, continue to wonder whether the Mayrian viewpoint might still hold some validity, and might now be subject to overly curt and confident dismissal.)

This situation creates a paradox for punctuated equilibrium. The pattern has been well documented and shown to predominate in many situations (Gould and Eldredge 1993), but its most obvious theoretical rationale has now fallen under strong skepticism. So either punctuated equilibrium is wrong—a proposition that this partisan views as unlikely (although obviously possible), especially in the face of such strong documentation—or we must identify another reason for the prominence of punctuated equilibrium as a pattern in the history of life. In our article on the "majority" (21st birthday!) of punctuated equilibrium, Eldredge and I expressed this dilemma in the following manner (Gould and Eldredge 1993, 226): "The pattern of punctuated equilibrium exists (at predominant relative frequency, we would argue) and is robust. *Eppur non si muove;* but why then? For the association of morphological change with speciation remains as a major pattern in the fossil record." (Our Italian parody, missed by many readers of the original article, alters Galileo's famous, but almost surely legendary, rebuke to the Inquisition, delivered secretly and sotto voce after he had been forced to recant his Copernican views in public: *Eppur si muove*—nevertheless it does move.

Our parody says, "Nevertheless it does *not* move"—a reference to the overwhelming evidence for predominant stasis in the history of species, even if our original evolutionary rationale, based on population size, must be reassessed.)

Such a paradox permits several approaches. One might simply argue that the pattern of punctuated equilibrium demonstrably exists, so the task falls to evolutionary theorists to find a proper explanation. The current absence of a satisfactory account does not threaten the empirical record, but rather directs inquiry by posing a problem. Or one might doubt that any single explanation can render the phenomenon, and suspect that many rationales will yield the observed pattern (including Mayrian genetic revolutions, even if we now regard their relative frequency as low). Thus, we need only list a set of enabling criteria from evolutionary theory and then argue that their combination may render the observed phenomena of the fossil record.

Most researchers would regard a third approach as preferable in science: an alternate general explanation of different form from the previous, but now rejected, leading candidate. I believe that such a resolution has been provided by Douglas Futuyma (1987, 1989), although his simple, yet profound, argument has not sunk into the consciousness of evolutionists because the implied and required hierarchical style of thinking remains so unfamiliar and elusive to most of us. (In fact, and with some shame, I am chagrined that I never recognized this evident and elegant resolution myself. After all, I am supposedly steeped in this alternative hierarchical mode of thinking—and I certainly have a strong stake in the problems of punctuated equilibrium.)

In short, Futuyma argues that we have been on the wrong track (and thinking at the wrong level) in trying to locate the reason for a correlation between paleontological punctuations and events of speciation in a direct mechanism of accelerated change promoted by the process of speciation itself. Yet Futuyma does agree that a strong correlation exists (and has been demonstrated, in large part by research and literature generated by debate about punctuated equilibrium). Since we all understand (but do not always put into practice!) the important logical principle that correlation does not imply causality (the so-called post hoc fallacy), an acknowledgment of the genuine link doesn't commit us to any particular causal scheme—especially, in this case, to the apparently false claim that mechanisms of speciation inherently enhance evolutionary rates.

Futuyma begins by arguing that morphological change may accumulate anywhere along the temporal trajectory of a species, and not exclusively (or even preferentially) during the geologic moment of its origin. What then could produce such a strong correlation between events of branching speciation and morphological change from an ancestor to the condition of subsequent stasis in a descendant? Futuyma—and I am somewhat rephrasing and extending his argument here—then draws an insightful and original analogy between macroevolution and the conventional Darwinism of natural selection in populations.

The operation of natural selection requires that Darwinian individuals interact with environments in such a manner that distinct features of these individuals (not held by all members in the varying population) bias their reproductive success relative to others in the population. As a defining criterion of Darwinian individuality, entities that interact with the environment must show sufficient stability—defined in terms of the theory and mechanism under discussion as enough coherence to perform as an interactor in the process of natural selection.

Darwin correctly recognized that organisms play this role as fundamental interactors for macroevolution within populations. (Gene selectionists make a crucial error in arguing that sexual organisms are not stable enough to be units of selection because they must disaggregate in forming the next generation. But units of selection are defined as interactors, and the "sufficient stability" required by the theory of natural selection only demands persistence through one episode [generational at this level] of interaction to bias reproductive success—as organisms do in the classical Darwinian "struggle for existence.") Organisms achieve this stability through ordinary mechanisms of bodily coherence (a protective skin, functional integration of parts, a regulated developmental program, etc.).

What, then, produces a corresponding stability for units of macroevolution? Species are constructed as complex units, composed of numerous local populations, each potentially separate (at any moment) due to limited gene flow, and each capable of adaptation to unique and immediate environments. Thus, in principle, substantial evolution can occur in any local population at any time during the geologic trajectory of a species. A large and developing literature, much beloved by popular sources of media and textbooks as a way of illustrating the efficacy of evolution in the flesh of immediacy (that is, within a time frame viscerally understood by human beings), has documented these rapid and adaptive changes in isolated local populations—substantial evolution of body size in guppies (Reznick et al. 1997), or of leg length in anolid lizards (Losos et al. 1997), for example (see Gould 1997).

But such changes in local populations cannot maintain any sustained macroevolutionary expression unless they become "locked up" in a Darwinian individual with sufficient stability to act as a unit of selection in geologic time. Local populations—as a primary feature of their definition—do not maintain such coherence. They can, in principle (and, ultimately, in the fullness of geologic time, almost invariably in practice), interbreed with other local populations of their species. Their distinctively evolved adaptations must therefore be ephemeral in geologic terms, unless these features can be stabilized by individuation—that is, by protection against amalgamation with other Darwinian individuals. Speciation—as the essence of its macroevolutionary meaning—provides such individuation by "locking up" evolved changes in reproductively isolated populations that can, thereafter, no longer amalgamate with others. The Darwinian individuation of organisms occurs by bodily coherence for structural and functional reasons. The Darwinian individuation of species occurs by reproductive coherence among parts (organisms), and prevention of intermingling between these parts and the parts of other macroevolutionary individuals (that is, organisms of other species).

Rapid evolution in local population of guppies and anoles illustrates a fascinating phenomenon that teaches us many important lessons about the general process of evolution. But such changes can only be ephemeral, because they do not become stabilized in coherent higher-level Darwinian individuals with sufficient stability to participate in macroevolutionary selection. They strut and fret their short hour on the geologic stage, and then disappear by death or amalgamation. They produce the ubiquitous and geologically momentary fluctuations that characterize the long-term stasis of species. They are, to use Mandelbrot's famous metaphor for fractals, the equivalent of calculating the coastline of Maine by measuring the distance around every boulder on every beach along the shore, and not at the scale properly enjoined when the entire state appears on a single page in an atlas. Macroevolution represents the page of the atlas; the distance around each boulder (marking substantial but ephemeral changes in local populations of guppies and lizards)—however important in the immediacy of an ecological moment—becomes invisible and irrelevant (as the jiggling fluctuations of stasis) in the study of sustained macroevolutionary change.

In other words, morphological change correlates so strongly with speciation not because cladogenesis accelerates evolutionary rates, but rather because such changes (which can occur at any time in the life of a local population) cannot be retained (and sufficiently stabilized to participate in selection) without the protection provided by individuation—and speciation, via reproductive isolation, produces macroevolutionary individuals. Speciation does not necessarily promote evolutionary change; rather, speciation "gathers in" and guards evolutionary change by locking it up for sufficient geologic time within a Darwinian individual of the appropriate scale. If a change in a local population does not gain such protection, it becomes—to borrow Dawkins's metaphor at a macroevolutionary scale—a transient dust storm in the desert of time, a passing cloud without borders, integrity, or even the capacity to act as a unit of selection, in the panorama of life's phylogeny.

To cite Futuyma's summary of his powerful idea (1987, 465): "I propose that because the spatial locations of habitats shift in time, extinction of and interbreeding among local populations makes much of the geographic differentiation of populations ephemeral, whereas reproductive isolation confers sufficient permanence on morphological changes for them to be discerned in the fossil record." Futuyma then directly follows this statement with the key implication of punctuated equilibrium for the explanation of evolutionary trends: "Long-term anagenetic change in some characters is then the consequence of a succession of speciation events."

Later in his article, Futuyma (467) explicitly links speciation with sufficient stability (individuation) for macroevolutionary expression:

> In the absence of reproductive isolation, differentiation is broken down by recombination. Given reproductive isolation, however, a species can retain its distinctive complex of characters as its spatial distribution changes along with that of its habitat or niche . . . Although speciation does not accelerate evolution within populations, it provides morphological changes with enough permanence to be registered in the fossil record. Thus, it is plausible to expect many evolutionary changes in the fossil record to be associated with speciation.

And, at the end of his article, Futuyma (470) notes the crucial link between punctuated equilibrium and the possibility of sustained evolutionary trends: "Each step has had a more than ephemeral existence only because reproductive isolation prevented the slippage consequent on interbreeding with other populations . . . Speciation may facilitate anagenesis by retaining, stepwise, the advances made in any one direction . . . Successive speciation events are the pitons affixed to the slopes of an adaptive peak."

I hope that Futuyma's simple yet profound insight may help to heal the remaining rifts and integrate punctuated equilibrium into an evolutionary theory hierarchically enriched in its light.

Coda. This article is an edited version of a key portion of a "monster" chapter on punctuated equilibrium that will appear in my forthcoming monograph *The Structure of Evolutionary Theory,* to be published by Harvard University Press in the maximally auspicious year of 2001, *Deo volente.* I dedicate this version with greatest pleasure to Alan H. Cheetham, who is not only the finest and most empirically meticulous of paleontologists—an inspiration to us all—but who has also provided the most important data ever assembled to test the notions of punctuated equilibrium in his classic and exemplary work on the bryozoan genus *Metrarabdotos*. His extensions of these studies, in collaboration with Jeremy Jackson for the genetics and morphometrics of living species as well, provides the major empirical support for the first part of this paper—yet another example for the centrality of careful empirics in any theoretical debate, and an illustration of the mastery of Alan H. Cheetham in this most worthy and enduring of all styles of science.

References

Arnold, A. J. 1983. Phyletic evolution in the *Globorotalia crassaformis* (Galloway and Wissler) lineage: A preliminary report. *Paleobiology* 9:390–97.

Cheetham, A. H. 1986. Tempo of evolution in a Neogene bryozoan: Rates of morphologic change within and across species boundaries. *Paleobiology* 12: 190–202.

———. 1987. Tempo of evolution in a Neogene bryozoan: Are trends in single morphologic characters misleading? *Paleobiology* 13:286–96.

Cope, J. C. W., and P. W. Skelton, eds. 1985. *Evolutionary case histories from the fossil record*. Special Papers in Palaeontology no. 33. London: Palaeontological Association.

Cronin, T. M. 1985. Speciation and stasis in marine Ostracoda: Climatic modulation of evolution. *Science* 227:60–63.

Eldredge, N., and S. J. Gould. 1972. Punctuated equilibria: An alternative to phyletic gradualism. In *Models in paleobiology,* ed. T. J. M. Schopf, 82–115. San Francisco: Freeman, Cooper and Company.

Erwin, D. H., and R. L. Anstey, eds. 1995. *New approaches to speciation in the fossil record*. New York: Columbia University Press.

Flynn, L. J. 1986. Species longevity, stasis, and stairsteps in rhizomyid rodents. *Contributions to Geology, University of Wyoming, Special Paper* 3:273–85.

Futuyma, D. J. 1987. On the role of species in anagenesis. *American Naturalist* 130:465–73.

———. 1989. Speciational trends and the role of species in macroevolution. *American Naturalist* 134:318–21.

Gould, S. J. 1982. The meaning of punctuated equilibrium and its role in validating a hierarchical approach to macroevolution. In *Perspectives on evolution,* ed. R. Milkman, 83–104. Sunderland, MA: Sinauer Associates.

———. 1989. Punctuated equilibrium in fact and theory. *Journal of Social Biological Structure* 12:117–36.

———. 1997. The paradox of the visibly irrelevant. *National History* 106:12–18, 60–66.

Gould, S. J., and N. Eldredge. 1977. Punctuated equilibria: The tempo and mode of evolution reconsidered. *Paleobiology* 3:115–51.

———. 1993. Punctuated equilibrium comes of age. *Nature* 366:223–27.

Hoffman, A. 1989. *Arguments on evolution*. New York: Oxford University Press.

Jackson, J. B. C., and A. H. Cheetham. 1990. Evolutionary significance of morphospecies: A test with cheilostome Bryozoa. *Science* 248:578–83.

———. 1994. Phylogeny reconstruction and the tempo of speciation in cheilostome Bryozoa. *Paleobiology* 20:407–23.

Levinton, J. 1988. *Genetics, paleontology, and macroevolution.* Cambridge: Cambridge University Press.

Lloyd, E. A., and S. J. Gould. 1993. Species selection on variability. *Proceedings of the National Academy of Sciences, USA* 90:595–99.

Losos, J. B., K. I. Warheit, and T. W. Schoener. 1997. Adaptive differentiation following experimental island colonization in *Anolis* lizards. *Nature* 387:70–73.

Malmgren, B. A., and J. P. Kennett. 1981. Phyletic gradualism in a late Cenozoic planktonic foraminiferal lineage DSDP site 284, south-west Pacific. *Paleobiology* 7:130–42.

Marshall, C. R. 1994. Confidence intervals on stratigraphic ranges: Partial relaxation of the assumption of randomly distributed fossil horizons. *Paleobiology* 20:459–69.

———. 1995. Distinguishing between sudden and gradual extinctions in the fossil record: Predicting the position of the Cretaceous-Tertiary iridium anomaly using the ammonite fossil record on Seymour Island, Antarctica. *Geology* 23: 731–34.

Mayr, E. 1954. Change of genetic environment and evolution. In *Evolution as a process,* ed. J. Huxley, 157–80. London: Allen and Unwin.

———. 1963. *Animal species and evolution.* Cambridge: Harvard University Press.

Meyer, C. J. A. 1878. Micrasters in the English chalk—two or more species? *Geological Magazine,* new series, 5:115–17.

McHenry, H. M. 1994. Tempo and mode in human evolution. *Proceedings of the National Academy of Sciences, USA* 91:6780–86.

Michaux, B. 1987. An analysis of allozymic characters of four species of New Zealand: *Amalda* (Gastropoda: Olividae: Ancillinae). *New Zealand Journal of Zoology* 14:359–66.

———. 1989. Morphological variation of species through time. *Biological Journal of the Linnean Society* 38:239–55.

Newell, N. D. 1949. Phyletic size increase: An important trend illustrated by fossil invertebrates. *International Journal of Organic Evolution* 3:103–24.

Prothero, D. R., and N. Shubin. 1989. The evolution of Oligocene horses. In *The evolution of Perissodactyles,* ed. D. R. Prothero and R. M. Schoch, 142–75. New York: Oxford University Press.

Raup, D. M. 1978. Cohort analysis of generic survivorship. *Paleobiology* 4:1–15.

Raup, D. M., and J. J. Sepkoski, Jr. 1986. Periodic extinction of families and genera. *Science* 231:833–36.

Reznick, D. N., F. H. Shaw, F. H. Rodd, and R. G. Shaw. 1997. Evaluation of the rate of evolution in natural populations of guppies *(Poecilia reticulata). Science* 275:1934–37.

Ridley, M. 1993. *Evolution.* Boston: Blackwell.

Sadler, P. H. 1981. Sediment accumulation rates and the completeness of stratigraphic sections. *Journal of Geology* 89:569–84.

Schindel, D. E. 1982. Resolution analysis: A new approach to the gaps in the fossil record. *Paleobiology* 8:340–53.

Sepkoski, J. J., Jr. 1982. A compilation of fossil marine families. *Milwaukee Public Museum Contribution to Biological Geology* 51:1–125.

Sylvester-Bradley, P. C., ed. 1956. *The species concept in palaeontology.* London: Systematics Association.

Turner, J. R. G. 1986. The genetics of adaptive radiation: A neo-Darwinian theory of punctuational evolution. In *Patterns and processes in the history of life,* ed. D. M. Raup and D. Jablonski, 183–207. Berlin: Springer-Verlag.

Williams, G. C. 1992. *Natural selection: Domains, levels, and challenges.* New York: Oxford University Press.

Williamson, P. G. 1981. Palaeontological documentation of speciation in Cenozoic molluscs from Turkana basin. *Nature* 293:437–43.

Part 3

Macroevolutionary Patterns and Trends

On the Ends of the Taxon Range Problem   **8**

LEE-ANN C. HAYEK AND EFSTATHIA BURA

Introduction

The fossil record left by a species in a dated horizon or sample can provide valuable information for assessing origination and extinction times (Marshall 1990). Although the intuitive estimates of origination and extinction are the observed ends of the local range, these almost always underestimate the true endpoints of the taxon range (Strauss and Sadler 1989). However, stratigraphic data can provide for interval or point estimates of the species origination or extinction dates (Shaw 1964; Paul 1982; Strauss and Sadler 1987, 1989; Springer and Lilje 1988; Springer 1990; Marshall 1990; Weiss and Marshall 1999) when probability-based models are fit to the observations.

Each fit of a probabilistic model with its attendant calculations for interval estimates relies upon a set of underlying assumptions for its accuracy of approximation and meaningfulness. These assumptions form a set of constraints on the observed situation. Lack of strict adherence or mistranslation of these assumptions can result in problematical estimates. If any one of the assumptions is conditioned on an extraneous force or is not met to a sufficient degree by the data, the resultant inferences will be in error. Current models for the taxon range problem suffer from both simplifying assumptions and conditioning, while not making use of all the sample evidence. For example, Shaw (1964), Paul (1982), Signor and Lipps (1982), Strauss and Sadler (1989), and others have made the simplifying assumption that fossils are distributed randomly within the interval (but see Marshall 1994), collecting intensity is constant, and species abundances have been assumed to be equal. Each of these authors has acknowledged that the current methods and models are in need of substantive improvement.

We have developed an assumption-free approach for modeling the observed data and quantifying strength of belief in the observed taxon range as a reliable estimate of true duration. Springer (1990) points out that

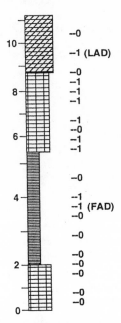

Fig. 8.1. Hypothetical stratigraphic section of three different lithologies, spanning approximately 11 million years (tick scale on left of column) and showing presence (1) and absence (0) data for a species. FAD is first observed find; LAD is last observed find in the sampled section.

the incompleteness of the fossil record needs to be disentangled from correlated factors before one can rely upon observed results. Alternatively, a statistical approach can be used to consider the incompleteness (Foote 1997). The importance of our empirical yet statistical approach is that we incorporate all of the field data and realistically take into account the existent incompleteness of the fossil record. We include the information from nonfinds or the absence data, as well as the observed pattern of presence/absence from the sampling process (and eliminate the assumption of randomness of sample occurrences). The fossiliferous horizons in such samples correspond to time points or intervals (fig. 8.1), and the observed data set consists of location markers with an accompanying indication of presence or absence. In addition, our method obviates the problem of inequality or rarity of species abundance by recognizing the interrelationship of the presence/absence pattern and species abundances and is applicable for analysis of continuous, binned, or discrete sampling data schemes. Using this approach we identify problems and shortcomings of alternative modeling efforts.

The Sample Situation—in Practice

Depending on the organism, the marine or terrestrial environment, and other factors, sampling methods vary for fossiliferous horizons within stratigraphic dated sections or samples. However, the resultant quantitative data for any group of interest consists of a sequence of occurrences (finds) and nonoccurrences (nonfinds), which we call presence/absence data, along with the stratigraphic position of the samples in the geologic section or dated material. The fossiliferous horizons in such samples correspond to time points or intervals. Finally, the observed data set consists of location markers with an accompanying indication of presence or absence.

Evidence of absence of a taxon in the interior of fossiliferous horizons can occur for a variety of reasons (Signor and Lipps 1982; Springer 1990; Foote and Raup 1996). Extinction is not one of them, however, when there is sound evidence that the species was present both immediately before and after the specified time. Abundance is an extraneous but vital factor affecting the overall pattern of presence and absence of any taxon.

Absences before the first or after the last observed presence are also possible when sampling, especially when interest lies in higher-order groupings (e.g., Cheetham and Jackson 1998). It is our belief that this total pattern of presence and absence provides valuable practical information for developing true duration estimates and should be incorporated into any inferential effort. We use all the data, not just some of it.

The Sample Situation—in Theory

The sample data of observed and recorded presence and absence indicators that correspond to the practical situation described above can be represented by a sequence of binary (1s and 0s) data. A value of 1 denotes a find, or presence of the taxon, and a 0 value denotes a nonfind, or absence of that taxon at the corresponding sampling position. Thus, an ordered vector of 1s and 0s can represent a set of n dated points, or fossiliferous horizons. That is, we have a statistical sample of size n. For example, the vector $(1, 0, 0, 1, 1, 1)$ represents a sample of size $n = 6$, with $k = 4$ presences, or finds, and $n - k = 2$ absences or nonfinds.

We could choose to model this data in any number of ways. As statisticians we could devise many answers to problems of little importance to the paleobiologist. The value of mathematical statistical application is to determine the problem of the paleobiologist and to solve that specific problem. To do this involves more than merely rewriting the data into numbers.

A probabilistic model relies for its usefulness totally on a set of underlying assumptions, which must match the field requirements. If we select a previously defined model for data description and inferential purposes, we must also be assured that the assumptions are upheld scrupulously, or suffer the inferential consequences. Foote (1997) noted that all attempts to account for bias in the fossil record involve some simplifying assumptions. Our method, without such assumptions, maximizes the information potential in the field data, and we use all the data.

Practical Limitations versus Theoretical Assumptions

In order to model a practical situation, we need consider the factors that affect sample results. The more practical information that can be incorporated into a model, the more reasonable and reliable the results are.

Clearly and predictably, stratigraphic gaps and variable preservation rates affect stratigraphic ranges (Foote 1997). Sampling effects, facies control, and sequencing each contribute their part to an uneven assortment of 1s and 0s in the resultant observation pattern from the field study.

Within the area to be sampled, the chance or probability of actually finding a species (obtaining a 1 as a data point) certainly is affected by a variety of factors, including taphonomic considerations. Weiss and Marshall (1999) constructed a Bayesian approach to the estimation of extinction when sampling data is unevenly spaced and discrete. They obtain estimates of the probability of a false negative or absence (0), which, in turn, yields a distribution for the description of the uncertainty in the true extinction date for a species of interest. The authors state that this modeling effort incorporates the assumption that the probability of finding a species in a given sample stays constant across stratigraphic position. That the relative abundance stays constant with time is a simplifying assumption but, of course, unrealistic, and its frequency of occurrence is not known.

The work of Strauss and Sadler (1989) presents a model of an underlying process that could have generated taxon range data. Their model considers the true endpoints of species' duration to be parameters in need of estimation. This is equivalent to allowing that the true range ends are unobserved but fixed evolutionary or migration events. Strauss and Sadler sought a correction to the proposed approach of Springer and Lilje (1988), who allowed for variable range ends. Strauss and Sadler state that their basic assumption, which simplifies the mathematics, is that fossil finds are distributed randomly between these endpoints. Shaw (1964), Paul (1982), and Signor and Lipps (1982) among others also have used a similar simplifying assumption that fossils are only randomly distributed, although all admit that such randomness is not probable. Holland (1995) sets out a hierarchical set of assumptions, which he incorporates into a series of test models. His third level in the hierarchy assumes facies-controlled taxa and parasequence-style cyclicity. These field conditions result in both randomly and nonrandomly distributed gaps between finds. Holland (1995) states that the abnormally long gaps he obtained with this level of assumption are frequently found in confidence interval study, which relies on an assumption of only randomly distributed gaps.

For Strauss and Sadler, randomness is then equated to modeling fossil finds with a Poisson process. The Poisson process that they invoke is obtained under the assumption that the preservation rate of a species across the sampled section is constant. Hence, the assumption of equal probability for both rare and common events, that is, species abundances, is imposed upon the data. Buzas et al. (1982) shows this to be a highly unlikely scenario, and Hayek and Buzas (1997) have described the substantial correlation that exists between abundance and occurrence across taxa. This means that these models assume the same constant probability of finding any fossil and this assumption is in use across the entire sampling interval, despite empirical field results to the contrary. Another way to say this is that fossil finds have a Uniform distribution between the two endpoints. That is, it is as likely for the researcher to find a rare species as a common or an abundant species. Marshall (1990, 3) states that this Uniform assumption is equivalent to claiming that "fossil horizons are randomly distributed, i.e., the sedimentation rates and fossilization potential were stochastically constant with stratigraphic position, and that the efficiency or intensity of collecting was uniform throughout the section." This can easily be shown implausible in many, if not most, cases. Sedimentation potential and pattern can be crucial to deciphering the information contained in the fossil record and should be incorporated or at least accounted for in modeling.

The so-called gaps or successive differences between finds have independent exponential distributions under some models, including that of Strauss and Sadler. According to the Poisson process, however, once we condition on the event (that is, once we have the observations in hand) that n observations are in the range between first and last find, the property of the exponential distribution is lost. The range is no longer infinite as needed for that distribution. The range used for modeling is $\theta_1 - \theta_2$ (that is, the true duration) like the length of a Broken Stick. When the interval $\theta_1 - \theta_2$ is mapped onto the interval (0, 1) as done for Strauss and Sadler's approach, the joint gap distribution of lengths for n points under the Broken Stick model is the Dirichlet distribution. Confidence intervals are computed for each endpoint as a fraction of the range.

In essence, Strauss and Sadler, as well as most other modeling efforts, ignore sample information except for the maximum and minimum presence indicators (FAD and LAD) and the total number of finds in the observed data. That is, the model, in effect, ignores all absences, both those within the sampled section, between the FAD and LAD, and those beyond the two observed endpoints, but within the range of observation. Evidence that a search was made for a given species but that it was not found is disregarded.

Assumptions are made also that the locations of the fossil horizons are stochastically independent, and that the collection process is unbiased.

Although the approach of Strauss and Sadler (1989) is mathematically rigorous and useful, it cannot be used as a basis for incorporating an additional set of conditions, like the inclusion of the pattern of sample results of presence/absence. Our approach shows that their model is in need at least of a mathematical statistical convolution because the total size is not merely k, the number of finds or presences, as the point of interest, but $k + (n - k)$, which is the number of presences and absences. So n should represent the total number of digs, tries, or sample locations. For these reasons, and others, we chose to let the data speak for themselves.

An Empirical Approach

The Observations

Given our observed pattern of 1s and 0s, we form a vector of binary observations. The sample points are not equidistant. For our purposes, the data form the entire sampling pattern or vector. That is, we use not only the first or last occurrence of the fossil taxon, but also the pattern of 1s and 0s over the entire sampled area. This pattern or vector will include the 0s that extend beyond the first or last observed presence. By incorporating this information, based upon the researcher's experience, into our modeling attempts we can obtain a more realistic probabilistic assessment of the likelihood of the true distance between the observed sample points and the true range ends.

For example, consider the vector or pattern (1, 0, 1, 1, 1, 1). It is widely accepted that the single, interior 0 should in no way be equated with evidence of extinction in time (or space). If no sampled material were obtained at an earlier time, then the first value of 1 is not only the first evidence of existence but also the outcome of the first sample we obtained. Signor and Lipps (1982) discuss artificial range truncation, which could be represented by a vector such as this. Foote and Raup (1996) note that wholesale truncation of stratigraphic ranges arise from a finite window of observation such as this vector can represent. Buzas and Culver (1998) provide explicit evidence of the ramifications of such sampling. Therefore, from a practical viewpoint, we have very little evidence, apart from possible statistical modeling, to give any weight to this first 1's being related to the origination point. If our observation vector is (0, 0, 0, 0, 1, 0, 1, 1, 1, 1), however, it seems reasonable to claim that the origination date is in some way closer to the observed first find. Because an experienced researcher will continue to search for a given taxon only where it has a high probability of being found, the absences before the FAD should increase our belief in the FAD as a realistic starting point for the taxon. Thus, intuitively the initial 0s play a part in an evidentiary picture.

As a second example, consider two distinct patterns of the form (0, 0, 0, 1, 1, 1, 1, 1) and (0, 0, 0, 1, 0, 0, 0, 1). Both represent statistical samples of the same size in our approach. In each of these vectors, the method of Strauss and Sadler (1989) would place symmetrical error bars around FAD and LAD (Marshall 1990). That is, the confidence interval on FAD would be of identical length to the confidence interval on LAD, even though this would appear to be rather unrealistic, considering the two patterns of presence/absence. The second vector or pattern could indicate that the species in question was of rare abundance or that the environment was inimical to it. Buzas et al (1982) noted and Koch and Morgan (1988) confirmed that in paleontological data rare species compose a sizable percentage of all observed species. Therefore, the existence of and the difference in sampling patterns appear to be vital considerations for correct inference.

The Observed Sampling Pattern Distribution

Here we develop an empirical method for a nonparametric approximation of the distribution of the observed patterns of presence/absence resulting from the field sampling process. Because our method is empirical, there will be no restrictive modeling assumptions imposed on the sedimentation, the sampling, the fossilization processes, or other potential processes involved. The only limitations are those of the experienced collector. We consider the data to be accurate, in the sense that they are not systematically biased due to collecting error and that they accurately represent the fossil record in the section being sampled.

We start with a pattern of finds and nonfinds of some taxon over a stratigraphic section, that is, the taxon's presence/absence sample pattern. The total number of samples taken, that is, the sum of both the finds and nonfinds (the number of elements in the vector of 1s and 0s), is denoted by n. The sequence of binary values does not represent equidistant samples. This reflects the fact that sampling may not be done at fixed or regular points equidistant from one another or in a column. Therefore, any given pattern can correspond to a wide variety of differing but ordered sample locations within a section or column.

In keeping with other model presentations, we map the sampled area to the interval [0, 1] on the x-axis. We choose to locate the earliest sampled date at 1 and the last at 0. Recall that this is done for all values in the pattern vector, even if the initial or final values were 0s.

Let the vector $\mathbf{b} = (b_1, b_2, ..., b_n)$ denote the presence/absence sequence, or the pattern for a sample of size n. For $i = 1, ..., n$, b_i equals either 0 or 1, depending on whether the ith observation is an absence or a presence, respectively.

For the given observed pattern, to incorporate into the distribution all possible locations at which fossils could have been observed, we construct another vector $\mathbf{u}$ of size n whose elements are random numbers generated from a Uniform distribution on the interval [0, 1], arranged in a decreasing order. That is, we use $\mathbf{u} = (u_1, u_2, ..., u_n)$ such that $u_1 > u_2 > ... > u_n$, where $u_1, u_2, ..., u_n$ are independent realizations of a Uniform random variable on the interval [0, 1]. The use of the Uniform distribution here merely generates location possibilities; it does not impose a restriction. The use of the Uniform is perfectly general since it specifies merely unequal distances or unequal-sized fossiliferous horizons along the length of the core, section, or sample. Thus, the elements of the vector $\mathbf{u}$ correspond to location measurements on the sedimentary rock or in the section, mapped to the interval [0, 1].

The vector $\mathbf{x}$ results from the element-wise multiplication of the vectors $\mathbf{b}$ and $\mathbf{u}$, that is,

$$\mathbf{x} = (x_1, x_2, ..., x_n) = (b_1 \times u_1, b_2 \times u_2, ..., b_n \times u_n)$$

Note that $x_i = u_i$ whenever $b_i = 1$, and $x_i = 0$ whenever $b_i = 0$. In other words, this vector represents the set of possible locations for pattern $\mathbf{b}$.

For example, suppose the sample size is 5 and the pattern is $\mathbf{b} = (0, 1, 1, 0, 1)$. We generate a vector of size 5 from the Uniform distribution on [0, 1] whose elements are placed in decreasing order, say $\mathbf{u} = (0.999, 0.878, 0.609, 0.218, 0.002)$. Then, the vector consisting of the element-wise products of $\mathbf{b}$ and $\mathbf{u}$ is $\mathbf{x} = (0, 0.878, 0.609, 0, 0.002)$.

If we repeat this generation, say, t times, we have a total of $N = \Sigma b_i \times t$ points from the observed pattern distribution. These represent N locations in which to find samples in the local section that could have originated from pattern $\mathbf{b}$, or whose results could be represented by pattern $\mathbf{b}$. Also, Σb_i represents the total number of finds.

Observed Pattern Density Function Estimation

Using the mathematical representation described above, we can generate data that conform to the observed pattern and, most important, that reflect its structure realistically. The ordered Uniform numbers correspond to possible locations of taxon presence, whereas the multiplication with the pattern vector makes sure that the nonfinds or absences beyond the observed range of first and last presence indicator, as well as the absences between the FAD and LAD, are also incorporated. The resultant data (from the vector multiplication) is data from the distribution of the observed taxon sampling pattern. A histogram of the N generated data points is formed. Histograms are the most widely used density estimators.

We use the technique of nonparametric density estimation on this data. Nonparametric density estimation methods are those methods for estimating (probability) density functions from data without assuming any functional form (model). A density function (or frequency function) is an expression for the frequency of each variate value. The density gives the likelihood of the possible values that the variable of interest can take. For us, this variable is the set of all possible distances, or dated markings for any fixed pattern. Therefore, the density function gives all the useful conclusions to be drawn from the data. To generate this density function estimate, we let f_b denote the density function for a given pattern **b**. A kernel estimator of f_b, denoted by the symbol f_b^*, is given by

$$f_b^*(y) = \frac{1}{Nh} \sum_{i=1}^{N} K\left(\frac{y - x_i}{h}\right),$$

where

$$K(t) = \frac{1}{\sqrt{2\pi}} \exp\left(\frac{-t^2}{2}\right)$$

is the Gaussian or Normal kernel, and h is an empirically chosen bandwidth or smoothing parameter. The probability density function at any point y in the unit interval is estimated by $f_b^*(y)$, which is a sum of "bumps" (amounts of area), $\{N^{-1}h^{-1}K\{(y - x_i)/h\}$, placed at the observations $x_1, x_2, ..., x_N$. The shape of these bumps is controlled by the kernel function, which is usually a density function itself (in our case a Normal), and the bandwidth determines the width (or window, similarly to the concept of a spectral window).

An illustration for pattern **b** = (0, 1, 1, 0, 1) in the last section is given in figure 8.2. We generated 400 (= t) ordered vectors of size 5 from the Uniform distribution on [0, 1], which in turn produced 1,200 (= N) data points in the unit interval via multiplication with **b,** as described above. The histogram of these 1,200 points from the distribution of the pattern **b** is the top plot in figure 8.2. The second and bottom plots are the kernel density estimates of the histogram for smoothing parameters 0.2 and 0.6, respectively. Note that, as the smoothing parameter increases, detail in the data is obscured. All calculations were carried out in S-PLUS (1998),

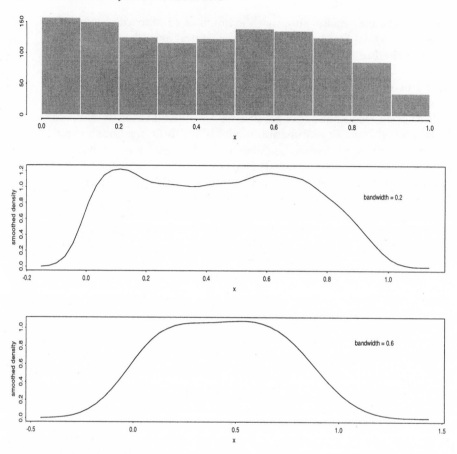

Fig. 8.2. Histogram of the kernel density estimates for the distribution of pattern **b** = (0, 1, 1, 0, 1), with bandwidths of 0.2 and 0.6, as indicated in the labels.

a statistical computing language. On the vertical axis of the histogram, frequencies are marked. For the estimated densities, the vertical axes may extend farther than 1 due to single-precision accuracy used in the calculations.

For the data we have generated, *h* was chosen to be twice the interquartile range of the data divided by 1.34. The interquartile range is the difference between the 75th and 25th percentile of the generated data. To obtain an optimal bandwidth, in the sense of minimizing the mean integrated square error, the interquartile range is divided by 1.34 (Silverman 1986). The resulting fitted curves were sufficiently smooth and followed the shape of the histogram. Also, with a sample size exceeding 10,000 for all data sets, the nonparametric density estimate is known to be a reliable approximation of the actual density function.

Rather than a nonparametric approach, we also tried to use the statistical package Bestfit to identify best-fitting parametric distributions (our smoothed estimates). The fit criterion was the Anderson-Darling test because it provides the most tail-sensitive test. For our simulated patterns used as test data, parametric fits were possible. In principle, as evidenced by the test data, we could use a parametric fit and extreme value theory to obtain the distribution of the extrema of the taxon range. However, many of the resultant histograms with existent field data were irregular with multi-modality and left skewness and were not amenable to computerized parametric fit. Therefore, we chose to use a nonparametric fit for the density estimation.

Simulation of the Distribution of the Endpoints of the Taxon Range Distribution

Once the density estimation of the observed pattern is completed, this density function can be used to estimate the distribution of the extrema, that is, the true duration limits. We accomplish this by resampling from our nonparametric density. Such resampling requires a specific algorithm appropriate for nonparametric density estimation (Silverman 1986, 143).

First, we let $(f_1^{(1)}, f_2^{(1)}, ..., f_k^{(1)})$, $(f_1^{(2)}, f_2^{(2)}, ..., f_k^{(2)})$, ..., $(f_1^{(m)}, f_2^{(m)}, ..., f_k^{(m)})$, be m independent samples of size k from the nonparametric density estimate. Second, we let $o_1 = \max(f_1^{(1)}, f_2^{(1)}, ..., f_k^{(1)})$, $o_2 = \max(f_1^{(2)}, f_2^{(2)}, ..., f_k^{(2)})$, ..., $o_m = \max(f_1^{(m)}, f_2^{(m)}, ..., f_k^{(m)})$, and $e_1 = \min(f_1^{(1)}, f_2^{(1)}, ..., f_k^{(1)})$, $e_2 = \min(f_1^{(2)}, f_2^{(2)}, ..., f_k^{(2)})$, ..., $e_m = \min(f_1^{(m)}, f_2^{(m)}, ..., f_k^{(m)})$. Then a resulting histogram of $(o_1, o_2, ..., o_m)$ is a reasonable estimate of the density function of the upper endpoint, maximum, or origination point. Likewise, a histogram of $(e_1, e_2, ..., e_m)$ is a reasonable estimate of the density of the lower endpoint, minimum, or extinction time of the species. These histograms or density function estimates contain all possible information concerning the values these extreme quantities can take on, given the observed pattern; in essence, they can be regarded as sampling distributions. In addition, the functions provide the likelihood of these values, given the observed presence/absence pattern.

To identify a parametric density distribution that could fit each of the two histograms of the nonparametric extrema estimates, normalized so that each will integrate to 1, we used the statistical package Bestfit (1995).

In consequence, we have modeled empirically the information about the true duration contained in the sample taxon range data. This can now be used for our inferential purposes.

Comparison of Probabilistic and Empirical Results

Table 8.1 contains the data for the 11 species of *Metrarabdotos*. There are 83 observations per species, but they do not appear to correspond to distinct time points. However, that is due to rounding error on the ages. It is possible with our method to use the observed species patterns with times as they are listed, that is, with a sample size of $n = 83$. Alternatively, to be able to establish a one-to-one relationship between time and measurements, we can drop the multiple observations. We follow the convention that, if at least one find has been observed at a specific time point, it corresponds to 1 in the pattern; if all of the observations are nonfinds, 0 is recorded. The length of the resulting vectors is 52. We shall use this vector length for comparisons. Similar results obtain with the larger vector.

Let us first consider n.sp. 10, which is the first column of data in table 8.1. Strauss and Sadler's calculations give a confidence interval on the FAD of 6.50 million years ago (Ma) that extends to 7.05 Ma. Our calculations for n.sp. 10 show that for our best-fitting distribution (Gamma (576, 0.00114)) the 95th percentile is 0.701272. To the right of this value lies 5% of the distribution, or 5% of the possible values that a random variable with this distribution can take on. We can transform this value to a score on our distribution of ages by the simple equation

$$y \times \text{range} + \text{last sample time point,} \tag{1}$$

where y is a value from a distribution, for example our 95th percentile. The "range" in formula (1) is merely (first sample time point − last sample time point). Figure 8.3 contains the histogram of the 1,000 data generated from the maximum of the distribution of the observed pattern for n.sp. 10, with the best-fitting curve of Gamma (576, 0.00114) overlaid. For comparison's sake, the density curve is overlaid to accommodate the frequency y-axis of the histogram.

For n.sp. 10, just as for each of the other ten species, the observed pattern ranges over $8.00 - 1.85 = 6.15$ Ma. Using formula (1), the 95th percentile of our best-fitting Gamma translates to a score of $0.701272 \times 6.15 + 1.85 = 6.16$ Ma. We base this calculation on the $n = 52$ unique time points, although similar results would have resulted from using the $n = 83$ data set.

Our time point of 6.16 Ma falls to the right of the FAD of 6.50 Ma, which observation translates to a score of 0.756098. Thus, our belief that, in fact, we have observed a FAD from our sample that is quite close to the true limit of this species' range is very (significantly) high. Of course, this statement is predicated upon the observed sampling pattern.

Table 8.1. Data for 11 species of *Metrarabdotos*

Age in Ma	n.sp. 10	n.sp. 9	auric.	n.sp. 8	collig.	n.sp. 5	n.sp. 6	n.sp. 7	lacrym.	n.sp. 3	n.sp. 4
1.85	0	0	1	0	0	0	0	0	0	0	0
2.50	0	0	0	0	0	0	0	0	1	0	0
3.30	0	0	1	0	0	0	0	0	0	0	0
3.35	0	0	1	0	0	0	0	0	0	0	0
3.40	0	0	0	0	0	0	0	0	1	0	0
3.85	1	1	0	1	0	0	0	0	1	1	0
4.55	0	1	0	0	0	0	0	0	1	0	0
4.80	0	0	1	0	0	0	0	0	1	0	0
4.80	0	0	1	0	0	0	0	0	0	0	0
5.00	1	1	1	0	0	0	0	0	1	0	0
5.10	0	1	0	0	0	0	0	0	0	0	0
5.15	0	1	1	0	0	0	0	0	0	0	0
5.25	0	0	0	0	0	0	0	1	1	0	1
5.30	1	0	0	0	0	0	0	1	1	0	0
5.35	1	0	0	0	0	0	0	1	1	0	1
5.40	1	1	1	0	0	0	0	1	1	0	1
5.40	1	0	0	0	0	0	0	0	0	0	0
5.45	0	1	1	0	0	0	0	1	0	0	0
5.55	1	1	1	0	0	0	0	0	1	0	0
5.55	0	1	1	0	0	0	0	0	0	0	0
5.55	0	1	1	0	0	0	0	0	0	0	0
5.60	0	1	0	0	0	0	0	1	1	0	1
5.65	1	1	1	0	0	0	0	1	1	0	1
5.65	0	1	0	0	0	0	0	0	0	0	0
5.70	0	0	0	0	0	0	0	0	0	1	1
5.70	0	0	0	0	0	0	0	0	0	1	0
5.75	1	1	0	0	0	0	0	0	0	1	1
5.90	1	1	0	0	0	0	0	0	1	0	1
5.95	1	1	0	0	0	0	0	0	1	0	0
5.95	0	1	0	0	0	0	0	0	0	0	0
6.20	0	1	0	0	0	0	0	1	0	1	1
6.25	1	0	1	0	0	0	0	0	1	0	0
6.25	1	0	0	0	0	0	0	0	0	0	0
6.35	1	0	1	0	0	0	0	0	1	0	0
6.40	1	0	1	0	0	0	0	0	1	0	0
6.45	1	1	1	0	0	1	0	0	1	1	1
6.45	0	0	1	0	0	0	1	0	1	0	0
6.45	0	0	0	0	0	0	0	0	1	0	0
6.50	1	0	1	0	1	1	0	0	1	1	1
6.50	0	0	1	0	0	1	0	0	1	0	0
6.50	0	0	0	0	0	1	0	0	1	0	0
6.50	0	0	0	0	0	0	0	0	1	0	0
6.55	0	0	0	0	0	0	0	0	1	0	1
6.55	0	0	0	0	0	0	0	0	1	0	0
6.65	0	0	0	1	0	0	0	0	0	1	1
6.80	0	0	0	1	0	0	0	0	1	1	1

Continued on next page

Table 8.1. *(continued)*

Age in Ma	n.sp. 10	n.sp. 9	auric.	n.sp. 8	collig.	n.sp. 5	n.sp. 6	n.sp. 7	lacrym.	n.sp. 3	n.sp. 4
6.90	0	0	0	1	1	0	0	0	1	1	1
6.90	0	0	0	1	0	0	0	0	1	0	1
7.00	0	0	0	1	0	0	0	0	1	0	0
7.05	0	0	1	0	1	0	0	0	0	0	0
7.15	0	0	1	0	1	0	1	0	1	0	0
7.15	0	0	1	0	1	0	0	0	1	0	0
7.20	0	0	1	0	1	0	1	1	1	0	0
7.25	0	0	1	0	1	0	0	0	0	0	0
7.30	0	0	1	0	1	0	0	0	1	0	0
7.35	0	1	1	0	0	0	0	0	0	0	0
7.40	0	0	0	0	1	0	0	1	1	1	1
7.40	0	0	0	0	0	0	0	0	0	0	1
7.45	0	0	0	0	1	0	0	0	1	0	0
7.50	0	0	0	0	1	0	1	0	1	0	0
7.50	0	0	0	0	1	0	0	0	0	0	0
7.55	0	0	0	0	1	0	0	0	0	0	0
7.60	0	0	0	0	1	1	1	0	0	0	0
7.60	0	0	0	0	1	0	0	0	0	0	0
7.60	0	0	0	0	1	0	0	0	0	0	0
7.60	0	0	0	0	1	0	0	0	0	0	0
7.65	0	0	0	0	1	0	1	0	0	0	0
7.70	0	0	0	0	1	1	1	0	0	1	0
7.70	0	0	0	0	1	1	1	0	0	1	0
7.70	0	0	0	0	1	0	0	0	0	0	0
7.75	0	0	0	0	1	1	0	0	0	0	0
7.75	0	0	0	0	1	1	0	0	0	0	0
7.75	0	0	0	0	1	0	0	0	0	0	0
7.80	0	0	1	0	1	1	0	0	1	0	0
7.80	0	0	0	0	1	0	0	0	0	0	0
7.85	0	0	0	0	1	0	0	0	1	0	0
7.85	0	0	0	0	1	0	0	0	1	0	0
7.85	0	0	0	0	1	0	0	0	0	0	0
7.90	0	0	0	0	1	0	0	0	1	0	0
7.90	0	0	0	0	1	0	0	0	0	0	0
7.90	0	0	0	0	1	0	0	0	0	0	0
8.00	0	0	1	0	1	0	0	0	0	0	0
8.00	0	0	0	0	1	0	0	0	0	0	0

Note: Data from A. H. Cheetham (1986 pers. comm.).

Figure 8.4 presents the histogram and kernel density estimate of the distribution of the observed pattern of species n.sp. 10. Apparently most of the mass concentrates around 0.45, translating to $0.45 \times 6.15 + 1.85 = 4.62$ Ma, with few observations falling to the right of 0.65 (giving a score of 5.85 Ma). As exhibited by the histogram of figure 8.4, there is pronounced evidence of lack of uniformity in the samples throughout the

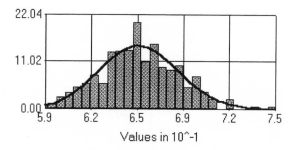

Fig. 8.3. Histogram and overlaid gamma (576, 0.00114) density estimate for the maximum of the distribution of n.sp. 10.

observed range with a peak at 0.5 (= 4.93 Ma). In fact, the distribution of the pattern for species n.sp. 10 appears to be unimodal with left skewness. We interpret our origination results for n.sp. 10 to be in keeping with Marshall's calculations of Strauss and Sadler's results, whose upper limit of 7.05 Ma, when compared to our Gamma description, translates to 0.845528. This value is beyond our 99th percentile. Thus, although 7.05 Ma could be probable overestimation, we still are in agreement that the observed range is close to the true range.

This same procedure and analogous statements can be made separately for each of the 11 species' origination and extinction times, or species' ranges. Results are listed in table 8.2, along with the accompanying species' results from the Strauss and Sadler method as calculated for these species by Marshall.

Let us now consider the LAD of n.sp. 10. From table 8.2 we see that the time of extinction is best described with a Normal density whose 5th percentile is 0.032727, translating to a score of 2.05 Ma. For this species the last observed occurrence is at 3.85 Ma. Thus, there is critical mass (a significant amount of density) of this distribution to the right of the LAD, and our strength of belief in this LAD's being a good estimate of the true extinction date is significantly small. In comparison, Marshall quotes a lower limit of 3.294, quite close to the observed 3.85 Ma for the LAD. Therefore, this is an apparent and considerable underestimate, whose magnitude appears to be related to the large number of absences after the LAD and to the sparseness of the finds at this end of the pattern.

Alternatively, consider *Metrarabdotos colligatum*, whose sample pattern is given in table 8.2. The data for *M. colligatum* is an example of truncation of the sampling near the FAD point, which is the terminus of a long series of finds. As is evident, the 95% confidence interval extends this observed range beyond the first observed occurrence of 8.00 Ma only to 8.143 Ma. The observed sample pattern for *M. colligatum*, in which

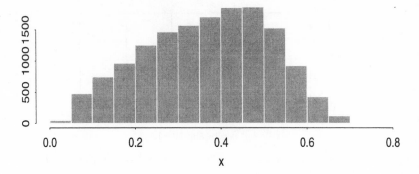

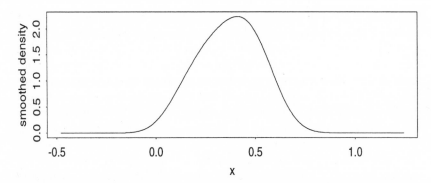

Fig. 8.4. Histogram and kernel density estimate for the distribution of n.sp. 10.

there is an extended series of 1s at the very beginning of its record, makes the estimated sampling distribution of the maximum excessively peaked at 1 and the area immediately to the left of 1, as exhibited in figure 8.5. Because there is no sample information to the right of the first sample location, the density of the maximum is truncated at the FAD. Thus, we are led to believe that there is critical density-function mass (or important but unknown fossil information) to the right of the first observed value of 8.00 Ma. A conservative estimate of the amount of this mass under the actual truncated curve fit is at least 36%. In essence, we have no solid basis upon which to construct any inference concerning the critical area, and our results do not agree with those of Marshall's (1995), which predicate a small distance between the FAD and the true point. Alternatively, we could say that because we can identify critical mass to the left of the FAD, in comparison, the confidence limit on this endpoint obtained with the Strauss and Sadler method appears unrealistically to underestimate the true duration.

Table 8.2. Comparison of estimated ranges for 11 species of *Metrarabdotos*

	95% Confidence Intervals				FAD Distribution	95th Percentile	95th Estimate	LAD Distribution	5th Percentile	5th Estimate
	F.O.	F.O. + 95%	L.O.	L.O. − 95%						
n.sp. 10	6.5	7.046	3.85	3.294	Gamma (576, 0.00114)	0.7013	6.1628	Normal (0.6320, 0.0185)	0.0327	2.0513
n.sp. 9	7.35	7.948	3.85	3.255	Logistic (0.82, 0.0198)	0.8757	7.2355	Gamma (16.6743, 0.0036)	0.0380	2.0838
auriculatum	8.0	8.694	1.85	1.174	Weibull (161, 1.00)	1.0030	8.0183[a]	Beta (0.7131, 132.2416) + 0.000026	0.0001	1.8506[b]
n.sp. 8	7.0	9.585	3.85	1.267	Extreme Value (0.7543, 0.0217)	0.8187	6.8852	Weibull (3.8854, 0.0517)	0.0241	1.9980
colligatum	8.0	8.143	6.50	6.350	Weibull (362, 1.00)	1.0008	8.0051[b]	Weibull (12.1213, 0.4834)	0.3784	4.1769
n.sp. 5	7.8	8.272	6.45	5.978	Normal (0.9776, 0.0085)	0.9915	7.9478	Weibull (12.8665, 0.4150)	0.3295	3.8762
n.sp. 6	7.7	8.056	7.15	6.793	Normal (0.9550, 0.0113)	0.9738	7.8388	Normal (0.5523, 0.0295)	0.5037	4.9475
n.sp. 7	7.4	8.249	5.25	4.412	Erlang (7.68, 0.14)	0.1648	7.0792	Weibull (7.68, 0.14)	0.0951	2.4348
lacrymosum	7.9	8.309	2.50	2.068	Weibull (126.4050, 0.9881)	0.9967	7.9798	Gamma (1.9909, 0.0107)	0.0038	1.8731[c]
n.sp. 3	7.7	8.792	3.85	2.772	Normal (0.9393, 0.0193)	0.9711	7.823	Normal (0.5744, 0.0177)	0.0284	2.0245
n.sp. 4	7.4	7.814	5.25	4.842	Logistic (0.8416, 0.0185)	0.8962	7.361	Normal (0.1487, 0.0277)	0.1031	2.4841

Note: 95% confidence interval for FAD and LAD from Marshall 1990. Species data from Cheetham 1986, pers. comm., table 8.1. First (F.O.) and last (L.O.) occurrences are in Ma. Confidence interval = (F.O. + 95%) for first occurrence and (L.O. − 95%) for last occurrence. FAD (LAD) distribution is the best fit distribution for the FAD (LAD), based upon the Hayek and Bura method.

[a] Possible truncation.

[b] Truncation.

[c] Significant mass exists to the left of the minimum value.

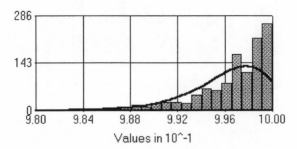

Fig. 8.5. Histogram and overlaid Weibull (362.3148, 0.9978) density estimate for the distribution of the maximum of *Metrarabdotos colligatum.*

The LAD for *M. colligatum* is 6.5 Ma (table 8.2). The 95% confidence limit is 6.35 Ma. Note the closeness of observed and estimate because of the symmetry of the modeling approach. However, table 8.1 shows a long run of 0s prior to the LAD, a distinctly different profile from the sampling at the FAD end of the range. By our method, using the Weibull description for extinction time, we find the 95th percentile to have a date of 4.18 Ma. It appears therefore that the 6.5 − 6.35 = 0.15 Ma extension to the range of *M. colligatum* is also a considerable underestimate.

M. lacrymosum, whose data are presented in table 8.1, is another informative example. The first observed occurrence of this species is at 7.9 Ma. With out empirical 95th percentile estimate score of 7.98 Ma and an interval estimate with upper limit of 8.309 Ma, the results of the two methods are in apparent agreement.

When we consider the last observed occurrence, however, we have a different profile. The limiting estimate of the true extinction date of 2.068 Ma is exceedingly close to the LAD of 2.5 Ma. By our method the best-fitting Gamma description is slightly truncated, with the fifth percentile estimate equal to 1.87. Since the estimated extinction point supplied by Marshall (1990) is much earlier than our 5% value, the interval estimate is apparently underestimating the true date. Our calculations conservatively show that at least 35% of the mass or area should be to the left of the observed value, or truncation point.

Finally, we examine *M. auriculatum,* whose data are in table 8.1. The sampling history or pattern for this species shows evidence of quite sparse sampling near the FAD and evidence of dense sampling near the LAD of this common species. Marshall (1990), using the Strauss and Sadler method, obtained an estimate of 8.694 Ma with the FAD of 8.00 Ma. Using our Weibull description, we obtain evidence of truncation at the FAD. Our table shows that the 95th percentile date score is 8.01 Ma. The amount of truncation conservatively is about 27%, and Marshall's results appear to place the confidence to the left, but relatively close, to the FAD. We take this to be probable underestimation due in part to the pattern of sparseness for this end of the species range.

The minimum has a Beta description evidencing truncation. The confidence limit estimate is 1.174 on a LAD of 1.85 Ma. There is clear indication of dense sampling at this end of the range for this common species. The amount of truncation is not noteworthy, and we have no basis upon which to disagree with previous results that place the confidence limit to the right of the LAD.

Discussion and Conclusions

The Empirical Approach

We have derived a method of empirical description of the field situation for extracting information for taxon duration. By introducing this empirical modeling we introduce a new way of thinking about the use of field data and practical observation results for problem solution. This method uses all potential and actual information in the observed samples and incorporates that information into modeling, without any attendant mathematical or statistical assumptions. That is, in no way do we constrain the field results, except to have the expectation that the observed samples will be representative of accurate and experienced fieldwork. We offer this as the necessary alternative to fitting probabilistic models, not only for the taxon range problem but also for other problems that make use of field data. The reasons for advocating an empirical approach are

1. A field sample is not a "random sample" in the statistical sense.
2. There are multiple meanings to the use of a dichotomous scoring system: a 1 always indicates presence of a taxon; a 0 has multiple meanings. That is, an interior 0, when taking a sample only in time, may in truth be a 1, or it may indicate sampling error or the result of other potential processes at work. Alternatively, when sampling in time and space, an interior 0 may be a true absence for a particular locality. In addition, 0s outside of the time of a LAD or FAD may similarly have multiple senses.
3. Most applications are of relatively unsophisticated probabilistic models without the necessary mathematical agility (or number of parameters) to describe field conditions completely realistically. Although some may result in realistic assessments under a limited set of conditions for which the assumptions are not too restrictive on the field situation, in general, the results are unrealistic assessments of the true situation, especially when applied indiscriminately.
4. An empirical approach, even one in its "infancy" such as the present

methodology, provides a utilitarian description of the desired paleobiological situation and can incorporate the true field characteristics. All the data is used.

5. An empirical approach switches the burden of proof for the results away from the statistician who devises or applies a model, onto the researcher who is responsible not only for the field results but for incorporating his or her expertise into the problem description.

In other words, we have developed a method that takes all the quantified field data, maximizes its information content, and uses statistical distributional theory to obtain meaningful results. Then we use this to evaluate the results of confidence limit calculations based upon a Poisson process and a probabilistic model. The estimated distribution of the endpoints (extrema) of the taxon range indicates where the true origination and extinction times would be expected to occur, given the observed pattern of the data. We mapped the observed range to the [0, 1] interval in keeping with other probabilistic models as an aid to comparison. This implicitly introduced upper and lower limits to the distribution of the extrema, but these limits are easily changed back to actual time dates by simple multiplication

The procedure can be summarized by saying that we derive the sampling distribution of all possible extreme values using only the observed sampling pattern or the contents of the field observations. Based solely upon this information concerning the values that the extrema can take on, and without imposing any assumptions on the underlying process that could have generated the data, we compute our quantified estimate of whether or not the local range is a good estimate of the true range.

This paper does not present a complete solution to the problem, but a new way to think about problems in paleobiology, which utilize field data. Our method can be improved in several ways, which would allow for immediate extinction and origination estimates. The present approach bears the limitation of the nonparametric estimation procedure we used and the accompanying unstructured consideration of error in the process. Alternatively, if we had available a computerized parametric fitting program, we could then capitalize upon extreme value theory and provide the desired estimates. The approach we are currently developing is to strengthen the empirical model with weighted functions summed over the sample section and to derive distribution theory for the empirical model directly. This should give immediate and precise duration limits.

Comparison of Empirical and Probabilistic Modeling Approaches

Koch and Morgan (1988) emphatically state that it is not the species-level fossil record that is hopelessly distorted, but that distortion is related to inattention to sampling bias and the quantitative aspects of the fossil record. Distortion can also result from inattention to coordination of practical and theoretical assumptions. These authors also provided evidence of the considerable influence of sample size in the absence of real differences in species' distribution. Buzas et al. (1982) established that in paleontological data rare species compose a sizable percentage of all observed species. Our thesis is that this well-established information must be incorporated into any efforts to model. Our results show that the present reliance on probabilistic non-Bayesian methods for estimating species durations for data from continuously sampled sections is useful yet flawed. A further powerful advantage of our method is that it can accommodate all practical sampling schemes: continuous, binned, or discrete sampling data. That is, an empirical approach when used wisely, in conjunction with field experience, can be the most expedient method by which to model complex, multifaceted situations like stratigraphic range sampling.

Philosophically, presence data is distinct from absence data. When a taxon is present in a sample, the meaning is clear. When we find no evidence of the existence of that taxon in a sample, there can be a variety of causes. However, translation of presence/absence into numerals (say 0, 1) gives little room for philosophical statements. For the taxon range problem in particular, when a Poisson process or other probabilistic model is used for problem solution, many restrictive assumptions, both implicit and explicit, must be made and the point of concentration is the LAD and/or FAD. For example, the Weiss and Marshall (1999) approach, whose focal point is estimation of absence, has implicit in it that there is a lack of information in the observations about the tail of the extinction time distribution. An empirical model is data-driven and therefore has the advantage over fitting probabilistic models for field sampling. We found that all the observations (including data in the tails) do contain extremely useful information for our problem solution.

With this empirical approach we have shown that (1) underlying assumptions may be unrealistic to the point of causing errors in estimating duration limits, and (2) the most serious problem is failure to incorporate the exterior and interior absences and the sampling pattern. This omission can be the cause of serious inferential errors.

When we compared our results with those calculations of Marshall (1990, based on Strauss and Sadler's model) in his application to Cheetham's data (1986, pers. comm.) on 11 species of *Metrarabdotos*, we found evidence of substantial overestimation and underestimation of extinction and origination date. Marshall (1990) discussed the possibility of underestimation, and Cheetham and Jackson (1998) gave concrete evidence of such bias. However, our work points out not merely existence but unrealistic assessment of magnitude of biased estimation.

When sampling extends beyond the time of a species' first or last observed occurrence, our method can make a quantitative evaluation of the likelihood of the LAD's or FAD's distance from the true duration limits. Then, without further sample evidence, we can say that calculated confidence limits will either be reasonable, overestimated, or underestimated.

When there was no evidence that sampling continued beyond the point of the LAD or FAD, our method shows that these confidence limits should not be used. It is reasonable to think of the observed range of a species as a truncated version of its true range. Our method gives further credence to this fact. In particular, however, if the sampling itself is also truncated at the point of last or first occurrence, the amount of information in the sample is lessened for inference. Estimation problems resulted from the Strauss and Sadler model when there was evidence of both dense and sparse sampling.

The Strauss and Sadler method gives a symmetric interval, since the number of finds used to calculate the extinction interval estimate is identical to that used to calculate the origination estimate. A glance at the distribution of 1s for many of the taxa in table 8.1 indicates that a symmetric interval is unrealistic. Our approach elucidates the nonsymmetry inherent in the sampling pattern and provides separate scores, or ages, for each end of the range. In addition, our results corroborate the influence of correlated factors in the sampling process.

Conclusions

1. If sampling is truncated, and the observed pattern is a series of 1s culminating in a final 1, the use of confidence limits is not recommended. There is no basis upon which to estimate true species duration, and there is evidence of severe underestimation when estimation is attempted.
2. When there is evidence that rare species are involved, this can lead to underestimation of duration time.
3. When there is evidence of species rarity in combination with nonfinds or absences outside of the observed range end, this can lead to overestimation. The pattern of rarity is then not being correctly weighted or considered.
4. Confidence estimates of true species duration time are best used when the sampling pattern shows (1) at least common abundance, (2) evidence of nonfinds outside of the observed range of FAD and LAD, and (3) dense sampling over a series of closely spaced horizons.

In summary, our work shows that the inclusion of nonfinds, as well as maximal sample information, is important evidence that should not be ignored if statistical inference is to give useful and accurate knowledge for phylogenetic or evolutionary study.

Acknowledgments

We thank M. Buzas, M. Foote, and an anonymous reviewer for their helpful suggestions and criticisms. The S-PLUS programs used to carry out the calculations in this article are available upon request from the authors. The first author would like to give a special acknowledgment to Alan Cheetham for his inspiration to her research. His logical and innovative approach to problem solution has been of major benefit over years of research and collaboration.

References

Bestfit: Distribution fitting for Windows. 1995. Newfield, NY: Palisade Corp.

Buzas, M. A., and S. J. Culver. 1998. Assembly, disassembly, and balance in marine paleocommunities. *Palaios* 13:263–75.

Buzas, M. A., C. F. Koch, S. J. Culver, and N. F. Sohl. 1982. On the distribution of species occurrence. *Paleobiology* 8 (2): 143–50.

Cheetham, A. H. 1986. Tempo of evolution in a Neogene bryozoan: Rates of morphologic change within and across species boundaries. *Paleobiology* 2: 190–222.

Cheetham, A. H., and J. B. C. Jackson. 1998. The fossil record of cheilostome Bryozoa in the Neogene and Quaternary of tropical America: Adequacy for phylogenetic and evolutionary studies. In *The adequacy of the fossil record*, ed. S. K. Donovan and C. R. C. Paul, 226–40. Chichester, United Kingdom: John Wiley and Sons.

Foote, M. 1997. Estimating taxonomic durations and preservation probability. *Paleobiology* 23 (3): 278–300.

Foote, M., and D. M. Raup. 1996. Fossil preservation and the stratigraphic ranges of taxa. *Paleobiology* 22 (2): 121–40.

Hayek, L. C., and M. A. Buzas. 1997. *Surveying natural populations.* New York: Columbia University Press.

Holland, S. M. 1995. The stratigraphic distribution of fossils. *Paleobiology* 21 (1): 92–109.

Koch, C. F., and J. P. Morgan. 1988. On the expected distribution of species' ranges. *Paleobiology* 14 (2): 126–38.

Marshall, C. R. 1990. Confidence intervals on stratigraphic ranges. *Paleobiology* 16 (1): 1–10.

———. 1994. Confidence intervals on stratigraphic ranges: Partial relaxation of the assumption of randomly distributed fossil horizons. *Paleobiology* 20 (4): 459–69.

Paul, C. R. C. 1982. The adequacy of the fossil record. In *Problems of phyloge-netic reconstruction,* ed. K. A. Joysey and A. E. Friday, Systematics Association Special Volume 21, 75–117. London: Academic Press.

Shaw, A. B. 1964. *Time in stratigraphy.* New York: McGraw-Hill.

Signor, P. W., and J. H. Lipps. 1982. Sampling bias, gradual extinction patterns, and catastrophes in the fossil record. *Geological Society of America, Special Paper* 190:291–96.

Silverman, B. W. 1986. *Density estimation for statistics and data analysis.* London: Chapman and Hall.

S-PLUS Guide to Statistics. 1998. Seattle: Data Analysis Products Division, MathSoft.

Springer, M. S. 1990. The effect of random range truncation on patterns of evolution in the fossil record. *Paleobiology* 16 (4): 512–20.

Springer, M. S., and A. Lilje. 1988. Biostratigraphy and gap analysis: The expected sequence of biostratigraphic events. *Journal of Geology* 96:228–36.

Strauss, D., and P. M. Sadler. 1987. *Confidence intervals for the ends of local taxon ranges.* Technical Report 158. Riverside, CA: Department of Statistics, University of California.

———. 1989. Classical confidence intervals and Bayesian probability estimates for ends of local taxon ranges. *Mathematical Geology* 21 (4): 411–27.

Weiss, R. E., and C. R. Marshall. 1999. The uncertainty in the true end point of a fossil's stratigraphic range when stratigraphic sections are sampled discretely. *Mathematical Geology* 31 (4): 435–53.

Evolutionary Rates and the Age Distributions of Living and Extinct Taxa

9

MIKE FOOTE

Estimating Origination and Extinction Rates

Any change in taxonomic diversity reflects the net difference between origination and extinction rates, but itself says nothing about the magnitude of these rates. In many cases, we would like to measure the two rates separately, not just their difference, for how the rates contribute to diversity changes has important implications for diversity dynamics. For example, global marine diversity showed little net change through much of the middle and late Paleozoic but increased substantially during the Mesozoic and Cenozoic (Sepkoski 1984). Thus the net difference between origination and extinction rates must have increased. In this case, there is evidence that both rates have declined since the early Paleozoic (Van Valen 1973, 1984, 1985; Van Valen and Maiorana 1985; Raup and Sepkoski 1982; Sepkoski 1986, 1987, 1998; Gilinsky and Bambach 1987; Gilinsky 1994); thus extinction rate must have declined more than origination rate. In general, however, an increase in diversity could occur by an increase in origination rate, by a decrease in extinction rate, or in some other way. Whether there is an overriding tendency for changes in diversity to be more strongly linked to changes in origination rate or to changes in extinction rate remains an open question.

In addition to the evolutionary issues involving the relationships among diversification, origination, and extinction, there are many methodological issues. Understanding the mechanisms of diversification requires independent estimates of origination and extinction rate, but consider just two problems with conventional origination and extinction metrics calculated over discrete stratigraphic intervals. (1) Single-interval taxa contribute to both rates for an interval, yet many of these taxa are artifacts of incomplete and variable preservation (Sepkoski 1993, 1998; Harper 1996; Foote and Raup 1996). (2) The lengths of intervals distort extinction and origination metrics in similar ways (Gilinsky 1991; Foote 1994). Certain origination and extinction metrics increase systematically with interval length while other metrics, notably many that attempt to normalize for interval length, decrease as interval length increases. Either of these factors can induce a spurious correlation between origination and extinction rates. Such interval-by-interval calculations are nonetheless useful if we compare two groups that coexist through the same stratigraphic intervals. Van Valen (1973) was thus able to suggest that Artiodactyla and Perissodactyla have had similar extinction rates, but that the former have diversified more as a result of higher origination rates.

A number of approaches to estimating evolutionary rates have focused on taxon-age distributions rather than interval-by-interval tabulations of origination and extinction events, but the analysis of taxon-age distributions is not without its problems (Raup 1975; Van Valen 1979). For example, cohort survivorship analysis of genera has been used to infer species origination and extinction rates, based on the theoretical relationship between species-level rates and genus-level survivorship (Raup 1978, 1991). The difference between species origination and extinction rates, however, is much better constrained than is either rate separately (Foote 1988; Brady et al. 1996).[1] In general, taxon-age distributions reflect the magnitude of taxonomic rates, with higher rates yielding steeper survivorship curves. Yet the kind of age distribution that is typically analyzed is based on the stratigraphic ranges of all taxa observed over some interval of time, and this distribution depends on both origination and extinction rates (Pease 1988; Foote 1997; appendix 9.1). It is therefore difficult to estimate the two rates independently.

In this chapter, I will survey properties of various taxon-age distributions (table 9.1, fig. 9.1), with the goal of obtaining independent estimates of origination and extinction rates.[2] This survey points to a number of pitfalls, some of them not generally appreciated, that complicate the estimation of rates. (1) Some taxon-age distributions depend on both origination and extinction rates, and therefore cannot be used to estimate these rates independently. (2) Other distributions depend on only one rate or the other, but it may be difficult to know which one. (3) And in some cases the distribution of ages is unaffected by interchanging origination and extinction rates. It is nevertheless possible to identify situations in which we should be able to follow cohorts of taxa forward in time to estimate extinction rate and backward in time to estimate origination rate. An attempt to do so for marine animal genera is thwarted by heterogeneity in the data which apparently reflects, at least in part, a dependence between the age of a genus at a given point in time and the probability that it will branch or become extinct at that time. The approaches outlined here should nevertheless be useful if there are cases where it can be shown that age-dependency is weak or absent.

Table 9.1. Summary of taxon-age distributions

Kinds of Taxon-Age Distribution

Distribution	Explanation	Examples from Fig. 9.1
General taxon-age distribution	Cumulative distribution of observed ranges	Fig. 9.1C
Cohort survivorship curve	Cumulative distribution of ages after time of reference	Fig. 9.1B, solid curves
Cohort prenascence curve	Cumulative distribution of ages before time of reference	Fig. 9.1B, dashed curves

Subsets of Taxa

Subset	Taxa Included	N	Applicable Distributions	Sections of Appendix 9.1
(t_1, t_2) birth cohort[a]	Taxa originating between t_1 and t_2	24[a]	Cohort survivorship	1, 11, 13, 14
(t_1, t_2) death cohort[a]	Taxa extinct between t_1 and t_2	22[a]	Cohort prenascence	1, 10, 12, 14
Boundary cohort (general)	Taxa extant at any moment in time	—	Cohort survivorship or prenascence	1, 2, 3, 14
$t = 0$ boundary cohort	Taxa extant at start of window of observation	10	Cohort survivorship	1, 2, 3, 14
$t = w$ boundary cohort[b]	Taxa extant at end of window of observation	24	Cohort prenascence	1, 2, 3, 14
(t_1, t_2) polycohort[c]	Taxa extant at any time between t_1 and t_2	37	Cohort survivorship or prenascence	1
Originating taxa	Taxa originating between $t = 0$ and $t = w$	104	General taxon age	1, 6, 7, 14
Extinct taxa[d]	Taxa extinct between $t = 0$ and $t = w$	90	General taxon age	1, 4, 5, 14
Combined	Taxa extant at any time between $t = 0$ and $t = w$	114	General taxon age	1, 8, 9, 14

[a] Eleven taxa that originate and become extinct between t_1 and t_2 are included in the tabulation of N in this table. These are excluded from analysis, however, and survivorship is followed from t_2 forward and prenascence from t_1 backward (see text).

[b] Prenascence curve of this boundary cohort is the classic "extant survivorship curve."

[c] The (t_1, t_2) polycohort or pseudocohort is equivalent to the t_1 boundary cohort plus the (t_1, t_2) birth cohort plus the (t_1, t_2) death cohort. Polycohorts are not considered in detail here (see text).

[d] Taxon-age distribution of these taxa gives the classic "extinct survivorship curve."

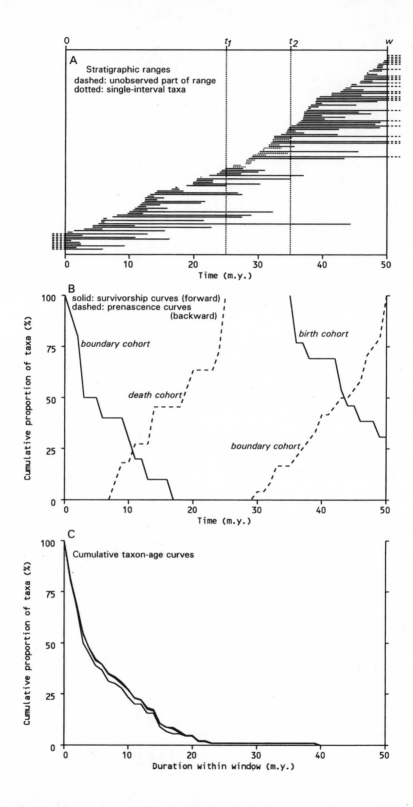

A Stratigraphic ranges
dashed: unobserved part of range
dotted: single-interval taxa

Time (m.y.)

B solid: survivorship curves (forward)
dashed: prenascence curves
(backward)

boundary cohort

birth cohort

death cohort

boundary cohort

Cumulative proportion of taxa (%)

Time (m.y.)

C

Cumulative taxon-age curves

Cumulative proportion of taxa (%)

Duration within window (m.y.)

Properties of Taxon-Age Distributions

One of the classic observations regarding taxon-age distribution is that extinct taxa have a different distribution of ages than do extant taxa. The age of an extinct taxon is the time between its first and last appearances, and the age of an extant taxon is the time between its first appearance and the Recent. The extinct taxon-age curve is generally steeper, as Van Valen (1973) showed most comprehensively. The difference is illustrated here with data from Sepkoski's unpublished compilation of marine animal genera (fig. 9.2). We generally see the same pattern if we treat an arbitrary time plane as the "Recent" and compare genera that cross this boundary with genera extinct at any time beforehand (fig. 9.3).

The difference between extinct and extant taxon-age distributions goes back to Simpson (1944, 1953).[3] Since longer-lived taxa preferentially extend forward to the Recent (or to any arbitrary time plane), the extinct and extant fractions are both biased with respect to the total duration distribution (Van Valen 1973, 1979; Raup 1975). As discussed below, however, the extant curve is not made too shallow by this bias; rather, the extinct curve is made too steep. By resampling from an estimate of the total duration distribution, itself based on a model of time-homogeneous rates and some additional assumptions, Gilinsky (1988) suggested that, for bivalve families and genera, the difference between extinct and extant taxon-age distributions is larger than would be expected from this

◁ **Fig. 9.1.** Hypothetical illustration of taxon-age distributions. *A,* Simulated ranges generated with origination rate equal to 0.12 per lineage-million-years (Lmy^{-1}), extinction rate equal to 0.1 Lmy$^{-1,}$ and window of observation (w) equal to 50 million years (m.y.). Refer to table 9.1. Parts of durations before and after window of observation, shown as *dashed lines,* are unknown and therefore cannot be tabulated as part of the observed taxon age. *Dotted vertical lines* bound an interval from $t_1 = 25$ m.y. to $t_2 = 35$ m.y. Ranges of taxa that originate and become extinct during this interval are shown as *dotted lines.* *B,* Cohort survivorship curves and prenascence curves derived from ranges in A. Survivorship curves *(solid)* show the proportion of taxa that survive at least until a given time after a time of reference. Curve going forward from $t = 0$ is the survivorship curve of taxa extant at the start of the window (the $t = 0$ boundary cohort). Curve going forward from t_2 is the survivorship curve of the cohort of taxa that originate between t_1 and t_2 (the $[t_1, t_2]$ birth cohort), excluding those that also become extinct before t_2. Prenascence curves *(dashed)* show the proportion of taxa that had already originated prior to a given time before a time of reference. Curve going backward from $t = 50$ is the prenascence curve of taxa extant at the end of the window (the $t = w$ boundary cohort). Curve going backward from t_1 is the prenascence curve of the cohort of taxa that become extinct between t_1 and t_2 (the $[t_1, t_2]$ death cohort), excluding those that also originate after t_1. *C,* Cumulative taxon-age distributions showing the proportion of taxa with durations greater than or equal to the value on the abscissa. The three distributions (all taxa that become extinct between $t = 0$ and $t = 50$ m.y.; all taxa that originate between $t = 0$ and $t = 50$ m.y.; and all taxa combined) are very similar in this case.

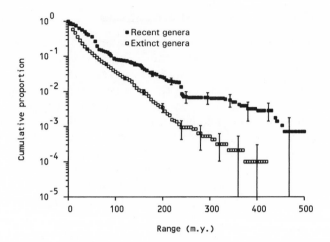

Fig. 9.2. Living and extinct taxon-age distributions for Phanerozoic marine animal genera, showing cumulative proportion (log scale) versus duration. The Phanerozoic was divided into 107 intervals with average duration 5.2 m.y. The timescale is mainly that of Harland et al. 1990, with some modification of the lower Paleozoic based on Tucker and McKerrow 1995. The data are 24,104 fossil genera whose first and last appearances can be properly resolved to one of these stratigraphic intervals. Age distribution for living genera shows the proportion of extant genera that had already originated before given times in the past, sampled at the boundaries between stratigraphic intervals (see text). Extinct distribution is the cumulative frequency of taxa having the given duration, with duration equal to the difference between the midpoint of the interval of first occurrence and the midpoint of the interval of last occurrence. Other conventions for placing first and last occurrences also yield conspicuous differences between living and extinct distributions. Durations are in 5-million-year bins. Single-interval genera are omitted. This follows the logic that, when preservation is incomplete, the age distribution excluding single-interval taxa approximates the true age distribution (Foote and Raup 1996; Foote 1997). This approach tends to make the living and extinct distributions more similar. *Error bars* at selected points in this and subsequent figures bracket the medial 90% of the binomial distribution corresponding to the total number of taxa and the observed proportion. *Error bars* near the left side of figure are generally smaller than the symbols themselves. Although points for a single age distribution are not independent, the two age distributions are independent and thus can be meaningfully compared statistically. See Sepkoski 1996 for description of data.

general bias. He discussed a number of possible explanations, somewhat inconclusively. Van Valen (1979) modeled the expected magnitude of this difference in the case of constant and equal origination and extinction rates. Appendix 9.1 modifies his approach slightly and extends it to the general case of unequal and variable rates.

At this point, it will be helpful to define some terms (table 9.1, fig. 9.1). Though some of these are unfortunately new, they are necessary in order to avoid confusion when comparing the age distributions that result from following taxa forward in time with those that result from following

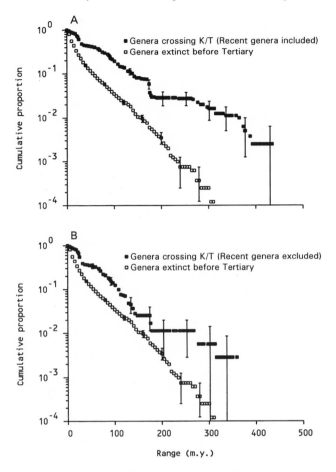

Fig. 9.3. Taxon-age distributions for genera that cross the Cretaceous-Tertiary (K/T) boundary versus all genera extinct before the Tertiary. Analyses are as in figure 9.2, except that the K/T boundary is treated as the Recent. *A*, Genera still extant today are included. *B*, Genera still extant today are excluded.

taxa backward in time. *Taxon-age distribution* is a general term reflecting the frequency distribution of times between origination and extinction, between origination and the last time of census, or between the first time of census and extinction. These are often portrayed as cumulative curves, referred to generally as *survivorship curves,* a term I will use in a more restricted way. A *birth cohort* (or simply a *cohort*) consists of all taxa that originate during a specified interval of time. A *death cohort* consists of all taxa that become extinct during a specified interval of time. Pease (1988) discussed cohorts as taxa originating or becoming extinct at exactly the same time. This is an impractical notion, so we generally combine the taxa over a longer span of time. A *boundary cohort* consists of all taxa that cross a specified time plane (not necessarily a conventional stratigraphic boundary). A *cohort survivorship curve* plots the proportion of a birth or boundary cohort still remaining as a function of elapsed, real time after a time of reference. A *cohort prenascence curve* is a "backward survivorship curve" (Pease 1988) that plots the proportion of taxa in a death or boundary cohort that had already originated as a function of real time before a time of reference.[4] Both types of cohort curve decline monotonically with time. Finally, a *polycohort* or *pseudocohort* (Hoffman and Kitchell 1984; Raup 1985) consists of all the taxa extant during some interval of time. Thus a boundary cohort along with the immediately following birth cohort constitutes a polycohort, as does a boundary cohort plus the immediately preceding death cohort. Because of their heterogeneous nature, polycohorts will not be considered in detail.

These different kinds of curves will be illustrated below. At this point it will help to consider how the terms apply to some previous studies. Each of Van Valen's comparisons (1973) between living and extinct survivorship curves is a comparison between a prenascence curve for the Recent boundary cohort and a general taxon-age distribution for all extinct taxa. Boyajian (1986, 1991, 1992) compared the ages of taxa that cross a particular boundary to those of taxa that became extinct during the last stratigraphic interval before the boundary in question. Thus he compared the noncumulative taxon-age distributions of boundary cohorts to those of death cohorts. If he had expressed the ages in a cumulative fashion, they would have yielded cohort prenascence curves. Foote (1988) compared the slopes of cohort survivorship curves for all Cambrian birth cohorts versus all Ordovician birth cohorts of trilobite genera. Lyellian percentages (Stanley 1979, 1985; Stanley et al. 1980; Pease 1987) represent the fraction of a boundary cohort, birth cohort (Simpson 1953), or more commonly, a polycohort (Stanley et al. 1980) still extant today; as such they are single points on survivorship curves. Hoffman and Kitchell's *accretion curve* (1984) is not a prenascence curve. Rather, it is a plot of the cumulative number of taxa, living and extinct, produced as a function of time; it is a plot of the *total progeny* over time (Kendall 1948; Raup 1985; appendix 9.1).

Theoretical Properties of Taxon-Age Distributions

Here I focus on obtaining independent estimates of origination and extinction rates. The results outlined in appendix 9.1 may be useful for other questions as well. For example, degrading the ideal distributions with a model of incomplete preservation may help to infer true durations from observed stratigraphic ranges (Sepkoski 1975; Pease 1988; Foote and Raup 1996; Foote 1997). Except where otherwise stated, I make the simplifying assumption that per-capita rates of origination and extinction at any arbitrary taxonomic level, while they may depend on real time, do not depend on taxonomic age. This is probably unrealistic at higher taxonomic levels, where extinction rates may depend on underlying species-level rates as well as the age of the taxon (Raup 1978, 1985). I also assume that rates are taxonomically homogeneous. These assumptions will eventually be relaxed below.

If we follow a boundary cohort forward, we track its extinction history (figs. 9.4, 9.5). If the extinction rate is constant, then future survivorship is unaffected by past history, and so the boundary cohort's survivorship curve is log-linear with a slope whose magnitude is equal to the extinction rate. The same is true of taxa that originate within an infinitesimal interval of time (Pease 1988). A number of taxon-age distributions are affected by the fact that there is a finite window of time over which taxa can be observed, and so durations are truncated, potentially at either end (Van Valen 1979; Foote 1997; appendix 9.1). The duration considered here is only the part that falls within the window. A birth cohort can be thought of as a sum of smaller cohorts, each originating within an infinitesimal interval of time. Because each of these has a different proportion still extant at the end of the window, the survivorship curve of the birth cohort is affected by the finite window of observation. A birth cohort nevertheless has the same form of survivorship curve as a boundary cohort if it starts out far enough before the end of the window of observation, so that there is a negligible probability that a taxon in the cohort will still be extant at the end of the window (figs. 9.4, 9.5; appendix 9.1). The slope of the survivorship curve of a boundary cohort or a birth cohort should therefore provide an estimate of the extinction rate.

It seems to be less widely appreciated that symmetrical results hold for prenascence curves and origination rates as for survivorship curves and extinction rates (Pease 1988; appendix 9.1). The prenascence curve of a boundary cohort is often discussed as reflecting both origination and extinction, even in the simple case of constant rates (e.g., Van Valen 1973). With constant rates, however, the classic extant survivorship curve (Van Valen 1973), that is, the prenascence curve of the Recent boundary cohort, depends only on origination rate (Pease 1988). This is true also for death cohorts, provided that the interval of extinction is long enough after the start of the window of observation that relatively few taxa in the death cohort were present at the start of the window. The dependence of prenascence curves on origination only and of survivorship curves on

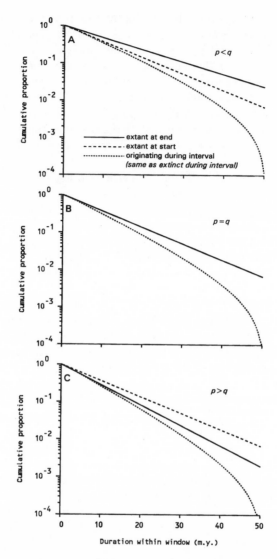

Fig. 9.4. Taxon-age distributions for time-homogeneous origination and extinction rates, showing effect of origination rate. Window of observation is 50 m.y., and extinction rate is 0.1 Lmy^{-1}. Origination rate equal to *(A)* 0.075 Lmy^{-1}, *(B)* 0.1 Lmy^{-1}, and *(C)* 0.125 Lmy^{-1}. *Solid line* is prenascence curve for taxa extant at the end of the window of observation; this is log-linear with a slope whose magnitude equals the origination rate. *Dashed line* is the survivorship curve for taxa extant at the start of the window; this is log-linear with a slope whose magnitude equals the extinction rate. *Dotted line* is the taxon-age curve for taxa that originate during the interval, which is the same as the curve for taxa that become extinct during the interval; this depends on both the origination and extinction rates and the length of the window (see text).

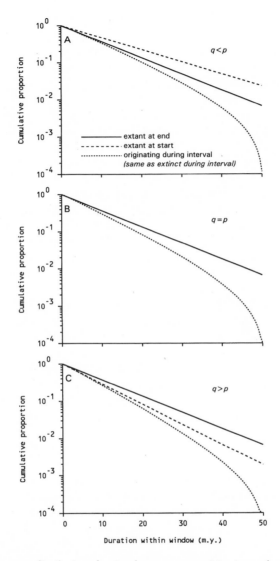

Fig. 9.5. Taxon-age distributions for time-homogeneous origination and extinction rates, showing effect of extinction rate. Window of observation is 50 m.y., and origination rate is 0.1 Lmy⁻¹. Extinction rate equal to *(A)* 0.075 Lmy⁻¹, *(B)* 0.1 Lmy⁻¹, and *(C)* 0.125 Lmy⁻¹. Curves are as in figure 9.4. Note that, for the taxon-age curve of taxa extinct during the window of observation (or of taxa originating during the window), origination and extinction rates are interchangeable. The same curve results, for example, when origination rate is 0.1 Lmy⁻¹ and extinction rate is 0.075 Lmy⁻¹ as when origination rate is 0.075 Lmy⁻¹ and extinction rate is 0.1 Lmy⁻¹ (fig. 9.4).

extinction only is not confined to the case of constant rates. It holds whenever rates are time-specific and taxonomically homogeneous. I will return to this below.

The symmetry between survivorship and prenascence is important and provides the key to estimating taxonomic rates from taxonomic age-frequency distributions. It is true that boundary crossers represent a biased part of the total duration distribution (Van Valen 1979). Provided that extinction and origination rate are independent of taxon age, however, the future extinction profile of boundary crossers is unaffected by the fact that they cross the boundary, as is the past origination profile (Pease 1987, 1988). Thus, if we use Van Valen's semilogarithmic coordinate scheme (1973), the slopes of survivorship curves and prenascence curves of boundary cohorts should provide unbiased, independent estimates of extinction rate and origination rate, respectively.

Taxonomic rates are often estimated from cumulative taxon-age curves that depend on edge effects imposed by the finite window of observation. The nature of edge effects should therefore be considered in more detail. Durations are artificially truncated, and the probability of entering the extinct versus the extant age distribution depends on the time of origination. Van Valen (1979) considered the case in which the origination rate is equal to the extinction rate, and he discussed how this rate can be estimated if we know the length of the window. More generally, the age distribution for all taxa that become extinct during the window of observation (including those extant at the start) depends on the origination rate, extinction rate, and length of the window (Pease 1988).

The expression for the extinct taxon-age distribution reveals an interesting symmetry (figs. 9.4, 9.5; appendix 9.1): The distribution is exactly the same if the origination and extinction rates are swapped. The extinct taxon-age distribution of a group that is increasing in diversity at a constant rate r is identical to that of a group that is declining in diversity at a constant rate $-r$, where the growth rate r is the difference between the origination and extinction rates. Thus extinct taxon-age distributions will not allow independent estimates of origination and extinction rates. They should, however, allow estimates of the general magnitude of these rates; this is useful, for example, in identifying groups with characteristically high or low rates.

Another interesting symmetry arises if we consider the taxon-age distribution of all taxa that originate during the window of observation (including those still extant at the end). This distribution also depends on the origination and extinction rates. In fact, it is exactly the same as the distribution of taxa that become extinct during the window, and it has the same property of swappability of origination and extinction rates. These various symmetries are intuitively appealing. Although there is a biological difference between origination and extinction, the events are mathematically interchangeable with a simple change in the direction in which time is followed.

The edge effect of the window of observation shrinks as the window becomes longer (Van Valen 1979; Pease 1988). The taxon-age distribution of extinct taxa (which is the same as that of taxa originating during the window) depends on both origination and extinction rates when the window is finite. What happens when the window is infinite or effectively so? It turns out that the distribution takes on the ideal form of an exponential curve that we would expect in the absence of edge effects (Van Valen 1979; Pease 1988), but the magnitude of the slope is equal to either the origination rate or the extinction rate, whichever is greater (figs. 9.6, 9.7; appendix 9.1).[5] This means that, even in the absence of edge effects, we do not know a priori whether an extinct taxon-age curve will reveal the origination rate or the extinction rate. Since these rates tend to be correlated, however (Stanley 1979, 1986, 1990), the slope of the taxon-age curve will at least tell us the general magnitude of the taxonomic rates. Moreover, if the fossil record is reliable enough to tell us whether the group in question is increasing or decreasing in diversity, this should indicate which of the two rates is estimated by the taxon-age distribution.

The combined distribution for all extinct and extant taxa (Raup 1975) shares the same symmetries as the distribution for extinct taxa. When the window of observation is finite, it depends on both origination and extinction rates, and the rates are interchangeable. As the window of observation becomes infinite, it depends only on the larger of the two rates (appendix 9.1; Foote 1997, appendix 1).

Taken together, these results suggest that, if we desire independent estimates of origination and extinction rates, we must focus on cohort prenascence curves and cohort survivorship curves (especially of boundary cohorts), and disregard the commonly used taxononomic age-frequency curves that reflect an accumulation of taxa originating and becoming extinct over some long span of time.

The need to focus on cohort prenascence and cohort survivorship curves in estimating origination and extinction rate raises two problems. First, if we focus on a single boundary cohort, we greatly reduce the sample of taxa. Second, if we use successive boundary cohorts, which are highly nonindependent, taxa that cross multiple boundaries enter the pool of data many times, contributing disproportionately to rate estimates (Hoffman and Kitchell 1984). A solution that allows the use of more data but counts each taxon only once for the estimation of origination rate and once for the estimation of extinction rate is to study a single boundary cohort as well as neighboring birth and death cohorts, provided that we choose the birth and death cohorts so that the edge effects are

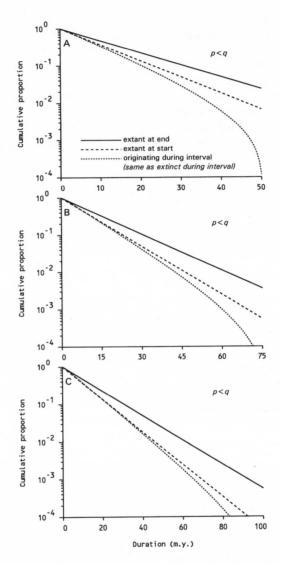

Fig. 9.6. Taxon-age distributions for time-homogeneous origination and extinction rates, showing effect of the length of the window of observation when extinction rate exceeds origination rate. Extinction rate is 0.1 Lmy^{-1}, and origination rate is 0.075 Lmy^{-1}. Window of observation equal to *(A)* 50 m.y., *(B)* 75 m.y., and *(C)* 100 m.y. Curves are as in figure 9.4. The curves for taxa extant at the start of the window and taxa extant at the end of the window have the same slope in each panel; they appear different because of the different temporal scales. As the window of observation increases, the taxon-age curve for taxa extinct during the window (or for taxa originating during the window) approaches the survivorship curve of taxa extant at the start; that is, it depends only on extinction rate.

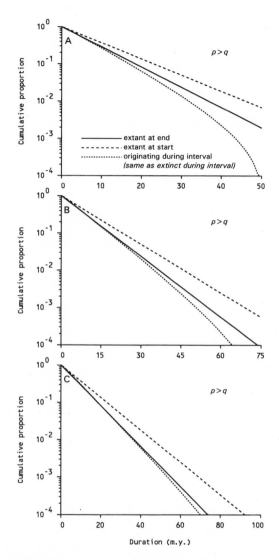

Fig. 9.7. Taxon-age distributions for time-homogeneous origination and extinction rates, showing effect of the length of the window of observation when origination rate exceeds extinction rate. Origination rate is 0.1 Lmy^{-1}, and extinction rate is 0.075 Lmy^{-1}. Window of observation equal to *(A)* 50 m.y., *(B)* 75 m.y., and *(C)* 100 m.y. Curves are as in figure 9.4. The curves for taxa extant at the start of the window and taxa extant at the end of the window have the same slope in each panel; they appear different because of the different temporal scales. As the window of observation increases, the taxon-age curve for taxa extinct during the window (or for taxa originating during the window) approaches the prenascence curve of taxa extant at the end; that is, it depends only on origination rate.

negligible. If the cohorts yield similar curves, we can calculate an average slope, perhaps weighted by the cohort size, in order to estimate taxonomic rate. As it turns out for the genus data studied here, the exercise is thwarted because the curves are not similar. Exploring this difference is nonetheless instructive.

Boundary Cohorts and the Birth and Death Cohorts Flanking Them

Consider prenascence data for marine animal genera alive today and for the Plio-Pleistocene death cohort (fig. 9.8). Contrary to the theoretical expectations for time-homogeneous or time-specific rates (figs. 9.4–9.7; appendix 9.1), the prenascence curve for the Recent boundary cohort has a substantially shallower slope than that for the immediately adjacent

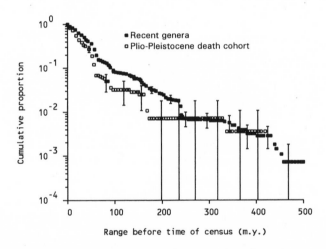

Fig. 9.8. Prenascence data for genera extant today (Recent boundary cohort) versus genera extinct during the Pliocene and Pleistocene (Plio-Pleistocene death cohort). All birth and death cohort curves in this study omit taxa that originate and become extinct during the same interval; durations are followed from the lower boundary of the death interval or the upper boundary of the birth interval. This approach circumvents the problem that the members of a birth (death) cohort originate (become extinct) at different times within the interval of origination (extinction), and so do not line up properly if we adopt some convention such as assigning all first or last appearances to the midpoint of the stratigraphic interval. Reference time for boundary cohorts is always the boundary itself. Contrary to expectations of the time-homogeneous model, the death cohort has a steeper slope than the boundary cohort. Note the early Tertiary radiation at about 60 Ma and the Triassic radiation at about 240 Ma. Prenascence curve for the boundary cohort is the same as the extant taxon-age curve of figure 9.2. In the area where the data overlap, the *longer error bars* belong to the Plio-Pleistocene death cohort.

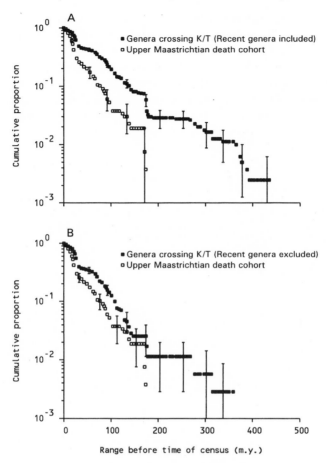

Fig. 9.9. Prenascence data for genera that cross the K/T boundary (K/T boundary cohort) versus genera extinct during the upper Maastrichtian (upper Maastrichtian death cohort). Compare to figure 9.8. Note the Triassic radiation at about 240 Ma (175 m.y. before the end of the Cretaceous), and the burst of origination in the Turonian, about 25 m.y. before the K/T. *A*, Genera still extant today are included. *B*, Genera still extant today are excluded. Prenascence curves for the boundary cohort are the same as extant taxon-age curves of figure 9.3.

death cohort. Since omitting single-interval taxa tends to increase the similarity of the curves, the discrepancy is not an artifact of this convention.

Similarly, if we treat the Cretaceous-Tertiary (K/T) boundary as the "Recent," we find a large discrepancy between the boundary cohort and the upper Maastrichtian death cohort (fig. 9.9). A considerable number of genera cross the K/T boundary and extend to the Recent. The observed difference does not seem to be an artifact of the pull of the Recent (Raup

1979), however, since the difference remains even when the Recent genera are omitted. The difference between boundary cohorts and the immediately preceding death cohorts proves to be rather general. We see it whether the boundary in question comes after a major extinction event (fig. 9.9), a time of intermediate turnover (fig. 9.10), or a rather quiescent period (fig. 9.11). We see the same kind of difference between cohort survivorship curves of boundary cohorts and of the birth cohorts that

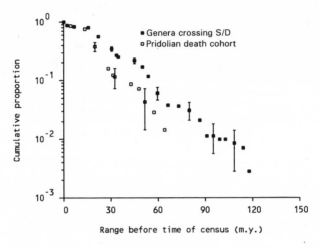

Fig. 9.10. Prenascence data for Siluro-Devonian (S/D) boundary cohort versus Pridolian death cohort.

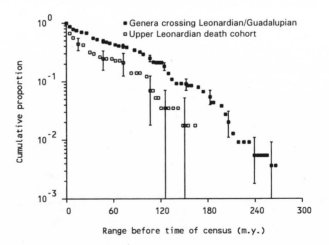

Fig. 9.11. Prenascence data for Leonardian-Guadalupian boundary cohort versus Upper Leonardian death cohort.

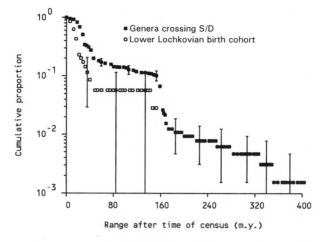

Fig. 9.12. Survivorship data for Siluro-Devonian (S/D) boundary cohort versus Lower Lochkovian birth cohort. Note the end-Permian extinction at about 160 m.y. after the time of census.

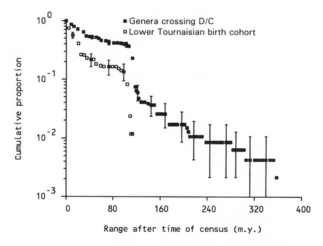

Fig. 9.13. Survivorship data for Devonian-Carboniferous (D/C) boundary cohort versus Lower Tournaisian birth cohort. Note the end-Permian extinction at about 115 m.y. after the time of census.

follow immediately (figs. 9.12–9.14). Additional analyses (not presented) like those in figures 9.8–9.14 suggest that this difference characterizes not just the immediately adjacent birth or death cohorts, but those in the vicinity of the boundary more generally. Except as indicated, all analyses here exclude Recent taxa from the boundary cohorts and birth cohorts,

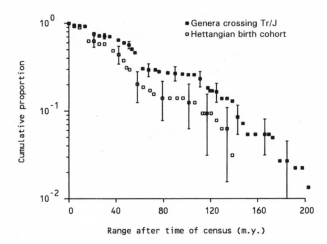

Fig. 9.14. Survivorship data for Triassic-Jurassic (Tr/J) boundary cohort versus Hettangian birth cohort.

in order to reduce the effects of the pull of the Recent. The difference is generally even greater if Recent genera are retained.

Explaining the Discrepancy between Boundary Cohorts and Nearby Birth and Death Cohorts

We may gain some insight into the discrepancy between boundary cohorts and nearby birth and death cohorts by relaxing the assumption of homogeneous origination and extinction rates. For simplicity, I will focus on the prenascence curves of a boundary cohort and the immediately preceding death cohort. The symmetries discussed above and in appendix 9.1 suggest that the results are applicable to the survivorship curves of a boundary cohort and the immediately following birth cohort.

Temporal Heterogeneity

Rates of origination and extinction are not constant over time, but instead show episodicity and a long-term decline during the Phanerozoic (Newell 1967; Van Valen 1973, 1984, 1985; Van Valen and Maiorana 1985; Raup and Sepkoski 1982; Sepkoski 1986, 1987, 1998; Gilinsky and Bambach 1987; Raup 1991; Gilinsky 1994; Foote 1994; see also Pearson [1992], who suggested ways of adjusting survivorship curves for temporal heterogeneity in order to test for age-dependency of rates). These heterogeneities must affect taxon-age distributions. It seems unlikely, however, that they produce large differences between boundary cohorts and adjacent birth and death cohorts, since the adjacent cohorts experience nearly the same history of origination and extinction rates as the boundary cohort. Heterogeneities are worth exploring nonetheless.

Figure 9.15 shows several cases in which rates decline linearly over time, with a twofold difference between the lowest and highest rates. The taxon-age distributions are clearly affected, but there is little difference between the boundary cohort and the adjacent death cohort. Nor does a striking difference result from a long-term increase in rates (fig. 9.16).

Figure 9.17 shows several cases of episodicity. If origination rate is constant while extinction is pulsed, the prenascence curves are unaffected (fig. 9.17A; appendix 9.1). The taxon-age curve for extinct taxa, though it depends on pulsed extinction, is smoothed out since the episodicity is spread throughout the window of observation. If extinction rate is constant while origination is pulsed, the prenascence curves, since they reflect the prior history of origination, are irregular (fig. 9.17B). The boundary cohort and the adjacent death cohort, however, have the same shape, offset by the length of the last interval of time before the boundary. The prenascence curve of the boundary cohort can be shifted upward and to the left to superimpose that of the death cohort.

In the foregoing examples of episodicity, extinction and origination are uncorrelated. A more realistic scenario may be one in which extinction is pulsed, and per-capita origination rates, being diversity-dependent, co-vary with the magnitude of prior extinction pulses (fig. 9.17C). In this case, the taxon-age curve for extinct taxa is somewhat but not completely smoothed. Since both rates are pulsed, there are large numbers of taxa that originate at a particular time and become extinct at a particular time, with relatively little evolutionary activity intervening. Again, however, since the death cohort and the boundary cohort experience nearly the same prior history of origination, we do not see a major discordance between them; they can be shifted and superimposed.

Taxonomic Heterogeneity

Van Valen (1973, 1979, 1994) discussed taxonomic heterogeneity, with particular emphasis on inferring the magnitude of heterogeneity from the shape of taxon-age curves. Here I am concerned with whether heterogeneity may cause the observed discordance between boundary cohorts and nearby birth and death cohorts. Imagine two groups of taxa, one characterized by higher origination and extinction rates, and one by lower rates. The taxa in the group with low extinction rate have longer durations and so are generally more likely to cross an arbitrary boundary from below. The low-rate group thus contributes to the boundary cohort more heavily, relative to its own diversity, than does the high-rate group, relative to its own diversity. Because the low-rate group has lower origination rates corresponding with its lower extinction rates, its prenascence curves

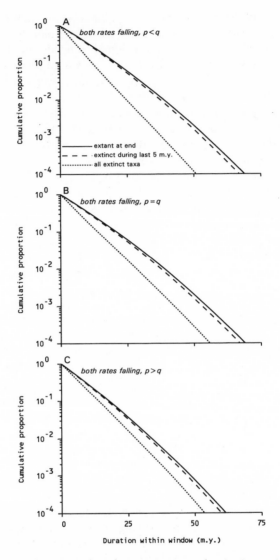

Fig. 9.15. Taxon-age curves resulting from origination and extinction rates that drop linearly with time. Window of observation is 100 m.y., but age distribution is shown only to 75 m.y. *Solid line* is prenascence curve for taxa extant at the end. *Dashed line* is prenascence curve for the death cohort consisting of taxa extinct during the last 5 m.y. of the window. *Dotted line* is the taxon-age distribution for all taxa extinct during the window. *A,* Extinction rate exceeds origination rate. Extinction rate drops from 0.22 Lmy^{-1} to 0.12 Lmy^{-1} and origination rate drops from 0.2 Lmy^{-1} to 0.1 Lmy^{-1}. *B,* Origination rate and extinction rate are equal. Both drop from 0.2 Lmy^{-1} to 0.1 Lmy^{-1}. *C,* Origination rate exceeds extinction rate. Origination rate drops from 0.22 Lmy^{-1} to 0.12 Lmy^{-1}, and extinction rate drops from 0.2 Lmy^{-1} to 0.1 Lmy^{-1}.

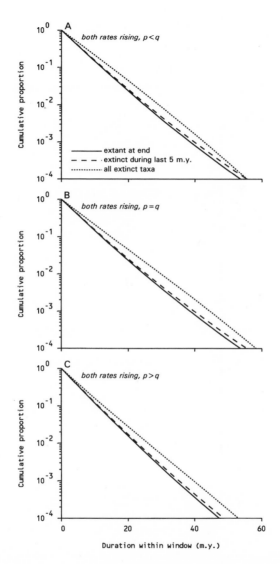

Fig. 9.16. Taxon-age curves resulting from origination and extinction rates that increase linearly with time. Window of observation is 100 m.y., but age distribution is shown only to 60 m.y. Curves as in figure 9.15. *A*, Extinction rate exceeds origination rate. Extinction rate increases from 0.12 Lmy⁻¹ to 0.22 Lmy⁻¹ and origination rate increases from 0.1 Lmy⁻¹ to 0.2 Lmy⁻¹. *B*, Origination rate and extinction rate are equal. Both increase from 0.1 Lmy⁻¹ to 0.2 Lmy⁻¹. *C*, Origination rate exceeds extinction rate. Origination rate increases from 0.12 Lmy⁻¹ to 0.22 Lmy⁻¹, and extinction rate increases from 0.1 Lmy⁻¹ to 0.2 Lmy⁻¹.

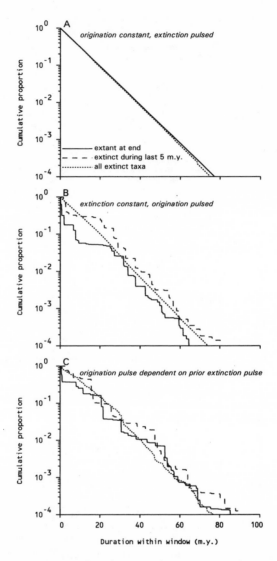

Fig. 9.17. Taxon-age curves resulting from episodic origination and extinction rates. Window of observation is 100 m.y. Curves as in figure 9.15. *A*, Origination rate is constant at 0.12 Lmy⁻¹, and extinction is pulsed, with long-term constant mean of 0.1 Lmy⁻¹. *B*, Extinction rate is constant at 0.1 Lmy⁻¹, and origination is pulsed, with long-term constant mean of 0.12 Lmy⁻¹. *C*, Origination and extinction are both pulsed, with constant long-term means, $\bar{p}$ and $\bar{q}$, of 0.12 Lmy⁻¹ and 0.1 Lmy⁻¹. Origination pulse, p_t, depends on the extinction pulse at the previous time step, q_{t-1}: $p_t = \bar{p}(q_{t-1}/\bar{q})$. Pulsed distributions follow the model of Foote (1994, equation [2]), using discrete time steps of 0.1 m.y. Pulses are pulled at random from these distributions, so each result shown here is just an example.

tend to be shallower. Its disproportionate contribution to the boundary cohort might therefore cause the prenascence curve of the boundary cohort to be shallower than that of the adjacent death cohort. There is reason to think that this sort of taxonomic heterogeneity does not contribute greatly to the discordance between boundary cohorts and the adjacent death cohorts. Both low- and high-rate taxa are originating throughout the window of observation, so the bias is not so great as it would be if we considered only the taxa produced during an early radiation.

Figures 9.18–9.20 show three cases in which there are two groups, one with about twice the taxonomic rates of the other. In one case, each group has constant diversity. In the other two cases, each group increases in diversity; in one of these, the high-rate group increases at twice the net rate of the low-rate group, while in the other case the roles are reversed. In each case, the boundary cohort and the adjacent death cohort are identical within each group. The curves for the mixture are of course between the end members, and they are dominated by the group that is diversifying more rapidly. The mixed curves tend to be nonlinear (Van Valen 1973, 1979), with the steeper, left side dominated by the high-rate group and the shallower, right side dominated by the low-rate group (Van Valen 1979). In no case, however, does the mixture show a great discordance between the boundary cohort and the adjacent death cohort.

Age-Specific Rates

The theoretical results up to this point all assumed that rates at any taxonomic level are independent of taxon age. This may be unrealistic, especially at higher taxonomic levels. Time-homogeneous origination and extinction of species, for example, predict that extinction risk of a genus will decline as the age of the genus increases (Kendall 1948; Raup 1978, 1985). Under some models, the probability that a genus will spawn a daughter genus increases with genus age (Patzkowsky 1995). Both age dependencies are related to species richness (Van Valen 1973; Flessa and Jablonski 1985), which tends to increase with genus age under the time-homogeneous model. They may also be related to the tendency for genera to increase in geographic extent as they age (Willis 1922; Miller 1997). There is empirical evidence consistent with age-dependent extinction risk of higher taxa. This includes the shape of cohort survivorship curves (Raup 1978) and the greater age of taxa that survive extinction events versus those that do not (Boyajian 1986, 1991, 1992; Van Valen and Boyajian 1987). This selectivity of taxon age is of course exactly what seems to be depicted by the discrepancy between boundary cohorts and death cohorts (figs. 9.8–9.11). If extinction risk declines with taxon age, then taxa that cross a boundary will have originated longer ago than taxa

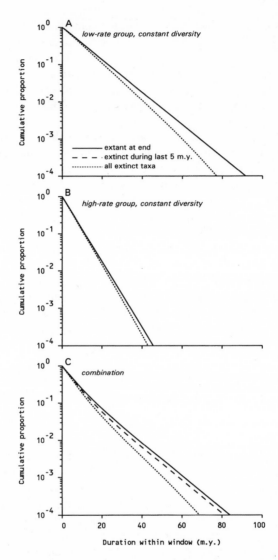

Fig. 9.18. Taxon-age curves resulting from combining two groups of taxa characterized by different origination and extinction rates. Window of observation is 100 m.y. Curves are as in figure 9.15. Both groups have constant diversity. *A*, Low-rate group, with origination rate equal to 0.1 Lmy^{-1} and extinction rate equal to 0.1 Lmy^{-1}. *B*, High-rate group, with origination rate equal to 0.2 Lmy^{-1} and extinction rate equal to 0.2 Lmy^{-1}. *C*, Combination of the two groups.

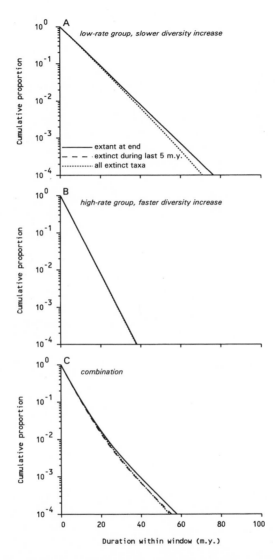

Fig. 9.19. Taxon-age curves resulting from combining two groups of taxa characterized by different origination and extinction rates. Window of observation is 100 m.y. Curves are as in figure 9.15. Both groups increase in diversity, with the high-rate group having a higher net rate of diversification. *A*, Low-rate group, with origination rate equal to 0.12 Lmy⁻¹ and extinction rate equal to 0.1 Lmy⁻¹. *B*, High-rate group, with origination rate equal to 0.24 Lmy⁻¹ and extinction rate equal to 0.2 Lmy⁻¹. *C*, Combination of the two groups.

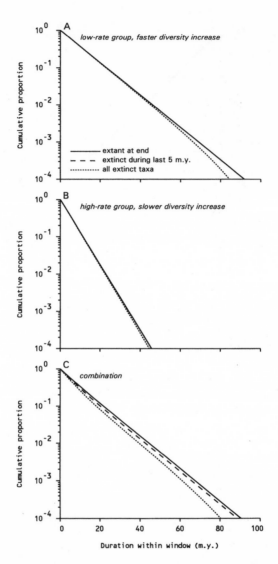

Fig. 9.20. Taxon-age curves resulting from combining two groups of taxa characterized by different origination and extinction rates. Window of observation is 100 m.y. Curves are as in figure 9.15. Both groups increase in diversity, with the low-rate group having a higher net rate of diversification. *A,* Low-rate group, with origination rate equal to 0.1 Lmy^{-1} and extinction rate equal to 0.06 Lmy^{-1}. *B,* High-rate group, with origination rate equal to 0.2 Lmy^{-1} and extinction rate equal to 0.18 Lmy^{-1}. *C,* Combination of the two groups.

that become extinct even shortly before the boundary; they will thus have shallower prenascence curves.

Since age dependence of higher taxonomic rates is expected theoretically as a simple consequence of time-homogeneous origination and extinction of constituent lower taxa, the crucial question is whether the effect of this age dependency is great enough to produce the observed discordances between boundary cohorts and adjacent death cohorts. Appendix 9.1 discusses a general model for age-specific rates, a special case of which is Patzkowsky's hierarchical branching model (1995) in which the total rate of genus origination depends directly on the number of species rather than the number of genera. Figure 9.21B shows taxon-age curves calculated with the rates of species origination, species extinction, and genus origination that Patzkowsky fit to data for Phanerozoic marine animal genera, essentially the same data studied here. Figure 9.21A depicts the corresponding curves calculated with rates equal to one-half the value of Patzkowsky's rates, and figure 9.21C shows the curves for twice Patzkowsky's values. The prenascence curves in this model have strikingly shallower slopes for the boundary cohort than for the adjacent death cohort, a difference comparable to that which we see in the genus data (figs. 9.8–9.11). If instantaneous extinction risk declines monotonically with taxon age, then future duration is not independent of past duration. Thus the hierarchical model also predicts a shallower survivorship curve for a boundary cohort than for the adjacent birth cohort, a result we see in the empirical data (figs. 9.12–9.14). We find similar theoretical results (not presented) if Patzkowsky's model is applied with a range of different parameter values and if other models are used, for example, if genus extinction risk is age-dependent but genus origination rate does not depend on species richness and therefore is not age-specific.

The Effect of Very Long-Lived Taxa

Age dependence of taxonomic rates may not be the sole reason for the empirical discordance between the prenascence and survivorship curves of boundary cohorts versus their adjacent death and birth cohorts. Leigh Van Valen (pers. comm.) has suggested that a similar discordance can arise if there is a mixture of "normal" taxa with time-homogeneous rates and taxa that are effectively "immortal." This situation is modeled in figure 9.22, in which immortal taxa are constrained to extend throughout the entire window of observation (i.e., neither to originate nor to become extinct during the window). The proportion of standing diversity that consists of immortal taxa is varied from 50% to 1% in this figure. This range of values is relevant to the case of Phanerozoic marine genera and windows of observation on the order of 100 million years, in which the taxa that range through the entire window represent on the order of 1%

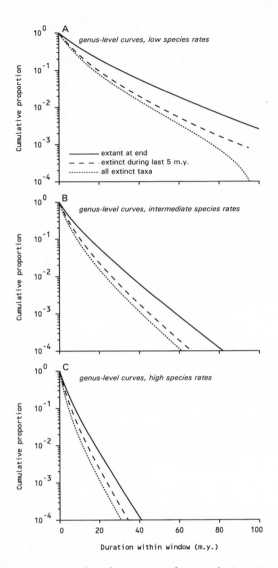

Fig. 9.21. Taxon-age curves resulting from age-specific rates of origination and extinction. Window of observation is 100 m.y. Curves are as in figure 9.15. Results are based on the hierarchical branching model of Patzkowsky (1995), in which species origination and extinction rates are constant and in which a constant proportion of speciation events found new genera. Taxon ages of genera are shown. *A*, Rate of species origination within existing genera (p) is 0.12 Lmy^{-1}, rate of species origination founding new genera (g) is 0.045 Lmy^{-1}, and species extinction rate (q) is 0.125 Lmy^{-1}. *B*, $p = 0.24$ Lmy^{-1}, $g = 0.09$ Lmy^{-1}, and $q = 0.25$ Lmy^{-1}. *C*, $p = 0.48$ Lmy^{-1}, $g = 0.18$ Lmy^{-1}, and $q = 0.5$ Lmy^{-1}. See text for explanation of choice of rates. The probability that a genus will spawn a new genus in a unit increment of time increases on average as the genus ages, and the probability that the genus will become extinct in a unit increment of time decreases as it ages. Genera that cross an arbitrary boundary are substantially longer lived than genera that die even shortly before the boundary; this produces a major discordance between the prenascence curves of boundary cohorts and adjacent death cohorts.

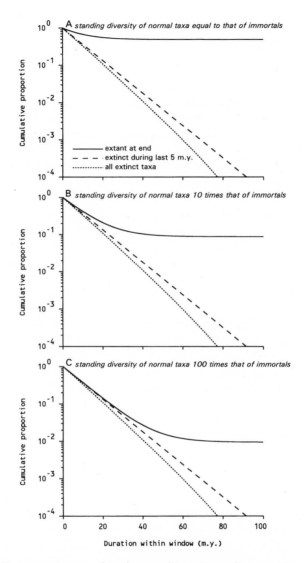

Fig. 9.22. Taxon-age curves resulting from combining "normal" taxa with constant, non-zero origination and extinction rate and taxa that are effectively "immortal," extending through the entire window of observation. Window of observation is 100 m.y., and diversity is constant. Curves are as in figure 9.15. Normal taxa have origination and extinction rates equal to 0.1 Lmy^{-1}. *A,* At any point in time there are as many normal taxa as immortal taxa. *B,* At any point in time there are ten times as many normal taxa as immortal taxa. *C,* At any point in time there are 100 times as many normal taxa as immortal taxa. Extant curves asymptotically approach the proportion of standing diversity consisting of immortal taxa. Substantial differences between the extant boundary cohort and the immediately preceding death cohort can result without age dependence of taxonomic rates.

to 10% of the taxa extant at either end of the window. For example, of about 800 genera that cross the K/T boundary, some 3% were also extant at the start of the Mesozoic, about 180 million years earlier.

If the number of immortal taxa is high relative to the number of vulnerable taxa, then there is a striking difference between the taxon-age curves of the boundary cohort and the adjacent death cohort, but the curve for the boundary cohort is unrealistically flat. If the number of immortal taxa is smaller, then the boundary cohort shows a more realistic curve, and there is still a difference between the boundary cohort and the death cohort. Depending on the relative numbers of immortal and vulnerable taxa, however, the boundary cohort and death cohort may be difficult to distinguish near the early part of the taxon-age curves. This contrasts with the empirical results for genera, in which the discordance between boundary cohorts and birth or death cohorts is generally evident throughout the taxon-age distributions.

The effect of "immortality" certainly deserves more detailed exploration, including more realistic modeling of immortality (e.g., taxa may be impervious to extinction during background times, times of mass extinction, or both) and empirical calibration. Although this study suggests that immortality of a subset of genera is not adequate by itself to yield results such as those in figures 9.8–9.14, this factor certainly can contribute substantially to the discordance between boundary cohorts and the flanking birth and death cohorts. Of the other likely contributions to this discordance, namely, temporal heterogeneity, taxonomic heterogeneity, and age dependency, the last seems to be the only one that, when modeled, produces an effect of the right magnitude. This certainly doesn't mean that age dependency is conclusively the cause of the observed discordance, but the possibility must be taken seriously.

Discussion and Conclusions

With some exceptions (Hoffman and Kitchell 1984; Pease 1987, 1988; Flessa and Jablonski 1996; Foote 1999), it seems that paleontologists do not generally follow lineages backward in time in order to study origination rates. Yet, with a simple reversal of the direction of time, the two processes are mathematically symmetrical. Whatever we can learn about extinction rates by following lineages forward in time we can hope to learn about origination rates by following them backward in time. Many taxon-age distributions depend on both extinction and origination rates in unknown or unknowable ways. In the case of homogeneous rates, however, survivorship curves and prenascence curves of boundary cohorts and adjacent birth and death cohorts allow these two rates to be estimated independently.

An attempt to apply this symmetrical approach to Sepkoski's genus data is complicated by heterogeneity of taxonomic rates. Boundary cohorts generally have shallower slopes than the adjacent birth and death cohorts, a result that is not expected if rates are homogeneous. Substantively, the observed discordance between boundary cohorts and birth and death cohorts seems to provide additional evidence for age dependency of taxonomic rates. The results also suggest that age dependency may provide a partial explanation for Gilinsky's conclusion (1988) that the difference between extinct and extant taxon-age distributions is greater than expected under the model of time-homogeneous rates. Of methodological importance, the discordance between boundary cohorts and birth or death cohorts may be large enough to preclude the analysis of prenascence curves and survivorship curves to obtain independent estimates of genus-level origination and extinction rates, at least for the data at hand and the approaches outlined here.

Although the goal of obtaining independent estimates of origination and extinction rates is largely defeated in this case, the results suggest some areas for future work. The comparison between boundary cohorts and birth and death cohorts may provide an effective way to test whether the nonlinearity of survivorship and prenascence curves reflects age dependence of rates versus taxonomic heterogeneity and other factors (figs. 9.18–9.22). The emphasis of this paper has been on some very general modeling of taxon-age distributions. There is certainly much room for making the models more realistic without overconstraining them and especially for scaling the models empirically.

The taxon-age distributions one would expect to provide the best means of estimating origination and extinction rates independently, namely prenascence and survivorship curves of boundary cohorts, may not meet their expectations if the rates are age dependent. Certain properties of other taxon-age distributions also hinder the independent estimation of rates even if the rates are strictly homogeneous. Most important, with respect to the extinct age distribution, (1) this distribution depends on both origination and extinction rates when the window of observation is short (i.e., when a considerable proportion of taxa intersect the edges), (2) the same distribution results if the rates are interchanged, and (3) the distribution depends only on the larger of the two rates when the window is long. Of course, we can still gain a general view of which groups have higher rates and which have lower rates, based on the steepness of taxon-age curves.

Many of the limitations of taxon-age distributions are quite general, but some problems depend on the case in question. Specifically, if the boundary cohorts and the flanking death (or birth) cohorts show similar prenascence (or survivorship) curves, thus suggesting an absence of age dependency or the presence of a compensating factor, then independent estimation of origination and extinction rates should be possible. Data other than those studied here may thus be more amenable to rate estimation. Moreover, if survivorship and prenascence approaches similar to those outlined here are used to estimate rates over shorter intervals of time such as single stratigraphic stages (Foote 1999), so that one seeks in effect an average rate over a brief interval rather than over the entire history of a group, rate heterogeneities may be less important. Understanding diversification in terms of its independent components of origination and extinction should still be seen as an attainable goal.

Acknowledgments

I thank the late J. J. Sepkoski, Jr., for providing his unpublished data. For discussion, advice, and reviews I thank R. H. Foote, D. Jablonski, C. F. Koch, A. I. Miller, D. M. Raup, and L. Van Valen. This work was supported by the National Science Foundation (EAR-9506568).

Appendix 9.1: Calculating Theoretical Taxon-Age Distributions

1. Introduction

The equations developed here largely build upon the foundations laid by Kendall (1948), Van Valen (1979), Raup (1985), Pease (1988), and Patzkowsky (1995). Let $p(T)$ and $q(T)$ be the time-specific, per-capita rates of origination and extinction at time T. Let $r(t)$ be the accumulated difference between origination and extinction from $T = 0$ until $T = t$.

$$r(t) = \int_0^t \{p(T) - q(T)\}dT.$$

Let $N(t)$ be the expected diversity at time t. Then

$$N(t) = N(0)e^{r(t)}.$$

In the derivations below, I will assume that $N(0) = 1$. Results differ only by a constant multiple if $N(0) > 1$. The total progeny at time t is the total number of taxa, living and extinct, produced up to time t, including taxa extant at $t = 0$ (Kendall 1948; Raup 1985). Let $dM(t)$ be the expected change in total progeny at time t and let $M(t)$ be the expected total progeny up to time t. Then

$$dM(0) = 1,$$

$$dM(t) = p(t)e^{r(t)},$$

and

$$M(t) = 1 + \int_0^t p(T)e^{r(T)}dT.$$

If p and q are constant, then

$$r(t) = (p - q)t,$$

$$N(t) = e^{(p-q)t},$$

$$dM(t) = pe^{(p-q)t},$$

and

$$M(t) = 1 + pt \quad \text{if} \quad p = q \quad \text{and} \quad M(t) = \frac{pe^{(p-q)t} - q}{p - q} \quad \text{if} \quad p \neq q.$$

Suppose a lineage exists at some arbitrary time t_1. Then the probability that it is still extant at least until some future time t_2 is equal to $e^{-\int_{t_1}^{t_2} q(T)dT}$. The density of extinction at time t_2 is equal to $q(t_2)e^{-\int_{t_1}^{t_2} q(T)dT}$. If q is constant, then this is equal to qe^{-qt}, and the probability of survival at least until time t is equal to e^{-qt}, where $t = t_2 - t_1$ and t_1 is any arbitrary point in the lineage's history.

Following lineages backward in time is symmetric with following them forward in time, except that origination rates rather than extinction rates are the relevant parameters (Pease 1988). Suppose a lineage exists at some arbitrary time t_1. Then the probability that it originated at or before some previous time t_2 is equal to $e^{-\int_{t_2}^{t_1} p(T)dT}$. The density of origination at time t_2 is equal to $p(t_2)e^{-\int_{t_2}^{t_1} p(T)dT}$. If p is constant, then this is equal to pe^{-pt}, and the probability of origination at or before time t is equal to e^{-pt}, where $t = t_1 - t_2$ and t_1 is any arbitrary point in the lineage's history.

In general, there will be a window of observation from $T = 0$ to $T = w$. Taxa extant at $T = 0$ have some part of the beginning of their duration truncated; taxa extant at $T = w$ have some part of the end of their duration truncated. The observed part of the duration is the *taxon age* that we have to work with. For example, in the case of constant rates, the probability that a taxon extant at $T = 0$ is still extant at $T = w$ is equal to e^{-qw}. Thus there is a discontinuity in the extinction density for taxa extant at $T = 0$; $f(t) = qe^{-qt}$ if $t < w$, and $f(t) = e^{-qt}$ if $t = w$. Similarly, for taxa extant at $T = w$, the origination density, where t increases backward in time from $T = w$, is given by $f(t) = pe^{-pt}$ if $t < w$, and $f(t) = e^{-pt}$ if $t = w$. I generally express taxon ages in terms of the probability densities; the taxon-age distributions are obtained by integrating the densities over the range of possible taxon ages. The value of $f(t)$ at a discontinuity is technically a probability rather than a density. But this difference matters little as long as we keep track of it in computing the cumulative distributions. I do not always note explicitly the range of durations over which various densities apply; this range is obvious from the length of the window and other stated constraints.

2. Age Distribution of Taxa Extant at $T = w$ or Taxa Extant at $T = 0$: p and q Constant

Arguing from the symmetry between forward and backward survivorship, Pease (1988) stated that the age distribution of living taxa is exponential with parameter p. Since the age distribution of living taxa is often discussed with respect to extinction rate, however, it is worth rederiving Pease's result from the perspective of forward survivorship. This derivation shows how extinction dependency cancels out. Let z be the age of an extant taxon, x its time of origin, and w the length of the window of observation; thus $z = w - x$. The total number of taxa produced is equal to $M(w)$, and the probability that a taxon is still extant at $T = w$ is equal to $N(w)/M(w)$. The probability that a taxon exists at $T = 0$ is equal to $dM(0)/M(w)$, which is equal to $1/M(w)$; the density of origination elsewhere on the interval is equal to $dM(x)/M(w)$, which is equal to $pe^{(p-q)x}/M(w)$. The probability that a taxon originating at $T = x$ is still extant at $T = w$ is equal to $e^{-q(w-x)}$, that is, e^{-qz}. Let $f(z)$ be the density of taxon ages of extant taxa. If $z = w$, the taxon must have been extant at $T = 0$, and $f(z)$ is given by

$$\frac{\dfrac{dM(0)}{M(w)} e^{-qz}}{N(w)/M(w)}. \tag{1a}$$

This is equal to

$$\frac{e^{-qz}/M(w)}{e^{(p-q)z}/M(w)}, \tag{1b}$$

which reduces to

$$e^{-pz}. \tag{1c}$$

If $z < w$, then the taxon originates at $T = w - z$, and $f(z)$ is given by

$$\frac{\dfrac{dM(w - z)}{M(w)} e^{-qz}}{N(w)/M(w)}. \tag{2a}$$

This is equal to

$$\frac{\dfrac{pe^{(p-q)(w-z)}}{M(w)} e^{-qz}}{e^{(p-q)w}/M(w)}, \tag{2b}$$

which reduces to

$$pe^{-pz}. \tag{2c}$$

Thus, although extant taxa are biased in the sense that they represent the longer-lived portion of taxa, their age distribution is independent of extinction rate and is an exponential function of origination rate (Pease 1988).

The corresponding taxon-age distribution for taxa extant at $T = 0$ was given in section 1:

$$f(z) = qe^{-qz} \text{ if } z < w \text{ and } f(z) = e^{-qz} \text{ if } z = w. \tag{3}$$

3. Age Distribution of Taxa Extant at $T = w$ or Taxa Extant at $T = 0$: $p(T)$ and $q(T)$ Time-Specific

To obtain the age distribution of taxa extant at $T = w$ when rates of origination and extinction vary, we modify (1a) and (2a). If $z = w$,

$$f(z) = \frac{\dfrac{dM(0)}{M(w)} e^{-\int_0^w q(T)\,dT}}{N(w)/M(w)}. \tag{4a}$$

This is equal to

$$\frac{e^{-\int_0^w q(T)\,dT}/M(w)}{e^{-\int_0^w \{p(T)-q(T)\}\,dT}/M(w)} \tag{4b}$$

which reduces to

$$e^{-\int_0^w p(T)\,dT}. \tag{4c}$$

If $z < w$,

$$f(z) = \frac{\dfrac{dM(w-z)}{M(w)} e^{-\int_{w-z}^w q(T)\,dT}}{N(w)/M(w)}. \tag{5a}$$

This is equal to

$$\frac{\dfrac{p(w-z)e^{\int_0^{w-z}\{p(T)-q(T)\}\,dT}}{M(w)} e^{\int_{w-z}^w q(T)\,dT}}{e^{\int_0^w \{p(T)-q(T)\}\,dT}/M(w)}, \tag{5b}$$

which reduces to

$$p(w-z)e^{-\int_{w-z}^w p(T)\,dT}. \tag{5c}$$

Thus, even when origination and extinction rates vary with time, the age distribution of extant taxa depends only on the prior history of origination rates.

The analogous age distribution of taxa extant at $T = 0$ follows directly from section 1:

$$f(z) = q(z)e^{-\int_0^z q(T)dT} \text{ if } z < w, \text{ and } f(z) = e^{-\int_0^w q(T)dT} \text{ if } z = w. \quad (6)$$

4. Age Distribution of Taxa That Become Extinct between $T = 0$ and $T = w$: p and q Constant

The probability that a taxon is extinct is the probability that it is not extant at $T = w$, which is equal to $[M(w) - N(w)]/M(w)$. The time of origination of a taxon with duration equal to z can be anywhere from $T = 0$ until $T = w - z$. Once the time of origination is specified, the duration density function is given by qe^{-qz}. Thus we have

$$f(z) = \frac{qe^{-qz}\left[\dfrac{dM(0)}{M(w)} + \dfrac{\int_0^{w-z} dM(T)dT}{M(w)}\right]}{[M(w) - N(w)]/M(w)}. \quad (7a)$$

Note that the numerator of this has two parts, corresponding to taxa extant at $T = 0$ and taxa that originate between $T = 0$ and $T = w$. Equation (7a) is equal to

$$\frac{qe^{-qz}\left[\dfrac{1}{M(w)} + \dfrac{\int_0^{w-z} pe^{(p-q)T}dT}{M(w)}\right]}{[M(w) - e^{(p-q)w}]/M(w)}, \quad (7b)$$

which reduces to

$$qe^{-qz} = pe^{-pz} \text{ if } p = q \quad (7c)$$

and

$$\frac{e^{-qz}[pe^{(p-q)(w-z)} - q]}{[e^{(p-q)w} - 1]} = \frac{e^{-pz}[qe^{(q-p)(w-z)} - p]}{[e^{(q-p)w} - 1]} \text{ if } p \neq q. \quad (7d)$$

Note that, for $p = q$, this result differs slightly from that of Van Valen (1979), apparently because I include taxa extant at $T = 0$. A remarkable property of the taxon-age distribution of extinct taxa is that it depends on both origination and extinction rates but that the rates are interchangeable. A group growing in diversity with $p - q = r$ has the same taxon-age distribution of extinct taxa as a group declining in diversity with $p - q = -r$. Also note that, as $w \to \infty$, $f(z) = pe^{-pz}$ if $p > q$, and $f(z) = qe^{-qz}$ if $p < q$.

5. Age Distribution of Taxa That Become Extinct between $T = 0$ and $T = w$: $p(T)$ and $q(T)$ Time-Specific

To obtain this distribution, we modify (7b), simultaneously canceling $M(w)$, to obtain

$$f(z) = \frac{q(z)e^{-\int_0^z q(T)dT} + \int_0^{w-z} [p(T)e^{r(T)}q(T+z)e^{-\int_T^{T+z} q(x)dx}]dT}{[M(w) - e^{r(w)}]}. \quad (8)$$

This expression must generally be solved numerically.

6. Age Distribution of Taxa That Originate between $T = 0$ and $T = w$: p and q Constant

The probability that a taxon originates between $T = 0$ and $T = w$ is the probability that it was not extant at $T = 0$, which is equal to $[M(w) - 1]/M(w)$. The time of origination of a taxon with duration equal to z can be anywhere from $T = 0$ until $T = w - z$. Once the time of origination is specified, the duration density function is given by qe^{-qz} if the taxon is extinct, and e^{-qz} if it is still extant at $T = w$. Thus we have

$$f(z) = \frac{\dfrac{qe^{-qz}\displaystyle\int_0^{w-z} dM(T)dT}{M(w)} + \dfrac{dM(w-z)e^{-qz}}{M(w)}}{[M(w) - 1]/M(w)}. \quad (9a)$$

Note that the numerator has two parts, corresponding to taxa that become extinct between $T = 0$ and $T = w$ and taxa still extant at $T = w$. Equation (9a) is equal to

$$\frac{\dfrac{qe^{-qz}\displaystyle\int_0^{w-z} pe^{(p-q)T}dT}{M(w)} + \dfrac{pe^{(p-q)(w-z)}e^{-qz}}{M(w)}}{[M(w) - 1]/M(w)}, \quad (9b)$$

which reduces to

$$qe^{-qz} = pe^{-pz} \quad \text{if} \quad p = q, \quad (9c)$$

and

$$\frac{e^{-qz}[pe^{(p-q)(w-z)} - q]}{[e^{(p-q)w} - 1]} = \frac{e^{-pz}[qe^{(q-p)(w-z)} - p]}{[e^{(q-p)w} - 1]} \quad \text{if} \quad p \neq q. \quad (9d)$$

The age distribution of taxa that become extinct between $T = 0$ and $T = w$ (including those extant at $T = 0$) is identical to the age distribution of taxa that originate between $T = 0$ and $T = w$ (including those taxa extant at $T = w$). For both distributions, p and q are interchangeable, and, as $w \to \infty$, $f(z) = pe^{-pz}$ if $p > q$, and $f(z) = qe^{-qz}$ if $p < q$.

7. Age Distribution of Taxa That Originate between $T = 0$ and $T = w$: $p(T)$ and $q(T)$ Age-Specific

To obtain this distribution, we modify (9b), simultaneously canceling $M(w)$, to obtain

$$f(z) = \frac{\int_0^{w-z} [p(T) e^{r(T)} q(T + z) e^{-\int_T^{T+z} q(x)dx}]dT + p(w - z)e^{r(w-z)}e^{-\int_{w-z}^{w} q(T)dT}}{[M(w) - 1]}. \tag{10}$$

This expression must generally be solved numerically.

8. Combined Age Distribution of All Taxa: p and q Constant

Raup (1975) suggested combining extinct and extant taxa into a single distribution. To do so, we combine equations (1c), (2c), and (7b), weighting each by the expected proportion of taxa that are extinct versus extant. This yields

$$f(z) = \frac{e^{-qz}}{1 + qw} \quad \text{if} \quad p = q \quad \text{and} \quad z = w, \tag{11a}$$

$$f(z) = \frac{e^{-qz}[2q + q^2(w - z)]}{1 + qw} \quad \text{if} \quad p = q \quad \text{and} \quad z < w, \tag{11b}$$

$$f(z) = \frac{(p - q)e^{-qz}}{pe^{(p-q)w} - q} = \frac{(q - p)e^{-pz}}{qe^{(q-p)w} - p} \quad \text{if} \quad p \neq q \quad \text{and} \quad z = w, \tag{11c}$$

and

$$f(z) = \frac{e^{-qz}[p^2 e^{(p-q)(w-z)} - q^2]}{pe^{(p-q)w} - q} = \frac{e^{-pz}(q^2 e^{(q-p)(w-z)} - p^2)}{qe^{(q-p)w} - p}$$
$$\text{if} \quad p \neq q \quad \text{and} \quad z < w. \tag{11d}$$

These equations are equivalent to those derived in a slightly different way by Foote (1997, appendix 1). Just as with the distribution of taxa extinct between $T = 0$ and $T = w$, the distribution of all taxa is the same if origination rate and extinction rate are swapped, and, as $w \to \infty$, $f(z) = pe^{-pz}$ if $p > q$, and $f(z) = qe^{-qz}$ if $p < q$.

9. Combined Age Distribution of All Taxa: $p(T)$ and $q(T)$ Time-Specific

To obtain this distribution we combine equations (4c), (5c), and (8), weighting each by the expected proportion of extinct versus extant taxa. We thus have, for $z = w$,

$$f(z) = \frac{N(w)e^{-\int_0^w p(T)dT}}{M(w)} = \frac{e^{-\int_0^w q(T)dT}}{M(w)}, \tag{12a}$$

and, for $z < w$,

$$f(z) = \frac{e^{r(w)}p(w-z)e^{-\int_{w-z}^w p(T)dT} + q(z)e^{-\int_0^z q(T)dT} + \int_0^{w-z} p(T)\,e^{r(T)}q(T+z)e^{-\int_T^{T+z} q(x)dx}dT}{M(w)}. \tag{12b}$$

These equations must generally be solved numerically.

10. Age Distribution of Taxa That Become Extinct During an Interval of Time (t_1, t_2): p and q Constant

Just as $dM(t) = pe^{(p-q)t}$ is the expected number of originations in some infinitesimal increment of time, the expected number of extinctions in some infinitesimal increment of time is equal to $qe^{(p-q)t}$. Thus, the overall probability of extinction between t_1 and t_2 is equal to

$$\int_{t_1}^{t_2} \frac{qe^{(p-q)T}dT}{M(w)}. \tag{13a}$$

This is equal to

$$\frac{q(t_2 - t_1)}{M(w)} \quad \text{if} \quad p = q, \tag{13b}$$

and

$$\frac{\frac{q}{(p-q)}[e^{(p-q)t_2} - e^{(p-q)t_1}]}{M(w)} \quad \text{if} \quad p \neq q. \tag{13c}$$

Once the time of extinction is specified, the taxon age depends on the time of origination, which is an exponential function of the origination rate (section 1). If the taxon age z is greater than or equal to t_1, so that it is possible for the taxon to have been extant at $T = 0$, then

$$f(z) = \frac{\dfrac{qe^{(p-q)z}e^{-pz}}{M(w)} + \dfrac{pe^{-pz}\displaystyle\int_{t_1}^{t_2}qe^{(p-q)T}dT}{M(w)}}{\dfrac{\displaystyle\int_{t_1}^{t_2}qe^{(p-q)T}dT}{M(w)}}. \tag{14a}$$

The numerator of equation (14a) has two parts, corresponding to taxa extant at $T = 0$ and taxa that originate between $T = 0$ and $T = t_2$. Equation (14a) is equal to

$$f(z) = \frac{e^{-pz}[1 + p(t_2 - t_1)]}{t_2 - t_1} \quad \text{if} \quad p = q \quad \text{and} \quad t_1 \leq z \leq t_2. \tag{14b}$$

and

$$f(z) = \frac{(p - q)e^{-qz} + pe^{-pz}[e^{(p-q)t_2} - e^{(p-q)t_1}]}{[e^{(p-q)t_2} - e^{(p-q)t_1}]}$$

$$\text{if} \quad p \neq q \quad \text{and} \quad t_1 \leq z \leq t_2. \tag{14c}$$

If the taxon age z is less than t_1, so that it is not possible for the taxon to have been extant at $T = 0$, then we lose the first term in the numerator of equation (14a), and we have, for any p and q,

$$f(z) = pe^{-pz} \quad \text{if} \quad z < t_1. \tag{14d}$$

Thus, if the interval of time is long enough after $T = 0$ that a member of the death cohort is unlikely to have been extant at $T = 0$, the age distribution of taxa extinct during the interval depends only on the origination rate.

In practice, limited stratigraphic resolution and the consequent lumping of taxa into cohorts imply that taxa that become extinct or originate over a range of times will be treated as if they had a common point of extinction or origination, a necessary practice that distorts the distributions derived in this and the following four sections. The amount of distortion depends on the length of the interval over which the cohort is lumped. This problem is partly circumvented in this paper by using interval boundaries as times of reference and tracking only the part of the duration that comes before the lower interval boundary for death cohorts, or after the upper interval boundary for birth cohorts.

11. Age Distribution of Taxa That Originate during an Interval of Time (t_1, t_2): p and q Constant

By analogy with (13a) and (13b), the overall probability of origination between t_1 and t_2 is equal to

$$\int_{t_1}^{t_2} \frac{pe^{(p-q)T}dT}{M(w)}, \tag{15a}$$

which is equal to

$$\frac{p(t_2 - t_1)}{M(w)} \quad \text{if} \quad p = q, \tag{15b}$$

and

$$\frac{\dfrac{p}{(p-q)}[e^{(p-q)t_2} - e^{(p-q)t_1}]}{M(w)} \quad \text{if} \quad p \neq q. \tag{15c}$$

Once the time of origination is specified, the taxon age depends on the time of extinction, which is an exponential function of the extinction rate (section 1). If the taxon age z is greater than or equal to $w - t_2$, so that it is possible for the taxon still to be extant at $T = w$, then

$$f(z) = \frac{\dfrac{pe^{(p-q)(w-z)}e^{-qz}}{M(w)} + \dfrac{qe^{-qz}\displaystyle\int_{t_1}^{t_2} pe^{(p-q)T}dT}{M(w)}}{\dfrac{\displaystyle\int_{t_1}^{t_2} pe^{(p-q)T}dT}{M(w)}}. \tag{16a}$$

The numerator of equation (16a) has two parts, corresponding to taxa still extant at $T = w$ and taxa that become extinct between $T = t_1$ and $T = w$. Equation (16a) is equal to

$$f(z) = \frac{e^{-qz}[e^{w-z} + q(t_2 - t_1)]}{t_2 - t_1}$$

$$\text{if} \quad p = q \quad \text{and} \quad w - t_2 \leq z \leq w - t_1, \tag{16b}$$

and

$$f(z) = \frac{(p - q)e^{(p-q)w}e^{-pz} + qe^{-qz}[e^{(p-q)t_2} - e^{(p-q)t_1}]}{[e^{(p-q)t_2} - e^{(p-q)t_1}]}$$

$$\text{if} \quad p \neq q \quad \text{and} \quad w - t_2 \leq z \leq w - t_1. \tag{16c}$$

If the taxon age z is less than $w - t_2$, so that it is not possible for the taxon still to be extant at $T = w$, then we lose the first term in the numerator of equation (16a), and we have, for any p and q,

$$f(z) = qe^{-qz} \quad \text{if} \quad z < w - t_2. \tag{16d}$$

Thus, if the interval of time is long enough before $T = w$ that a member of the birth cohort is unlikely still to be extant at $T = w$, the age distribution of taxa originating during the interval depends only on the extinction rate.

12. Age Distribution of Taxa That Become Extinct during an Interval of Time (t_1, t_2): $p(T)$ and $q(T)$ Time-Specific

To obtain this distribution, we modify equation (14a), simultaneously canceling $M(w)$, to obtain

$$f(z) = \frac{q(z)e^{r(z)}e^{-\int_0^z p(T)dT} + \int_{t_1}^{t_2} [q(T)e^{r(T)}p(T-z)e^{-\int_{T-z}^T p(x)dx}]dT}{\int_{t_1}^{t_2} q(T)e^{r(T)}dT}$$

$$\text{if} \quad t_1 \leq z \leq t_2, \tag{17a}$$

and

$$f(z) = \frac{\int_{t_1}^{t_2} [q(T)e^{r(T)}p(T-z)e^{-\int_{T-z}^T p(x)dx}]dT}{\int_{t_1}^{t_2} q(T)e^{r(T)}dT} \quad \text{if} \quad z < t_1. \tag{17b}$$

This generally must be solved numerically.

13. Age Distribution of Taxa That Originate during an Interval of Time (t_1, t_2): $p(T)$ and $q(T)$ Time-Specific

To obtain this distribution, we modify equation (16a), simultaneously canceling $M(w)$, to obtain

$$f(z) = \frac{p(w-z)e^{r(w-z)}e^{-\int_{w-z}^w q(T)dT} + \int_{t_1}^{t_2} [p(T)e^{r(T)}q(T+z)e^{-\int_T^{T+z} q(x)dx}]dT}{\int_{t_1}^{t_2} p(T)e^{r(T)}dT}$$

$$\text{if} \quad w - t_2 \leq z \leq w - t_1, \tag{18a}$$

and

$$f(z) = \frac{\int_{t_1}^{t_2} [p(T)e^{r(T)}q(T+z)e^{-\int_T^{T+z} q(x)dx}]dT}{\int_{t_1}^{t_2} p(T)e^{r(T)}dT} \quad \text{if} \quad z < w - t_2.$$

(18b)

This generally must be solved numerically.

14. Age Distributions with Age-Specific Rates

Here I focus on prenascence curves and on age distributions of all taxa extinct between $T = 0$ and $T = w$. Results for survivorship curves and for all taxa that originate between $T = 0$ and $T = w$ either exist already (e.g., Raup 1985) or can be derived in an analogous way. Suppose that the per-capita rates of origination and extinction depend not on absolute time but on the age of a taxon, that is, the probability that a taxon will leave a descendant in some increment of time and the probability that it will become extinct in some increment of time are functions of the age of that taxon. Age dependency of per-capita rates is to be expected, for example, if the taxa in question consist of subtaxa that are governed by constant rates (Raup 1978; Patzkowsky 1995).

To illustrate the effect of age dependency, consider the hierarchical branching model of Patzkowsky (1995). Three rates are relevant, all of which are constant: (1) the rate of origination of species within existing genera, p; (2) the rate of origination of species that found new genera, g; and (3) the rate of species extinction, q. Because the probability of genus origination depends on the number of species, and the number of species within a genus is expected to increase over time, the genus-level origination rate *per genus* actually increases on average during the age of a genus. The age dependencey of genus-level origination rate is implicit in the function $dM(t)$.

For a population of genera under Patzkowsky's model, $dM(0) = 1$ and $dM(t) = ge^{(p+g-q)t}$. The total number of genera produced is given by

$$M(w) = 1 + \int_0^w ge^{(p+g-q)T}dT,$$

(19a)

which is equal to

$$1 + gw \quad \text{if} \quad (p+g) = q \quad \text{and} \quad \frac{[p - q + ge^{(p+g-q)w}]}{(p + g - q)}$$

$$\text{if} \quad (p + g) \neq q. \quad (19b)$$

The probability of genus survival at least until time t beyond its time of origin is given by

$$P_{s,t} = \frac{1}{(1+pt)} \quad \text{if} \quad p = q \quad \text{and} \quad P_{s,t} = \frac{(p-q)e^{(p-q)t}}{[pe^{(p-q)t} - q]} \quad \text{if} \quad p \neq q. \quad (20)$$

Thus the expected standing diversity of genera at time t is given by

$$N(t) = dM(0)P_{s,t} + \int_0^t dM(T)P_{s,t-T}dT, \qquad (21a)$$

which is equal to

$$P_{s,t} + \int_0^t ge^{(p+g-q)T}P_{s,t-T}dT. \qquad (21b)$$

The probability that a genus is extant at $T = w$ is equal to $N(w)/M(w)$ and the probability that it is extinct is equal to $[M(w) - N(w)]/M(w)$.

First consider genera extant at $T = w$. If a genus is extant also at $T = 0$, then we have, by analogy with equation (1a),

$$f(z) = \frac{\dfrac{1}{M(w)}P_{s,z}}{N(w)/M(w)}. \qquad (22)$$

This implicitly assumes that genera are monotypic at $T = 0$. $P_{s,z}$ can be modified if a distribution of genus sizes is assumed (Raup 1985; Patzkowsky 1995). If the genus originates after $T = 0$, then we have, by analogy with (2a),

$$f(z) = \frac{\dfrac{dM(w-z)}{M(w)}P_{s,z}}{N(w)/M(w)}. \qquad (23)$$

Now consider genera that become extinct between $T = 0$ and $T = w$. By analogy with equation (7a) we have

$$f(z) = \frac{\dfrac{1}{M(w)}P_{s,z} + \dfrac{\displaystyle\int_0^{w-z} dM(T)P_{s,z}dT}{M(w)}}{[M(w) - N(w)]/M(w)}. \qquad (24)$$

Finally, consider genera that become extinct during some interval of time $T = t_1$ to $T = t_2$. Let $P_{d,t}$ denote the probability that a genus becomes extinct during some infinitesimal increment of time between t and $t + \delta$, where time is measured from the time of origin of the genus. Then $P_{d,t} = P_{s,t} - P_{s,t+\delta}$. The expected number of genera extinct during the interval, including those that originate and become extinct during the interval, is given by

$$E(t_1, t_2) = \left[dM(0) \int_{t_1}^{t_2} P_{d,T} dT \right]$$
$$+ \left\{ \int_0^{t_2} dM(x) \left[\int_{\max(x,t_1)}^{t_2} P_{d,T-x} dT \right] dx \right\}. \tag{25}$$

Here x is the time of origin. The first term in equation (25) is for genera extant at $T = 0$, and the second term is for genera that originate between $T = 0$ and $T = t_2$. Thus the probability of extinction between t_1 and t_2 is equal to $E(t_1, t_2)/M(w)$. For a genus to become extinct between t_1 and t_2 and to have a taxon age z, it must have originated between $T = t_1 - z$ and $T = t_2 - z$. For simplicity, assume the interval (t_1, t_2) is far enough above $T = 0$ that genera extant at $T = 0$ can be ignored. Then we have

$$f(z) = \frac{\int_{t_1-z}^{t_2-z} dM(x) P_{d,z}}{E(t_1, t_2)/M(w)}. \tag{26}$$

The foregoing equations generally need to be solved numerically (fig. 9.21). This approach can be generalized to any case of age-specific rates.

Notes

1. This result stems primarily from the fact that, in mathematical models, the probability of genus survival at least to time t, given species-level origination and extinction rates p and q, is dominated by an exponential function of the difference between the species-level rates. The relevant probability is equal to $(p - q)e^{(p-q)t}/pe^{(p-q)t} - q$ (Kendall 1948; Raup 1985; appendix 9.1, section 14).

2. A previous study (Foote 1999) attempted to estimate these rates independently on an interval-by-interval basis rather than as a long-term average. If N_b and N_t are the numbers of lineages crossing the bottom and top boundaries of the stratigraphic interval in question, N_{bt} is the number of lineages ranging through the entire interval, t is the interval duration, and p and q are the origination and extinction rates per lineage-million-years, then p and q are estimated as $\ln(N_t/N_{bt})/t$ and $\ln(N_b/N_{bt})/t$ (see appendix 9.1). By focusing on boundary crossers and ignoring single-interval taxa, many problems of interval duration and incomplete preservation are circumvented.

3. In contrast to Van Valen's distributions for living taxa, Simpson's extant curves show Lyellian percentages, in this case the proportion of taxa first appearing at some time in the past that are still extant today. There is a slight inaccuracy in Simpson's classic "clam vs. carnivore" figure (1953, table 10 and fig. 5), which I believe has not been pointed out before. Lyellian percentages need not decline monotonically with age before the Recent, although they tend to do so (Stanley et al. 1980; Stanley 1985). Simpson's figure shows a monotonic decline in Lyellian percentages, even though the data in his table, on which the figure is based, do not. The points that deviate from monotonicity have been omitted.

4. This term (*pre-*, before, + *nascence*, birth) refers to the tabulation of genera that had already originated before some specified time. I suggest the new term to replace Pease's only because I have found that discussing forward survivorship and backward survivorship sometimes generates confusion.

5. This result may seem counterintuitive, since we are used to thinking about longevity in association with extinction rates of taxa or failure rates of manufactured objects, but it does make sense. If diversity grows exponentially over a long period of time, then nearly all extinct taxa will have their last appearances near the end of the window of observation; they will in effect constitute a death cohort. Conversely, if diversity declines exponentially over a long period of time, then most extinct taxa will have their first appearances near the beginning of the window; they will in effect constitute a birth cohort.

References

Boyajian, G. E. 1986. Phanerozoic trends in background extinction: Consequence of an aging fauna. *Geology* 14:955–58.

———. 1991. Taxon age and selectivity of extinction. *Paleobiology* 17:49–57.

———. 1992. Taxon age, origination, and extinction through geologic time. *Historical Biology* 6:281–91.

Brady, M. S., H. J. Sims, and A. K. Grant. 1996. Extinction and speciation rates: Tests of the sensitivity of cohort survivorship analysis. *Geological Society of America Abstracts with Programs* 28:A107.

Flessa, K. W., and D. Jablonski. 1985. Declining Phanerozoic background extinction rates: Effect of taxonomic structure? *Nature* 313:216–18.

———. 1996. The geography of evolutionary turnover: A global analysis of extant bivalves. In *Evolutionary paleobiology,* ed. D. Jablonski, D. H. Erwin, and J. H. Lipps, 376–97. Chicago: University of Chicago Press.

Foote, M. 1988. Survivorship analysis of Cambrian and Ordovician trilobites. *Paleobiology* 14:258–71.

———. 1994. Temporal variation in extinction risk and temporal scaling of extinction metrics. *Paleobiology* 20:424–44.

———. 1997. Estimating taxonomic durations and preservation probability. *Paleobiology* 23:278–300.

———. 1999. Morphological diversity in the evolutionary radiation of Paleozoic and post-Paleozoic crinoids. *Paleobiology Memoir* 1 (supplement to *Paleobiology* 25[2]).

Foote, M., and D. M. Raup. 1996. Fossil preservation and the stratigraphic ranges of taxa. *Paleobiology* 22:121–40.

Gilinsky, N. L. 1988. Survivorship in the Bivalvia: Comparing living and extinct genera and families. *Paleobiology* 14:370–86.

———. 1991. The pace of taxonomic evolution. In *Analytical paleobiology,* ed. N. L. Gilinsky and P. W. Signor, 157–74. Knoxville: University of Tennessee and Paleontological Society.

———. 1994. Volatility and the Phanerozoic decline of background extinction intensity. *Paleobiology* 20:445–58.

Gilinsky, N. L., and R. K. Bambach. 1987. Asymmetrical patterns of origination and extinction in higher taxa. *Paleobiology* 13:427–45.

Harland, W. B., R. L. Armstrong, A. V. Cox, L. E. Craig, A. G. Smith, and D. G. Smith. 1990. *A geologic time scale 1989*. Cambridge: Cambridge University Press.

Harper, C. W., Jr. 1996. Patterns of diversity, extinction, and origination in the Ordovician-Devonian Stropheodontacea. *Historical Biology* 11:267–88.

Hoffman, A., and J. A. Kitchell. 1984. Evolution in a pelagic planktic system: A paleobiologic test of models of multispecies evolution. *Paleobiology* 10:9–33.

Kendall, D. G. 1948. On the generalized "birth-and-death" process. *Annals of Mathematical Statistics* 19:1–15.

Miller, A. I. 1997. A new look at age and area: The geographic and environmental expansion of genera during the Ordovician Radiation. *Paleobiology* 23:410–19.

Newell, N. D. 1967. Revolutions in the history of life. *Geological Society of America Special Paper* 89:63–91.

Patzkowsky, M. E. 1995. A hierarchical branching model of evolutionary radiations. *Paleobiology* 21:440–60.

Pearson, P. N. 1992. Survivorship analysis of fossil taxa when real-time extinction rates vary: The Paleogene planktonic foraminifera. *Paleobiology* 18:115–31.

Pease, C. M. 1987. Lyellian curves and mean taxonomic durations. *Paleobiology* 13:484–87.

———. 1988. Biases in the survivorship curves of fossil taxa. *Journal of Theoretical Biology* 130:31–48.

Raup, D. M. 1975. Taxonomic survivorship curves and Van Valen's law. *Paleobiology* 1:82–96.

———. 1978. Cohort analysis of generic survivorship. *Paleobiology* 4:1–15.

———. 1979. Biases in the fossil record of species and genera. *Bulletin of the Carnegie Museum of Natural History* 13:85–91.

———. 1985. Mathematical models of cladogenesis. *Paleobiology* 11:42–52.

———. 1991. A kill curve for Phanerozoic marine species. *Paleobiology* 17:37–48.

Raup, D. M., and J. J. Sepkoski, Jr. 1982. Mass extinctions in the marine fossil record. *Science* 215:1501–3.

Sepkoski, J. J., Jr. 1975. Stratigraphic biases in the analysis of taxonomic survivorship. *Paleobiology* 1:343–35.

———. 1984. A kinetic model of Phanerozoic taxonomic diversity: 3, Post-Paleozoic families and mass extinctions. *Paleobiology* 10:246–67.

———. 1986. Phanerozoic overview of mass extinction. In *Patterns and processes in the history of life*, ed. D. M. Raup and D. Jablonski, 277–95. Berlin: Springer.

———. 1987. Environmental trends in extinction during the Paleozoic. *Science* 235:64–66.

———. 1993. Phanerozoic diversity at the genus level: Problems and prospects. *Geological Society of America Abstracts with Programs* 25:A50.

———. 1996. Patterns of Phanerozoic extinctions: A perspective from global databases. In *Global events and event stratigraphy,* ed. O. H. Walliser, 35–52. Berlin: Springer.

———. 1998. Rates of speciation in the fossil record. *Philosophical Transactions of the Royal Society, London,* series B 353:315–26.

Simpson, G. G. 1944. *Tempo and mode in evolution.* New York: Columbia University Press.

———. 1953. *The major features of evolution.* New York: Columbia University Press.

Stanley, S. M. 1979. *Macroevolution.* San Francisco: Freeman.

———. 1985. Rates of evolution. *Paleobiology* 11:13–26.

———. 1986. Population size, extinction, and speciation: The fission effect in Neogene Bivalvia. *Paleobiology* 12:89–110.

———. 1990. The general correlation between rate of speciation and rate of extinction: Fortuitous causal linkages. In *Causes of evolution,* ed. R. M. Ross and W. D. Allmon, 103–27. Chicago: University of Chicago Press.

Stanley, S. M., W. O. Addicott, and K. Chinzei. 1980. Lyellian curves in paleontology: Possibilities and limitations. *Geology* 8:422–26.

Tucker, R. D., and W. S. McKerrow. 1995. Early Paleozoic chronology: A review in light of new U-Pb zircon ages from Newfoundland and Britain. *Canadian Journal of Earth Sciences* 32:368–79.

Van Valen, L. 1973. A new evolutionary law. *Evolutionary Theory* 1:1–30.

———. 1979. Taxonomic survivorship curves. *Evolutionary Theory* 4:129–42.

———. 1984. A resetting of Phanerozoic community evolution. *Nature* 307:50–52.

———. 1985. How constant is extinction? *Evolutionary Theory* 7:93–106.

———. 1994. Concepts and the nature of selection by extinction: Is generalization possible? In *The mass-extinction debates: How science works in a crisis,* ed. W. Glen, 200–16. Stanford: Stanford University Press.

Van Valen, L., and G. E. Boyajian. 1987. Comment and reply on "Phanerozoic trends in background extinction: Consequence of an aging fauna." *Geology* 15:875–76.

Van Valen. L., and V. C. Maiorana. 1985. Patterns of origination. *Evolutionary Theory* 7:107–25.

Willis, J. C. 1922. *Age and area.* Cambridge: Cambridge University Press.

Contrasting Patterns in Rare and Abundant Species During Evolutionary Turnover 　**10**

Ann F. Budd and Kenneth G. Johnson

Introduction

One of the more controversial evolutionary patterns in paleoecology today is the empirical pattern of "coordinated stasis," in which long periods of biotic stasis are interrupted by short episodes of upheaval and biotic change (Brett et al. 1996; Miller 1997). Brett and Baird (1995) recently described the pattern in early Silurian to middle Devonian benthic marine invertebrate assemblages of the northern Appalachian Basin, stimulating others to consider whether the pattern prevails in different community types elsewhere in the geologic record (e.g., Ivany and Schopf 1996). One approach involves testing whether accelerated pulses of origination and extinction exist within a given biota and, if so, whether they occur simultaneously at regular 3–7 million-year intervals in association with external environmental events (Patzkowsky and Holland 1997). In Cenozoic marine and terrestrial biotas, the results of such tests have been equivocal, most likely because deteriorating climates acting on more regional scales played a key role driving biotic change. For example, studies of the mammals of the North American High Plains have shown that, during the early Oligocene, a high rate of extinction occurred a million years after the most extreme drop in temperature, and was preceded by a minor increase in the rate of origination (Prothero 1994; Prothero and Heaton 1996). Similarly, although Vrba (1985a, 1985b) detected coordinated pulses of origination and extinction in mid-Pliocene subtropical African antelopes, new data suggest that these peaks are actually spread over a considerably wider time interval when viewed on a broader geographic scale (Behrensmeyer et al. 1996).

Our recent investigations of Caribbean reef corals have yielded equally complex evolutionary patterns (Budd and Johnson 1997, 1999). Namely, Plio-Pleistocene faunal change involved a prolonged period of increased species origination that preceded increased rates of extinction by 4–6 million years, thereby resulting in a protracted period of biotic change. The addition of new species proceeded at a much slower rate than the early Pleistocene decimation of the late Pliocene fauna. In this case, a unique series of environmental events may have been primarily responsible for the observed patterns. Origination appears to have been stimulated by changes in oceanic circulation caused by emergence of the Central American isthmus, whereas extinction seems to have been more closely linked to the onset of Northern Hemisphere glaciation (Jackson and Budd 1996; Budd et al. 1996).

Metapopulation theory predicts that climatically induced habitat fragmentation and patch isolation would cause simultaneous increases in both speciation and extinction at intermediate disturbance levels (Stanley 1986, 1990; McKinney and Allmon 1995; Hanski and Gilpin 1997). High speciation would be expected when both patch connectivity is low and disturbance is local in scale and low to intermediate in frequency, wavelength, and/or amplitude. High extinction would occur when both patch connectivity is low and disturbance is regional in scale and intermediate to high in frequency, wavelength, and/or amplitude. The observed asynchroneity in origination and extinction of corals during reef turnover could therefore possibly be explained by increasing the level and geographic scale of disturbance through geologic time (e.g., climatic deterioration). Still, it remains unclear if origination and extinction events during biotic turnover were the result of independent responses to changes in the external environment, or if they were the result of local biotic interactions such as those described by Jackson (1983).

In the present chapter, in the spirit of Cheetham and Jackson (1996), we attempt further to explore the processes causing late Cenozoic faunal turnover in Caribbean reefs by separating species with different ecological properties in our analyses. Specifically, we compare evolutionary patterns in rare and abundant species to determine if noisy evolutionary patterns in rare species may have masked a more uniform response in abundant species that took place over a more constrained time interval as predicted by "coordinated stasis" (see McKinney et al. 1996). We do this by addressing two questions: do long-term patterns of evolution of rare species differ fundamentally from those of abundant or widespread species, and could these differences explain the complex patterns observed in the fauna as a whole? Because of their small population sizes, rare species would be expected to experience generally higher rates of origination and extinction (Pimm 1991; Gaston 1994), whereas abundant species would be expected to be affected mainly by mass extinctions (Stanley 1990; Eldredge 1992). Rare species are also more likely to be irregularly preserved in the fossil record and thus are subject to greater taphonomic bias (Buzas et al. 1982). We make our comparisons by examining rarity on three different scales: abundance within local populations, geographic range, and niche breadth or environmental tolerance.

Plio-Pleistocene Caribbean Reef Corals

Analysis of newly revised occurrence data for Caribbean reef corals shows that, over the past 10 million years, overall origination was high across the region between 8.5 and 3.5 million years ago (Ma), whereas overall extinction peaked between 2 and 1 Ma (fig. 10.1). As a consequence, Caribbean species richness increased from ~60 to >120 species between 8.5 and 1.5 Ma and subsequently dropped back to 60 species. The data for this and subsequent analyses in this chapter consist of occurrences of 166 species (appendix 10.1) in 141 Miocene to Recent Caribbean localities; 166 species occur in the 121 localities that are younger

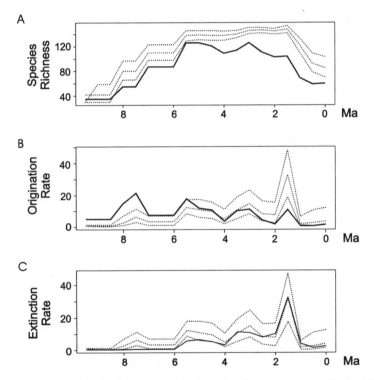

Fig. 10.1. Overall species richness and origination and extinction rates in the Caribbean reef-coral fauna over the past 10 million years. Species richness is estimated by study of stratigraphic ranges of species. In evolutionary-rate calculations, first and last occurrences are weighted relative to the age range assigned to the locality in which they took place. The numbers of first and last occurrences are divided by the total species richness in each interval. *Solid lines* indicate observed values; *dotted lines* indicate the median and 10th and 90th percentile of a permutation distribution obtained by repeated random shuffling of sample age assignments. Where the observed values exceed the randomization values, there is evidence that the observed pattern is not solely a result of uneven sampling.

than 10 Ma (table 10.1). The occurrences in 115 of the 141 localities are based on large collections that were made over the past 18 years by extracting individual specimens from outcrop exposures and cores (Budd and McNeill 1998; Budd et al. 1998, 1999). During collection, the locations of samples were plotted on detailed stratigraphic sections, and samples were grouped into 3–5-m-thick lithostratigraphic units (= localities). Highest possible resolution age dates were determined for each stratigraphic unit by integrating up-to-date microfossil, paleomagnetic, and strontium isotope data (Saunders et al. 1986; Coates et al. 1992; McNeill et al. 1997; Budd and McNeill 1998; Budd et al. 1998, 1999).

The specimens in the collections were identified to species using standardized sets of morphological characters, established in part by compar-

Table 10.1. List of 121 localities that are younger than 10 Ma

Region	Locality Name	Oldest (Ma)	Youngest (Ma)	Environment*	Collected Specimens	89 Localities in Analyses of Abundance
Bahamas	Unda_7	5.4	5.3	Intermediate		
Bahamas	Unda_6	5.3	5.2	Intermediate		
Bahamas	Clino_6	2.2	2.1	Deep		
Bahamas	Unda_5	2.2	2.1	Intermediate		
Bahamas	Unda_4	2.1	1.9	Intermediate		
Bahamas	Clino_5	2.0	1.9	Deep		
Bahamas	Clino_4	1.9	1.8	Intermediate		
Bahamas	Unda_3	1.9	1.8	Intermediate		
Bahamas	Unda_2	1.8	1.7	Intermediate		
Bahamas	Clino_3	1.8	1.6	Shallow		
Bahamas	Clino_2	1.6	1.3	Intermediate		
Bahamas	Clino_1	0.9	0.2	Shallow		
Bahamas	Unda_1	0.7	0.5	Shallow		
Costa Rica	Brazo_Seco	5.2	4.3	Intermediate	56	x
Costa Rica	Q_Choco	3.5	3.2	Intermediate	59	x
Costa Rica .	Buenos_4	3.1	2.9	Shallow	16	
Costa Rica	Buenos_1	3.1	2.9	Intermediate	20	x
Costa Rica	Buenos_8	3.1	2.9	Shallow	21	x
Costa Rica	Buenos_3	3.1	2.9	Intermediate	30	x
Costa Rica	Buenos_2	3.1	2.9	Intermediate	86	x
Costa Rica	Buenos_6	3.1	2.9	Intermediate	114	x
Costa Rica	Buenos_7	3.1	2.9	Shallow	160	x
Costa Rica	Empalme_3	2.9	1.9	Shallow	18	
Costa Rica	St_Rosa_4	2.9	1.9	Shallow	23	x
Costa Rica	St_Rosa_1	2.9	1.9	Shallow	30	x
Costa Rica	St_Rosa_3	2.9	1.9	Shallow	56	x
Costa Rica	St_Rosa_2	2.9	1.9	Shallow	74	x
Costa Rica	Lomas_w_3	1.9	1.5	Deep	22	x
Costa Rica	Lomas_e_3	1.9	1.5	Deep	28	x
Costa Rica	Lomas_e_1	1.9	1.5	Shallow	41	x

Continued on next page

Table 10.1. *(continued)*

Region	Locality Name	Oldest (Ma)	Youngest (Ma)	Environment*	Collected Specimens	89 Localities in Analyses of Abundance
Costa Rica	Lomas_w_2	1.9	1.5	Shallow	49	x
Costa Rica	Portete_2	1.9	1.5	Shallow	52	x
Costa Rica	Lomas_e_8	1.9	1.5	Shallow	79	x
Costa Rica	Lomas_e_7	1.9	1.5	Shallow	126	x
Costa Rica	Lomas_e_2	1.9	1.5	Deep	135	x
Costa Rica	Portete_1	1.9	1.5	Shallow	139	x
Costa Rica	Lomas_e_5	1.9	1.5	Shallow	168	x
Costa Rica	Lomas_e_9	1.9	1.5	Shallow	175	x
Costa Rica	Lomas_e_4	1.9	1.5	Deep	242	x
Costa Rica	Lomas_e_10	1.9	1.5	Shallow	320	x
Cuba	La_Cruz	3.5	1.6	Shallow		
Cuba	Matanzas	3.5	1.0	Intermediate		
Curaçao	salinaE_67	10.3	7.8	Shallow	25	x
Curaçao	salinaE_53	10.3	7.8	Intermediate	46	x
Curaçao	salina_51	5.9	4.6	Intermediate	30	x
Curaçao	salina_50	5.9	4.6	Intermediate	41	x
Curaçao	salina_55	5.9	4.6	Intermediate	47	x
Curaçao	salina_35	5.9	4.6	Shallow	53	x
Curaçao	salina_57	5.9	4.6	Intermediate	79	x
Curaçao	ridges_33	5.6	3.0	Shallow	35	x
Curaçao	ridges_65	5.6	3.0	Shallow	37	x
Curaçao	ridges_59	5.6	3.0	Intermediate	40	x
Curaçao	ridges_63	5.6	3.0	Intermediate	43	x
Curaçao	ridges_32	5.6	3.0	Intermediate	45	x
Curaçao	ridges_71	5.6	3.0	Intermediate	48	x
Curaçao	ridges_31	5.6	3.0	Intermediate	63	x
Curaçao	ridges_60	5.6	3.0	Intermediate	70	x
Curaçao	ridges_74	5.6	3.0	Intermediate	70	x
Curaçao	ridges_77	5.6	3.0	Shallow	72	x
Curaçao	ridges_70	5.6	3.0	Intermediate	92	x
Curaçao	terr_hs39	3.0	2.5	Intermediate	38	x
Curaçao	seacliff_17	2.6	2.0	Intermediate	33	x
Curaçao	seacliff_21	2.6	2.0	Shallow	49	x
Curaçao	terr_h38	2.6	2.0	Shallow	49	x
Curaçao	seacliff_26	2.6	2.0	Intermediate	64	x
Curaçao	seacliff_23	2.6	2.0	Intermediate	77	x
Curaçao	terr_m80	0.6	0.5	Shallow	34	x
Curaçao	terr_l40	0.2	0.1	Shallow	25	x
Dom. Rep.	Gurabo_1	8.3	7.5	Margin	13	
Dom. Rep.	Yaque_4	8.3	7.5	Shallow	26	x
Dom. Rep.	Gurabo_2	8.3	7.5	Intermediate	44	x
Dom. Rep.	Gurabo_3	8.3	7.5	Margin	49	x
Dom. Rep.	Cana_1	8.3	7.5	Shallow	67	x
Dom. Rep.	Mao_3	7.5	5.6	Margin	15	
Dom. Rep.	Yaque_5	7.5	5.6	Shallow	18	

Continued on next page

Table 10.1. *(continued)*

Region	Locality Name	Oldest (Ma)	Youngest (Ma)	Environment*	Collected Specimens	89 Localities in Analyses of Abundance
Dom. Rep.	Mao_1	7.5	5.6	Margin	33	x
Dom. Rep.	Cana_3	7.5	5.6	Margin	48	x
Dom. Rep.	Cana_4	7.5	5.6	Margin	51	x
Dom. Rep.	Mao_2	7.5	5.6	Intermediate	65	x
Dom. Rep.	Gurabo_5	7.5	5.6	Margin	118	x
Dom. Rep.	Gurabo_8	7.5	5.6	Intermediate	298	x
Dom. Rep.	Gurabo_6	7.5	5.6	Intermediate	467	x
Dom. Rep.	Yaque_6	5.6	4.5	Margin	11	
Dom. Rep.	Amina_1	5.6	4.5	Margin	38	x
Dom. Rep.	Gurabo_9	5.6	4.5	Intermediate	53	x
Dom. Rep.	Cana_5	5.6	4.5	Intermediate	306	x
Dom. Rep.	Cana_7	4.5	4.0	Margin	22	x
Dom. Rep.	Cana_6	4.5	4.0	Margin	43	x
Dom. Rep.	Cana_9	4.0	3.7	Shallow	35	x
Dom. Rep.	Gurabo_11	4.0	3.7	Intermediate	63	x
Dom. Rep.	Gurabo_12	3.7	3.4	Intermediate	45	x
Dom. Rep.	Cana_11	3.7	3.4	Intermediate	88	x
Dom. Rep.	Santo_Domingo	0.5	0.1	Shallow		
Florida	Pinecrest	3.5	3.0	Shallow		
Florida	Caloosahatchee	1.8	1.6	Shallow		
Florida	Glades	1.6	1.0	Shallow		
Florida	Key_Largo	0.5	0.1	Shallow		
Jamaica	Bowden_1	3.8	2.7	Margin	67	x
Jamaica	Bowden_2	3.3	3.0	Margin	71	x
Jamaica	OldPera_1	3.0	2.0	Intermediate	86	x
Jamaica	OldPera_2	2.0	1.8	Intermediate	70	x
Jamaica	TropicW	2.0	1.4	Shallow	33	x
Jamaica	HectorsR	2.0	1.4	Shallow	38	x
Jamaica	FollyPt_2	2.0	1.4	Shallow	44	x
Jamaica	FollyPt_1	2.0	1.4	Shallow	49	x
Jamaica	FollyPt_3	2.0	1.4	Deep	71	x
Jamaica	Navyls_1	1.9	1.6	Shallow	30	x
Jamaica	Navyls_2	1.9	1.6	Shallow	54	x
Jamaica	HopeGate_3	1.8	1.0	Shallow	27	x
Jamaica	HopeGate_1	1.8	1.0	Shallow	41	x
Jamaica	HopeGate_4	1.8	1.0	Intermediate	45	x
Jamaica	HopeGate_2	1.8	1.0	Intermediate	77	x
Jamaica	Falmouth	0.2	0.1	Shallow		
Jamaica	Rec_Deep	0.0	0.0	Shallow		
Jamaica	Rec_Intermediate	0.0	0.0	Shallow		
Jamaica	Rec_Shallow	0.0	0.0	Shallow		
Panama	Hill_Pt	3.5	1.7	Intermediate	29	x
Panama	Paunch	3.5	1.7	Shallow	32	x
Panama	Fish_Hole	3.0	2.2	Shallow	21	x
San Andrés	San_Andres	0.5	0.1	Shallow		
Trinidad	Manzanilla	10.5	5.3	Deep		

* Environments are defined on pp. 307–8.

ing morphometric and molecular data (Budd et al. 1994; Budd and John-son 1996; http://nmita.geology.uiowa.edu), and are currently deposited at the U.S. National Museum of Natural History (NMNH) and the Natural History Museum in Basel, Switzerland (NMB). To gain a more complete representation of the Caribbean region, we examined museum collections from 13 additional fossil localities (Budd et al. 1996; Budd and Johnson 1997), and published faunal lists of modern reef corals at three depth intervals along the north coast of Jamaica (Goreau 1959; Wells and Lang 1973). All of the species that were identified are known only from the Caribbean region; therefore, first occurrences in the compilation estimate species originations and last occurrences estimate extinctions.

In our analyses of evolutionary patterns, we divided the past 10 million years into half-million-year intervals and made three calculations using STATPOD (Johnson and McCormick 1998): species richness (a count of the total number of species whose ranges extended through or into a given time interval); numbers of originations within each time interval, and origination rate (relative proportion of first occurrences of species in a given time interval); and numbers of extinctions within each time interval, and extinction rate (relative proportion of last occurrences of species in a given time interval). We calculated evolutionary rates from these data by weighting occurrences relative to the total time range assigned to the locality in which they took place (Budd et al. 1996). This weighting not only corrected for variability in the assigned durations of localities, but also dampened artificial peaks caused by different interval lengths and cutpoint locations.

A stratigraphic permutation test was used to assess the influence of differential sampling on turnover estimates (Johnson and McCormick 1999). In this procedure, observed patterns of turnover are compared with a randomization distribution obtained by repeated shuffling of fossil assemblage among samples. Throughout all iterations of the resampling, the stratigraphic distribution of samples is constant and the co-occurrence of taxa within assemblages remains constant. Only the association of a particular assemblage with a particular sample is randomly shuffled. Therefore, any patterns that are only a function of the distribution of samples will appear in the randomization distribution.

As shown in figure 10.1, the origination highs at 8.5–7 Ma and at 6–5 Ma and the extinction peak at 2–1.5 Ma lie above the median predicted by randomization. However, the extinction peak lies below the 90th percentile predicted by randomization, and is therefore likely to be influenced by sampling bias. The observed ranges of early Pleistocene species appear truncated because there are few early Pleistocene localities in the data set. Regardless, a very large number of species became extinct between 2 and 1.5 Ma. Of a total of 106 extinctions documented in the compilation, approximately one-third occurred in the interval between

2.0 and 1.5 Ma ($n = 33$), and nearly half of the extinctions ($n = 52$) were recorded in the three subintervals between 2.5 and 1.0 Ma. In contrast, only four extinctions are recognized after 1.0 Ma. Therefore, although the rates are likely to be overestimated, the true values are likely to be higher than at any time during the Neogene. In general, the observed asynchroneity in the pattern of origination and extinction and the resulting prolonged period of turnover indicate that turnover was not coordinated and that reef assemblages were not tightly integrated during faunal change. At face value, this suggests that local ecological interactions may have had less influence on the overall pattern of faunal change.

Richness values were significantly lower for the past 10 million years than would be predicted by chance, and were highest between 5.5 and 3 Ma (fig. 10.1). The increase in richness of the entire species pool during turnover is also reflected in counts of species whose occurrences were actually observed within each time interval (fig. 10.2); however, statistical tests comparing species richnesses within preturnover assemblages (>5.5 Ma) and richness within turnover assemblages (5.5–1.5 Ma) reveal no significant differences (Kruskal-Wallis chi-square = 2.73, df = 1, p = 0.099). This result suggests that richnesses within assemblages may have remained constant during faunal change and that, contrary to interpreta-

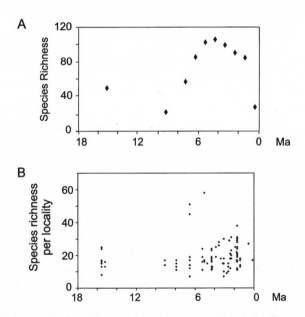

Fig. 10.2. Richness of reef-coral species observed to occur within half-million-year intervals over the past 18 million years. *A*, Total richness for the entire Caribbean species pool. *B*, Richness within assemblages.

tions of evolutionary rates, biological factors at the metapopulation level may have somehow been responsible for limiting community membership (Jackson et al. 1996). In general, the patterns in evolutionary rates and richness for the species pool as a whole are complex and suggest a variety of seemingly contradictory interpretations of the role of biological factors during evolutionary turnover.

Comparisons between Locally Rare and Abundant Species

Our first analysis of the effects of rarity on evolutionary patterns in Caribbean reef corals compares locally rare and abundant species. To perform these comparisons, we used occurrence data in 89 of the 121 late Miocene to Recent lithostratigraphic units that had adequate sampling (table 10.1). To make this assessment, we examined scatter plots of numbers of specimens per collection versus numbers of species per collection for each major reef sequence and found that species numbers began to level off at collections with >20 specimens (fig. 10.3, top). Only collections with >20 specimens (89 in total) were therefore used in the analysis. As found in most faunal surveys (Buzas et al. 1982), histograms of numbers of localities per species revealed that most species in our samples were uncommon or rare (fig. 10.3, bottom).

Within each of the 89 selected localities, species were coded as rare, common, abundant, or superabundant using a modified version of the "proportion of species" method described by Gaston (1994). This approach assigns relative abundance to species based on quartiles of the distribution of absolute abundance within an assemblage. Species with abundance lower than the first quartile of the abundance distribution are considered "rare," species with abundance between the first and the third quartile are designated "common," and species that are more abundant than 75% of the others are considered "abundant." In cases where the highest count exceeded the next highest count by two times, we designated species as "superabundant." Study of variation in relative abundance within species shows that all species are sometimes rare or common; however, 90 species (54%) were never abundant, and 76 species (46%) were sometimes abundant (table 10.2). Following Hanski (1982), we designated the 90 species that were never abundant as "satellite species," and the 76 species that were sometimes abundant as "core species." Exploratory analyses of our data further show that taxa can be abundant (core species) even if they are found only in a few localities (less than ten), but nonabundant taxa (satellite species) are usually found only in a few localities (table 10.2). Therefore, local abundance and number of localities are correlated, indicating that sampling may affect some of the patterns we observe.

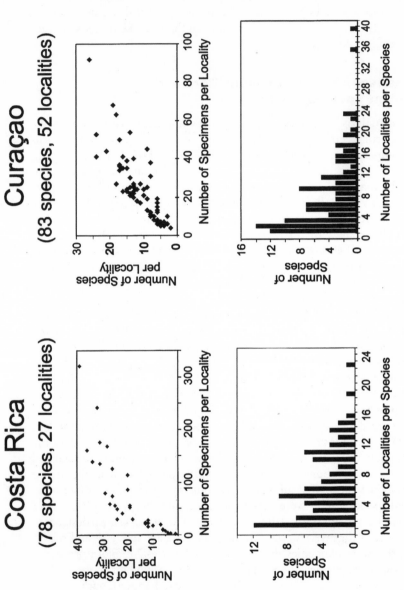

Fig. 10.3. Graphs summarizing specimen, species, and locality data for sequences collected in Costa Rica and Curaçao. *Top*, scatter plots of numbers of specimens versus species for each locality. *Bottom*, frequency histograms for numbers of localities per species.

Table 10.2. Frequencies of satellite and core species per number of localities

Species Type	Relative Abundance	Total Species	Localities						
			<10	11–20	21–30	31–40	41–50	51–60	>61
Satellite	Rare only	28	28	0	0	0	0	0	0
	Common only	4	4	0	0	0	0	0	0
	Rare or common	58	46	10	2	0	0	0	0
Core	Abundant only	0	0	0	0	0	0	0	0
	Superabundant only	0	0	0	0	0	0	0	0
	Rare to abundant	59	21	17	9	6	4	2	0
	Rare to superabundant	17	0	4	5	3	3	0	2
Total		166	99	31	16	9	7	2	2

Table 10.3. Contingency tables comparing maximum colony size and extinction susceptibility between satellite and core species

Species Type	Colony-Size Category			Survivorship	
	Small (<10 cm in Diameter)	Medium (10–30 cm in Diameter)	Large (>30 cm in Diameter)	Extinct	Extant
Satellite	24	54	12	60	30
Core	16	29	31	47	29
Chi-square		16.462 (df = 2)		0.418 (df = 1)	
p-value		0.0003		0.52	

One important trait that distinguishes persistent coral species that dominate reef ecosystems is their generally large (>30 cm in diameter) colony size (Endean and Cameron 1990). Statistical comparisons between core and satellite species indicate that satellite species have significantly smaller maximum colony sizes (table 10.3), confirming the interpretation of core species as reef dominants (Johnson et al. 1995). Because of their small colony size, however, satellite species are also less likely to be sampled, again introducing possible sampling bias into our analyses.

Quantitative comparisons of species richness and rates of origination and extinction between satellite and core species (fig. 10.4) suggest that, first, more core than satellite species existed within each time interval. This result can be attributed to the fact that core species have longer durations than satellite species (fig. 10.5), and were therefore sampled in more time intervals. Second, satellite species generally originated later during turnover than core species. Stratigraphic permutation tests suggest that satellite species show peaks of origination at 4–3 Ma and 2–1.5 Ma; whereas core species show peaks at only 6–5.5 Ma. Third, the two species types did not differ in times of extinction. Extinction peaks occur in both

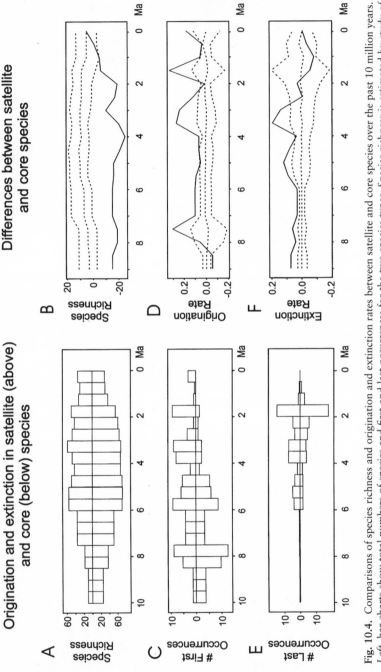

Fig. 10.4. Comparisons of species richness and origination and extinction rates between satellite and core species over the past 10 million years. *Left*, bar charts show total numbers of species and first and last occurrences for the two species types. Species richness is estimated by study of stratigraphic ranges of species. In evolutionary-rate calculations, first and last occurrences are weighted relative to the age range assigned to the locality in which they took place. *Right*, line graphs show differences between the two species types. *Solid lines* indicate observed values. *Dashed lines* indicate the median and 10th and 90th percentiles of a permutation distribution obtained by repeated shuffling of group membership prior to calculating differences in turnover pattern. If the observed value is not different than most of the permutation distribution, then there is no evidence that the observed pattern is generated by group differences.

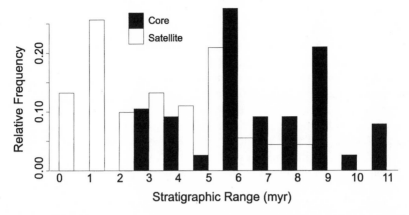

Fig. 10.5. Distribution of the stratigraphic ranges of core and satellite species. The median range for core species is 6.15 million years (myr) and the median range for satellite species is 3.8 million years.

species types at 2–1.5 Ma. The results of permutation tests for significant differences between satellite and core species indicate that richness was lower and origination rates were higher in satellite species between 8 and 1 Ma. Extinction in satellite species was higher at times >3 Ma, although not during the extinction pulse at 2–1 Ma, which equally affected core species. The permutation test was performed by comparing the observed differences in taxonomic turnover between core and satellite species with a large number (1,000) of sets of differences generated after random shuffling of group membership (see Johnson et al. 1995). Contingency tables (table 10.3) confirm that core and satellite species were equally susceptible to extinction during the final extinction pulse.

In sum, satellite species had higher overall background rates of origination and extinction than core species, and they originated later during turnover. Core species were equally susceptible to extinction during the extinction pulse at the end of the turnover interval.

Comparisons of Narrowly and Widely Distributed Species

Two other ways of assessing the effects of rarity on evolutionary patterns in Caribbean reef corals involve examining regional and environmental distributions. To perform these analyses, we assigned a region code and an environmental code to each of the 121 late Miocene to Recent localities that were younger than 10 Ma (table 10.1). The five categories for regions consisted of Bahamas, Costa Rica, Curaçao, Dominican Republic, and Jamaica. The four categories for environment consisted of shallow reef assemblage, defined as containing >50% massive species; intermediate depth or protected reef assemblage, defined as containing >35% branching species; deep reef assemblage, defined as containing >20% platy species; reef margin or soft substrate assemblage, defined as containing >40% free-living species. We then counted the number of different types of regions and environments in which each species occurred. Geographically restricted species were defined as species occurring in only one region; environmentally restricted (stenotopic) species were defined as species occurring in only one environment. Thirteen species did not occur in the target regions and were excluded from the geographic analysis.

Analyses of evolutionary rates show that, as observed for satellite species, species that occurred in more geographic regions had higher richnesses within each time interval, again probably because of their longer durations. Moreover, restricted species originated later during turnover than species with broad geographic distributions (fig. 10.6). Restricted species show peaks of origination at 8–7.5 Ma, 6–5.5 Ma, and 2–1.5 Ma, whereas widespread species show peaks of origination at 8–7.5 Ma, 6–5.5 Ma, and 4–3 Ma. Rates of extinction were generally higher for restricted species, except at 2–1.5 Ma, when extinction of widespread species peaked. Randomization tests confirm that origination was higher in restricted species from 3 to 1 Ma and from 8 to 4 Ma (fig. 10.6). Extinction was generally higher in more restricted species between 6 and 2 Ma, but not during the extinction pulse at 2–1.5 Ma.

Similarly, as in satellite and geographically restricted species, species that occurred in fewer reef environments were less numerous within time intervals and originated later during turnover than species with broad environmental distributions (fig. 10.7). Stenotopic species show minor peaks of origination at 6–5.5 Ma, 4–3 Ma, and 2–1.5 Ma, whereas eurytopic species show peaks of origination at 8–7.5 Ma and 6–5.5 Ma. Rates of extinction were generally higher for stenotopic species, except at 2–1.5 Ma when extinction of eurytopic species peaked. Randomization tests confirm that origination was high in stenotopic species from 7.5 to 1 Ma (fig. 10.7). Extinction was generally higher in stenotopic species from 6 to 3 Ma.

Contingency tables further indicate that restricted species, both geographically and environmentally, were generally more susceptible to extinction (table 10.4). They also show that abundances at three spatial scales (within assemblages, among environments, and among regions) are correlated (table 10.5), and thus, as theoretically predicted (see review by Gaston 1994), core species were more likely to be geographically and environmentally widespread. Further scrutiny of our data (table 10.6) indicates, however, that satellite species can be widespread or restricted (50% are widespread) and that they can also be either eurytopic or stenotopic. The final pulse of extinction is unusual in that it included large numbers of core species, so that core and satellite species were equally affected (table 10.3).

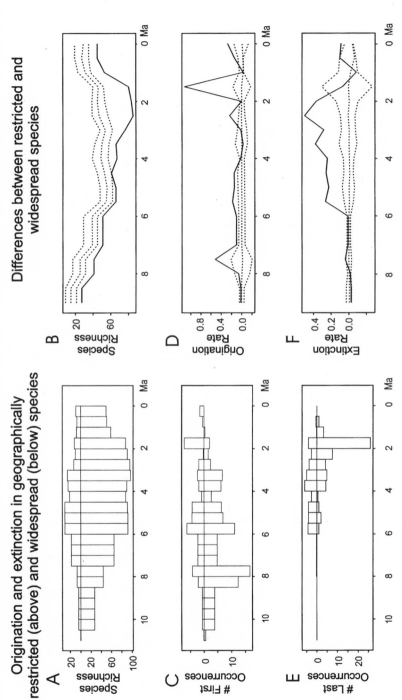

Fig. 10.6. Comparisons of species richness and origination and extinction rates between geographically restricted and widespread species over the past 10 million years. *Left,* bar charts show total numbers of species and first and last occurrences for the two species types. Species richness is estimated by study of stratigraphic ranges of species. In evolutionary-rate calculations, first and last occurrences are weighted relative to the age range assigned to the locality in which they took place. *Right,* line graphs show differences between the two species types. *Solid lines* indicate observed turnover patterns, and *dashed lines* indicate the median and 10th and 90th percentile of the permutation distribution obtained by repeatedly recalculating new patterns of group difference after shuffling group membership. One thousand iterations were performed to generate the permutation distribution.

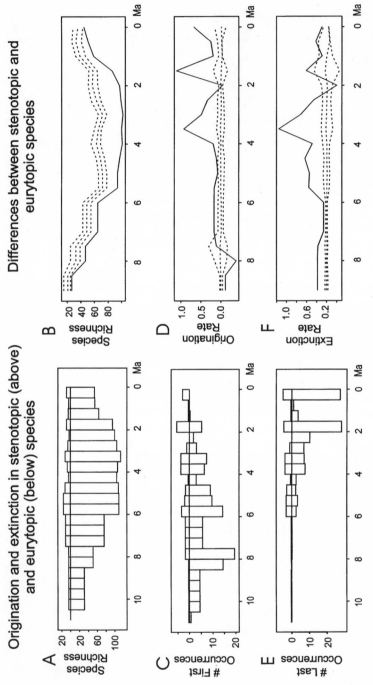

Fig. 10.7. Comparisons of species richness and origination and extinction rates between environmentally restricted (stenotopic) and widespread (eurytopic) species over the past 10 million years. *Left*, bar charts show total numbers of species and first and last occurrences for the two species types. Species richness is estimated by study of stratigraphic ranges of species. In evolutionary-rate calculations, first and last occurrences are weighted relative to the age range assigned to the locality in which they took place. *Right*, line graphs show differences between the two species types. *Solid lines* indicate observed values; *dashed lines* indicate medians and 10th and 90th percentiles of a permutation distribution. If the observed pattern is different than the majority of patterns calculated after the random shuffling of group membership, then there is evidence that the differences are not a result of random error.

Table 10.4. Contingency tables comparing extinction susceptibility among species with differing geographic and environmental distributions

Number of Regions	Extinct	Extant	Number of Environments	Extinct	Extant
1	39	8	1	31	5
2	26	14	2	35	21
3	14	9	3	33	26
4	8	15	4	8	7
5	8	12			
Chi-square	20.30 (df = 4)			10.143 (df = 3)	
p-value	0.0004			0.017	

Table 10.5. Contingency tables comparing geographic and environmental distributions between core and satellite species

Number of Regions	Satellite	Core	Number of Environments	Satellite	Core
1	35	12	1	34	2
2	27	13	2	19	37
3	9	14	3	18	41
4	6	17	4	1	14
5	0	20			
Chi-square	42.498 (df = 4)			53.664 (df = 3)	
p-value	≪0.0001			≪0.0001	

Table 10.6. The frequency of occurrences for each relative abundance

Species Type	Relative Abundance				Occurrences
	Rare	Common	Abundant	Superabundant	
Satellite	0.67	0.33	0	0	538
Core	0.42	0.40	0.17	0.01	1,660

However, because of the variable distribution patterns of satellite species and the differential extinction of restricted satellite species, restricted species in general were still found to be more likely to become extinct than widespread species during the final extinction pulse (table 10.4).

Discussion and Conclusions

Our analyses suggest that, during late Cenozoic time, rare species in Caribbean reef communities exhibited many different evolutionary patterns from abundant or widespread species. Most important, rare species had generally higher evolutionary rates, and their peaks of origination and extinction were less well developed. These results indicate that rare species exhibited more evolutionary flux, and responded in a less coordinated fashion to changes in the external environment. They therefore can be interpreted as less tightly integrated into reef ecosystems, as has been argued in general for many other ecosystems by McKinney et al. (1996). Indeed, except during the five great mass extinctions in earth history, abundant and/or widespread species have long been observed to persist longer in the fossil record and be less susceptible to extinction than rare species (Stanley 1986; Hansen 1978, 1980; Jablonski 1986, 1987).

Our analyses also indicate that locally rare species (satellite species) had small to medium colony sizes, whereas abundant species (core species) had medium to large colony sizes. In most reef-building corals, colony growth is indeterminate, although growth rates vary among species and in response to environmental conditions (Buddemeier et al. 1974; Buddemeier and Kinzie 1976). As a consequence, colony size coarsely reflects colony longevity. The observed correlation between local abundance and colony size, therefore, can be interpreted to indicate that abundant species were often longer-lived and thus more persistent members of reef communities. They thus may have played a more important role in structuring reef ecosystems. The apparent correlation between local abundance and persistence contradicts the generalizations made by Endean and Cameron (1990), who reported that persistent species in Great Barrier Reef communities are often rare, but are completely consistent with the metapopulation models described by McKinney et al. (1996).

Our analyses indicate that satellite and core species were equally susceptible to extinction during the extinction pulse at the end of turnover. Because core species generally had larger colony sizes, the results appear at first to contradict earlier findings (Johnson et al. 1995) that small free-living species were more susceptible to extinction. However, because of their often high local abundances, many of the most important small, free-living species are categorized as core species in the present study (e.g., *Manicina grandis, Thysanus excentricus, Placocyathus variabilis, Trachyphyllia bilobata,* and *Antillia dentata*). The present results, therefore, enhance rather than contradict earlier findings.

Our analyses further show a correlation between abundances at three spatial scales (within assemblages, among environments, and among regions), indicating that abundant species tend to have greater niche breadths and are more likely to disperse widely. This result agrees with generalizations reviewed by Jackson (1974) and more recently by Brown (1995). Thus our results reveal another contradiction. Environmentally and geographically restricted species were more susceptible to extinction during the extinction pulse at the end of turnover, but satellite and core species were equally susceptible. Interestingly, this result suggests that spatial distribution may be more important than local abundance in controlling susceptibility to extinction during the final extinction pulse, thus indicating that patch connectivity within metapopulations may play a greater role during extinctions than local biotic interactions. Indeed, the effects of extinction on core species are manifested by the rise in dominance of *Acropora palmata* and *A. cervicornis* at shallow and mid-fore-reef depths following turnover (Jackson 1994; Jackson and Budd 1996; Jackson et al. 1996).

Despite the differences between rare and abundant species, removal of rare species from the data set does not simplify the messy patterns of extinction and origination associated with turnover, and our results argue against "coordinated stasis." The late Cenozoic history of Caribbean reef communities cannot be characterized by long periods of biotic stasis interrupted by short episodes of upheaval and biotic change. Although peaks of origination and extinction are better defined in core species, indicating a more coordinated response, the gap between major peaks of origination and extinction in core species is enhanced rather than reduced. Origination appears to have preceded extinction by more than 3–4 million years in core species, indicating that members of the pre- and postturnover fauna belonged to the same Caribbean-wide species pool. In fact, observations of individual stratigraphic units suggest that pre- and postturnover species lived together at the same locations. Although evolutionary responses were more coordinated in core species, local communities were not so tightly integrated that individual members of either the pre- or postturnover faunas were excluded.

Instead, our results indicate that exclusion may have tied to colonization dynamics operating on regional spatial scale, as predicted by metapopulation theory (McKinney and Allmon 1995; Jackson et al. 1996). In metapopulation theory, evolutionary events, such as speciation and extinction, result from the interplay between processes operating locally within populations and processes operating regionally to connect populations at different locations. If disturbance is primarily local in scale (e.g., caused by sedimentation, runoff, or local temperature/salinity extremes), extinctions of restricted species result. If disturbance is regional in scale (e.g., caused by change in current patterns or in sea level) and thus reduces connectivity among patches, high speciation results at low disturbance levels and high extinction results at high disturbance levels. The observed pattern of speciation followed by selective extinction of restricted species can therefore be interpreted as the result of initial regional disturbance at low levels followed by more localized disturbance at high levels. This pattern suggests two more or less independent processes, and thus confirms earlier interpretations (Jackson and Budd 1996; Budd et al. 1996) that the observed evolutionary patterns may have been a response to an unusual sequence of environmental events (i.e., changes in oceanic circulation caused by emergence of the Central American isthmus followed by changes in temperature associated with the onset of Northern Hemisphere glaciation) rather than a gradual increase in disturbance level.

More refined ecological and evolutionary data are needed to better understand this balance between local and regional processes during turnover and thus its ultimate cause. Relative abundances of species and their variation need to be more rigorously assessed within and among individual stratigraphic units throughout the region, and species evolution needs to be more carefully examined within the context of refined morphometric and phylogenetic analyses. The morphometric and phylogenetic analyses of Cheetham (Cheetham 1986; Cheetham and Jackson 1995; Jackson and Cheetham 1990, 1994) serve as a model.

In summary, our analyses have shown that

1. In late Cenozoic Caribbean reef assemblages, relative abundance varies along a continuum. All species were sometimes rare; however, 54% were never abundant (satellite), and 46% were sometimes abundant (core).
2. Core species occur in more samples than satellite species, and had longer durations and larger colony sizes. They thus were more persistent and more important in structuring reef ecosystems.
3. Although satellite species originated later during turnover than core species, satellite and core species did not differ in timing of extinction, and they were equally susceptible to extinction during the final pulse of extinction associated with faunal turnover. Satellite species thus exhibited greater evolutionary flux during normal times.
4. Like satellite species, species that occurred in fewer reef environments and in fewer geographic regions also originated later during turnover, and were more susceptible to extinction. Local abundance appears to be generally correlated with niche breadth and geographic range. Core species tend to be widespread and eurytopic, however, whereas satellite species can be either widespread or restricted as well as eurytopic or stenotopic.
5. Peaks of origination and extinction were better developed in core species, suggesting that core species were more strongly integrated within local ecosystems. A 3–4-million-year time gap still exists between peaks of origination and extinction in core species, however, indicating that the processes involved in turnover were complex. These patterns may be explained by variation in patch connectivity within metapopulations.

Acknowledgments

We thank Daniel Miller and Martin Buzas for reviews of the manuscript. This research was supported by a U.S. National Science Foundation Grant (EAR-9219138) to AFB and by a U.K. Natural Environment Research Council Advanced Postdoctoral Fellowship in Taxonomy to KGJ.

Appendix 10.1

Table 10A.1. List of species in the compilation, their age ranges, and their documented pre-Pleistocene local, regional, and environmental distributions

Species	Total Samples	Oldest (Ma)	Youngest (Ma)	Local Abundance	Bahamas	Costa Rica	Curaçao	Dominican Republic	Jamaica	Shallow	Interm	Deep	Margin
					Geographic Regions					Reef Environments			
Stephanocoenia intersepta	26	5.6	Extant	Core	1	1	1	1	1	1	1	1	0
Stephanocoenia duncani	30	10.3	1.5	Core	1	1	1	1	1	1	1	1	1
Stephanocoenia spongiformis	7	7.5	1.5	Core	0	1	0	1	0	1	1	1	0
Stylophora affinis	29	10.3	1.6	Core	0	1	1	1	1	1	1	0	0
Stylophora affinis-2	3	7.5	3.0	Satellite	0	0	0	0	0	1	1	0	0
Stylophora granulata	47	10.3	1.0	Core	1	1	1	1	1	1	1	0	1
Stylophora imperatoris	12	10.5	3.0	Satellite	0	0	0	0	0	1	1	0	0
Stylophora minor	30	8.3	1.0	Core	1	1	1	1	1	1	1	1	1
Stylophora monticulosa	33	10.3	1.7	Core	1	1	1	1	1	1	1	0	1
Stylophora panamensis	13	10.3	2.0	Core	0	0	1	0	0	1	0	0	0
Stylophora aff. portobellensis	11	5.9	3.0	Core	0	0	1	0	0	1	1	0	0
Pocillopora arnoldi	4	5.9	3.0	Satellite	0	0	1	0	0	0	1	0	0
Pocillopora baracoaensis	3	7.5	1.0	Satellite	0	0	0	1	0	0	1	0	0
Pocillopora crassoramosa	18	8.3	1.8	Core	0	1	1	1	1	1	1	1	1
Pocillopora palmata	2	0.5	0.1	Satellite	0	0	0	0	0	1	0	0	0
Madracis asperula	6	1.9	Extant	Core	1	1	0	0	0	0	0	1	0
Madracis decactis	35	10.3	Extant	Core	1	1	1	1	0	1	0	1	0
Madracis decaseptata	4	7.5	3.4	Satellite	0	0	0	1	0	0	0	0	0
Madracis formosa	1	0.1	Extant	Satellite	0	0	0	0	0	1	1	0	0
Madracis mirabilis	27	7.5	Extant	Core	0	1	1	1	1	1	1	1	1
Madracis pharensis	2	1.9	Extant	Satellite	0	1	0	0	0	1	0	0	0
Madracis cf. herricki	6	7.5	3.4	Satellite	0	0	0	1	0	0	1	0	0

Continued on next page

Table 10A.1. *(continued)*

Species	Total Samples	Oldest (Ma)	Youngest (Ma)	Local Abundance	Geographic Regions					Reef Environments			
					Bahamas	Costa Rica	Curaçao	Dominican Republic	Jamaica	Shallow	Interm	Deep	Margin
Madracis sp. A	5	7.5	3.4	Satellite	0	1	0	1	0	0	1	0	0
Madracis sp. Z	6	5.9	2.0	Satellite	0	0	1	0	0	1	1	0	0
Acropora cervicornis	62	5.9	Extant	Core	1	1	1	1	1	1	1	1	1
Acropora palmata	42	5.6	Extant	Core	1	1	1	1	1	1	1	1	1
Acropora panamensis	3	10.3	1.6	Satellite	1	0	1	0	0	1	1	0	0
Acropora prolifera	3	0.1	Extant	Satellite	0	0	0	0	0	1	0	0	0
Acropora saludensis	10	10.3	1.7	Core	1	0	1	1	0	0	0	0	1
Acropora sp. Y	1	3.8	2.7	Satellite	0	0	0	0	1	0	0	0	0
Acropora sp. A	9	5.9	2.0	Core	0	0	1	0	0	1	0	0	0
Acropora sp. B	16	5.9	2.0	Core	0	0	1	0	0	1	1	0	0
Acropora sp. C	9	5.9	3.0	Core	0	0	1	0	0	1	0	0	0
Agaricia grahamae	15	5.6	Extant	Core	0	1	1	0	0	1	1	1	1
Agaricia lamarcki	33	10.3	Extant	Core	1	1	1	1	1	1	1	1	1
Agaricia sp. A	1	4.0	3.7	Satellite	0	0	0	1	0	0	0	0	0
Agaricia undata	23	10.3	Extant	Core	0	1	1	1	1	1	1	1	1
Undaria agaricites	61	8.3	Extant	Core	1	1	1	1	1	1	1	1	1
Undaria crassa	28	7.5	Extant	Core	1	1	0	1	1	1	1	1	0
Undaria pusilla	8	5.6	Extant	Satellite	0	1	1	0	0	1	1	0	0
Undaria sp. A	2	7.5	4.5	Satellite	0	1	0	1	0	1	0	0	0
Gardineroseris planulata	6	7.5	1.9	Satellite	1	0	0	1	0	1	1	0	1
Pavona sp. A	8	7.5	2.1	Satellite	1	0	0	1	0	1	1	0	0
Pavona sp. B	7	7.5	3.7	Core	0	0	0	1	0	1	1	0	1
Pavona pennyi	1	10.5	5.3	Satellite	0	0	0	0	0	0	0	1	0
Pavona trinitatis	1	10.5	5.3	Satellite	0	0	1	0	0	0	0	1	1
Helioseris cucullata	24	5.6	Extant	Core	0	1	1	0	1	1	1	1	0

Species											
Leptoseris cailleti	6	7.5	Extant	Satellite	0	0	0	0	1	1	0
Leptoseris gardineri	6	7.5	4.5	Core	1	1	0	0	1	1	0
Leptoseris glabra	10	7.5	1.9	Core	1	1	0	0	1	1	0
Leptoseris sp. A	4	7.5	4.5	Satellite	0	1	0	0	1	1	0
Leptoseris sp. B	3	7.5	4.5	Satellite	0	1	0	0	1	1	0
Leptoseris walli	1	10.5	5.3	Satellite	0	0	0	0	0	0	0
Psammocora trinitatis	4	10.5	5.2	Satellite	1	1	1	0	1	1	1
Siderastrea mendenhalli	4	8.3	3.4	Satellite	0	1	1	0	1	1	0
Siderastrea pliocenica	3	3.5	1.0	Satellite	0	1	1	0	1	0	0
Siderastrea radians	8	3.5	Extant	Satellite	0	1	0	1	0	0	0
Siderastrea silecensis	2	3.5	1.6	Satellite	0	0	0	0	0	0	0
Siderastrea siderea	49	8.3	Extant	Core	1	1	1	1	1	1	1
Pironastrea anguillensis	1	5.9	4.6	Satellite	0	0	1	1	1	1	0
Porites astreoides	52	5.6	Extant	Core	1	1	1	1	1	1	1
Porites aff. astreoides	6	10.3	3.0	Satellite	0	0	1	0	1	0	0
Porites macdonaldi	9	7.5	3.7	Core	1	1	0	1	0	1	1
Porites portoricensis	25	10.3	1.5	Core	1	1	1	1	1	1	1
Porites trinitatis	1	10.5	5.3	Satellite	0	0	0	0	0	0	0
Porites waylandi	17	10.5	1.5	Core	1	1	1	1	1	1	1
Porites baracoaensis	32	10.5	1.0	Core	1	1	1	1	1	1	1
Porites branneri	10	3.5	Extant	Satellite	1	1	1	0	0	1	0
Porites colonensis	5	1.9	Extant	Satellite	0	1	0	1	0	0	0
Porites convivatoris	3	7.5	4.5	Core	0	0	0	0	0	1	0
Porites divaricata	11	5.6	Extant	Satellite	1	1	1	1	1	1	1
Porites furcata	33	5.6	Extant	Core	1	1	1	1	1	1	1
Porites porites	17	3.5	Extant	Core	0	1	1	1	1	1	0
Goniopora calhounensis	2	7.5	3.7	Satellite	0	1	1	1	1	1	1
Goniopora hilli	7	8.3	1.6	Satellite	1	1	0	1	0	1	1
Goniopora imperatoris	13	10.3	1.6	Core	1	1	1	1	1	1	1
Alveopora sp. A	1	5.9	4.6	Satellite	0	0	0	0	1	0	0

Continued on next page

Table 10A.1. (continued)

Species	Total Samples	Oldest (Ma)	Youngest (Ma)	Local Abundance	Geographic Regions					Reef Environments			
					Bahamas	Costa Rica	Curaçao	Dominican Republic	Jamaica	Shallow	Interm	Deep	Margin
Caulastraea portoricensis	31	5.9	0.1	Core	1	1	1	1	1	1	1	1	1
Caulastraea sp. A	5	5.9	3.0	Satellite	0	0	1	0	0	1	1	0	0
Cladocora arbuscula	2	1.6	Extant	Satellite	0	0	0	0	0	1	0	0	0
Cladocora johnsoni	1	1.8	1.6	Satellite	0	0	0	0	0	1	0	0	0
Favia fragum	13	5.6	Extant	Satellite	0	1	1	1	1	1	1	0	0
Favia vokesae	3	8.3	1.8	Satellite	1	0	1	1	0	1	1	0	0
Diploria clivosa	31	10.3	Extant	Core	0	1	1	1	1	1	1	0	0
Diploria labyrinthiformis	20	3.5	Extant	Core	1	1	1	1	1	1	1	0	0
Diploria sarasotana	1	3.5	3.0	Satellite	0	0	0	0	0	1	0	0	0
Diploria strigosa	55	5.9	Extant	Core	1	1	1	1	1	1	1	1	1
Diploria adentrensis	2	8.3	3.4	Core	0	0	0	1	0	1	0	0	0
Diploria zambensis	8	7.5	2.7	Satellite	0	1	1	1	1	1	0	0	1
Manicina areolata	23	5.6	Extant	Satellite	1	1	1	0	1	0	1	1	0
Manicina aff. *areolata*	6	5.6	1.8	Satellite	0	0	0	1	1	1	0	0	1
Manicina mayori	13	3.1	Extant	Satellite	0	1	1	0	1	1	1	0	0
Manicina aff. *mayori*	1	5.6	4.5	Satellite	0	1	0	0	0	0	0	0	0
Manicina puntagordensis	22	10.3	1.5	Core	0	1	1	1	1	0	1	1	1
Manicina grandis	20	8.3	2.7	Core	0	0	1	1	1	1	1	0	1
Thysanus navicula	9	7.5	2.0	Satellite	0	0	0	1	1	1	0	0	1
Manicina geisteri	9	7.5	2.7	Satellite	0	0	0	1	1	1	0	0	1
Manicina jungi	11	8.3	2.7	Core	0	0	1	1	1	0	1	0	1
Hadrophyllia saundersi	8	7.5	1.0	Satellite	0	0	1	1	1	1	1	0	1
Manicina sp. E	9	10.3	2.0	Core	0	0	1	0	0	1	0	0	0
Thysanus sp. A	6	3.8	1.9	Satellite	0	1	0	0	1	1	1	0	1
Thysanus excentricus	12	7.5	1.0	Core	1	0	0	1	1	1	0	0	1

Species														
Thysanus corbicula	6	7.5	1.6	Core	0	0	0	1	1	1	1	1	0	1
Colpophyllia amaranthus	4	3.1	Extant	Satellite	0	1	1	0	0	0	1	1	1	0
Colpophyllia breviserialis	6	5.6	Extant	Satellite	1	0	1	0	0	0	1	1	1	0
Colpophyllia natans	43	5.6	Extant	Core	0	1	1	1	1	1	1	1	1	0
Colpophyllia sp. A	9	5.6	Extant	Satellite	0	1	0	1	0	0	0	1	1	0
Montastraea annularis	14	2.6	Extant	Core	1	1	1	0	0	1	1	1	1	0
Montastraea faveolata	41	5.6	Extant	Core	1	1	1	0	1	1	1	1	1	0
Montastraea franksi	28	5.6	Extant	Core	1	1	1	0	1	1	1	1	1	0
Montastraea brevis	3	7.5	4.5	Core	0	0	0	1	0	0	0	0	1	0
Montastraea limbata-1	11	10.3	1.5	Core	0	1	1	1	0	1	1	1	0	1
Montastraea limbata-2	35	10.3	1.5	Core	1	1	1	1	1	0	1	1	1	1
Montastraea limbata-3	18	10.3	2.0	Core	0	1	1	1	1	1	1	1	0	0
Montastraea sp. A	8	5.2	1.5	Satellite	0	0	1	0	1	0	1	1	1	0
Montastraea trinitatis	3	10.5	4.6	Core	0	1	0	1	0	1	0	0	0	0
Montastraea canalis	13	5.6	1.5	Core	1	0	1	1	1	0	1	1	1	0
Montastraea cavernosa-1	1	0.1	Extant	Satellite	0	0	0	0	1	0	1	1	0	0
Montastraea cavernosa-2	41	10.3	Extant	Core	1	1	1	1	0	1	1	1	1	0
Montastraea cavernosa-3	16	5.6	Extant	Satellite	0	0	1	1	1	0	1	1	0	0
Montastraea cylindrica	23	7.5	1.0	Core	1	1	1	1	0	1	1	1	1	0
Montastraea endothecata	5	10.3	4.5	Satellite	0	0	1	1	1	0	1	1	0	0
Solenastrea bournoni	37	10.3	Extant	Core	0	1	1	1	0	1	1	1	0	1
Solenastrea hyades	9	8.3	Extant	Satellite	0	0	0	0	0	1	1	1	0	0
Trachyphyllia bilobata	29	8.3	1.0	Core	1	1	1	0	1	1	0	1	1	1
Trachyphyllia sp. A	5	8.3	1.9	Satellite	1	1	1	0	1	1	1	1	0	1
Trachyphyllia sp. B	6	5.9	3.0	Core	0	0	0	0	1	1	1	1	1	1
Antillophyllia sawkinsi	8	10.5	1.5	Core	0	0	1	1	0	1	1	1	1	0
Meandrina braziliensis	14	8.3	1.8	Core	0	1	1	1	1	0	0	1	1	1
Meandrina meandrites	23	5.6	Extant	Satellite	1	1	1	1	1	1	1	1	1	1
Meandrina sp. A	9	5.9	Extant	Core	1	1	0	1	0	1	1	1	1	0
Placocyathus alveolus	11	7.5	1.8	Core	1	0	0	1	1	1	1	1	1	1

Continued on next page

Table 10A.1. (continued)

Species	Total Samples	Oldest (Ma)	Youngest (Ma)	Local Abundance	Geographic Regions					Reef Environments			
					Bahamas	Costa Rica	Curaçao	Dominican Republic	Jamaica	Shallow	Interm	Deep	Margin
Placocyathus barretti	6	5.9	1.8	Satellite	0	0	1	0	1	1	1	0	1
Placocyathus costatus	16	8.3	2.0	Core	0	0	0	1	1	1	1	0	1
Placocyathus trinitatis	10	10.5	2.7	Core	0	1	0	1	1	0	1	1	1
Placocyathus variabilis	44	8.3	1.0	Core	1	1	1	1	1	1	1	1	1
Dichocoenia caloosahatcheensis	9	5.9	1.6	Satellite	0	1	1	1	0	1	1	0	0
Dichocoenia eminens	4	3.5	1.6	Satellite	0	1	0	0	0	1	1	0	0
Dichocoenia stokesi	27	5.9	Extant	Core	1	1	1	1	1	1	1	1	0
Dichocoenia stellaris	17	2.9	Extant	Satellite	0	1	1	0	1	1	0	1	0
Dichocoenia tuberosa	19	10.3	1.5	Satellite	0	1	1	1	1	1	1	0	0
Dendrogyra cylindricus	8	5.6	Extant	Satellite	1	0	1	1	1	1	1	1	0
Galaxea excelsa	2	8.3	5.6	Satellite	0	0	0	1	0	1	1	0	0
Archohelia limonensis	1	1.9	1.5	Satellite	0	1	0	0	0	1	0	0	0
Antillia dentata	11	8.3	2.0	Core	0	0	0	1	1	1	1	0	1
Antillia gregorii	2	3.8	2.7	Satellite	0	0	0	0	1	0	0	0	1

Species													
Scolymia cubensis	18	8.3	Extant	Satellite	0	1	0	1	1	1	1	1	1
Scolymia lacera	6	3.8	Extant	Satellite	0	1	1	0	1	1	1	0	1
Mussa angulosa	18	7.5	Extant	Satellite	1	1	1	1	1	1	1	1	0
Mussismilia hartti	1	0.2	Extant	Satellite	0	0	0	0	0	1	0	0	0
Mussismilia aff. *hartti*	9	7.5	0.5	Core	0	1	1	1	1	1	1	1	0
Mussismilia hispida	1	3.5	3.0	Satellite	0	0	0	0	0	1	0	0	0
Mussismilia sp. A	1	5.9	4.6	Satellite	0	0	1	0	1	1	1	1	0
Isophyllia sp. A	6	8.3	3.4	Satellite	0	0	1	1	0	1	1	0	0
Isophyllia sinuosa	6	3.5	Extant	Satellite	0	0	0	1	0	1	0	0	0
Isophyllastrea sp. B	1	3.1	2.9	Satellite	0	1	0	0	0	1	0	0	0
Isophyllastrea rigida	4	1.9	Extant	Satellite	0	0	1	0	0	1	0	0	0
Mycetophyllia aliciae	3	1.9	Extant	Satellite	0	1	0	0	0	1	1	1	1
Mycetophyllia bullbrooki	4	10.5	3.4	Core	1	1	1	1	1	1	1	1	1
Mycetophyllia danaana	16	3.5	Extant	Satellite	0	1	1	0	0	1	1	0	0
Mycetophyllia ferox	9	3.5	Extant	Satellite	0	1	0	0	0	1	0	1	0
Mycetophyllia lamarckiana	8	2.0	Extant	Satellite	0	1	0	0	0	1	0	0	0
Mycetophyllia reesi	10	1.9	Extant	Satellite	0	1	1	0	0	1	1	1	0
Mycetophyllia sp. A	2	1.9	1.5	Satellite	0	1	0	0	0	1	0	0	0
Eusmilia fastigiata	15	3.5	Extant	Satellite	0	1	1	1	1	1	1	1	1
Eusmilia sp. A	6	7.5	1.5	Satellite	1	1	1	1	1	1	1	0	0

References

Behrensmeyer, A. K., N. E. Todd, R. Potts, and G. E. McBrinn. 1996. Environmental change and faunal turnover in late Pliocene terrestrial faunas of Africa. *Geological Society of America Abstracts with Programs* 28 (7): A177.

Brett, C. E., and G. C. Baird. 1995. "Coordinated stasis" and evolutionary ecology of Silurian and middle Devonian faunas in the Appalachian Basin. In *New approaches to speciation in the fossil record*, ed. D. H. Erwin and R. L. Anstey, 285–315. New York: Columbia University Press.

Brett, C. E., L. C. Ivany, and K. M. Schopf. 1996. "Coordinated stasis": An overview. *Palaeogeography, Palaeoclimatology, Palaeoecology* 127:1–20.

Brown, J. H. 1995. *Macroecology.* Chicago: University of Chicago Press.

Budd, A. F., and K. G. Johnson. 1996. Recognizing species of late Cenozoic Scleractinia and their evolutionary patterns. *Paleontological Society Papers* 1:59–79.

———. 1997. Coral reef community dynamics over 8 million years of evolutionary time: Stasis and turnover. In *Proceedings of the Eighth International Coral Reef Symposium,* ed. H. A. Lessios and I. G. Macintyre, 1:423–28. Balboa, Republic of Panamá: Smithsonian Tropical Research Institute.

———. 1999. Origination preceding extinction during late Cenozoic turnover of Caribbean reefs. *Paleobiology* 25:188–200.

Budd, A. F., K. G. Johnson, and T. A. Stemann. 1996. Plio-Pleistocene turnover and extinctions in the Caribbean reef coral fauna. In *Evolution and environment in tropical America,* ed. J. B. C. Jackson, A. F. Budd, and A. G. Coates, 168–204. Chicago: University of Chicago Press.

Budd, A. F., K. G. Johnson, T. A. Stemann, and B. H. Tompkins. 1999. Pliocene to Pleistocene reef coral assemblages in the Limon Group of Costa Rica. In *A paleobiotic survey of the Caribbean faunas from the Neogene of the Isthmus of Panama,* ed. L. S. Collins and A. G. Coates, Bulletins of American Paleontology, no. 357, 119–58. Ithaca, NY: Paleontological Research Institution.

Budd, A. F., and D. F. McNeill. 1998. Zooxanthellate scleractinian corals of the Bowden Shell Bed, southeast Jamaica. *Contributions to Tertiary and Quaternary Geology* 35:49–65.

Budd, A. F., R. A. Petersen, and D. F. McNeill. 1998. Stepwise faunal change during evolutionary turnover: A case study from the Neogene of Curaçao, Netherlands Antilles. *Palaios* 13:167–85.

Budd, A. F., T. A. Stemann, and K. G. Johnson. 1994. Stratigraphic distributions of genera and species of Neogene to Recent Caribbean reef corals. *Journal of Paleontology* 68:951–77.

Buddemeier, R. W., and R. A. Kinzie, III. 1976. Coral growth. *Oceanography and Marine Biology* 14:183–225.

Buddemeier, R. W., J. E. Maragos, and D. W. Knutson. 1974. Radiographic studies of reef coral exoskeletons: Rates and patterns of coral growth. *Journal of Experimental Marine Biology and Ecology* 14:179–200.

Buzas, M. A., C. F. Koch, S. J. Culver, and N. F. Sohl. 1982. On the distribution of species occurrence. *Paleobiology* 8:143–50.

Cheetham, A. H. 1986. Tempo of evolution in a Neogene bryozoan: Rates of morphologic change within and across species boundaries. *Paleobiology* 12: 190–202.

Cheetham, A. H., and J. B. C. Jackson. 1995. Process from pattern: Tests for selection versus random change in punctuated bryozoan speciation. In *New approaches to speciation in the fossil record*, ed. D. H. Erwin and R. L. Anstey, 184–207. New York: Columbia University Press.

———. 1996. Speciation, extinction, and the decline of arborescent growth in Neogene and Quaternary Bryozoa of tropical America. In *Evolution and environment in tropical America*, ed. J. B. C. Jackson, A. F. Budd, and A. G. Coates, 205–33. Chicago: University of Chicago Press.

Coates, A. G., J. B. C. Jackson, L. S. Collins, T. M. Cronin, H. J. Dowsett, L. M. Bybell, P. Jung, and J. A. Obando. 1992. Closure of the Isthmus of Panama: The nearshore record of Costa Rica and western Panama. *Geological Society of America Bulletin* 104:814–28.

Eldredge, N. 1992. Where the twain meet: Causal intersections between the genealogical and ecological realms. In *Systematics, ecology, and the biodiversity crisis*, ed. N. Eldredge, 1–14. New York: Columbia University Press.

Endean, R., and A. M. Cameron. 1990. Trends and new perspectives in coral-reef ecology. In *Coral Reefs,* ed. Z. Dubinsky, 469–92. New York: Elsevier.

Gaston, K. J. 1994. *Rarity*. London: Chapman and Hall.

Goreau, T. F. 1959. The ecology of Jamaican coral reefs: 1, Species composition and zonation. *Ecology* 40:67–90.

Hansen, T. A. 1978. Larval dispersal and species longevity in Lower Tertiary gastropods. *Science* 199:855–87.

———. 1980. Influence of larval dispersal and geographic distribution on species longevity in neogastropods. *Paleobiology* 6:194–209.

Hanski, I. 1982. Dynamics of regional distribution: The core and satellite species hypothesis. *Oikos* 38:210–21.

Hanski, I., and M. E. Gilpin. 1997. *Metapopulation biology: Ecology, genetics, and evolution.* San Diego: Academic Press.

Ivany, L. C., and K. M. Schopf, eds. 1996. New perspectives on faunal stability in the fossil record. *Palaeogeography, Palaeoclimatology, Palaeoecology* 127: 1–361.

Jablonksi, D. 1986. Background and mass extinctions: The alternation of macroevolutionary regimes. *Science* 231:129–33.

———. 1987. Heritability at the species level: Analysis of geographic ranges of Cretaceous mollusks. *Science* 238:360–63.

Jackson, J. B. C. 1974. Biogeographic consequences of eurytopy and stenotopy among marine bivalves and their evolutionary significance. *American Naturalist* 104:541–60.

———. 1983. Biological determinants of present and past sessile animal distributions. In *Biotic interactions in Recent and fossil sessile benthic communities*, ed. M. J. S. Tevesz and P. L. McCall, 39–120. New York: Plenum.

———. 1994. Community unity? *Science* 264:1412–13.

Jackson, J. B. C., and A. F. Budd. 1996. Evolution and environment: Introduction and overview. In *Evolution and environment in tropical America*, ed. J. B. C.

Jackson, A. F. Budd, and A. G. Coates, 1–20. Chicago: University of Chicago Press.

Jackson, J. B. C., A. F. Budd, and J. M. Pandolfi. 1996. The shifting balance of natural communities? In *Evolutionary paleobiology,* ed. D. Jablonski, D. H. Erwin, and J. Lipps, 89–122. Chicago: University of Chicago Press.

Jackson, J. B. C., and A. H. Cheetham. 1990. Evolutionary significance of morphospecies: A test with cheilostome Bryozoa. *Science* 248:579–83.

———. 1994. Phylogeny reconstruction and the tempo of speciation in cheilostome Bryozoa. *Paleobiology* 20:407–23.

Johnson, K. G., A. F. Budd, and T. A. Stemann. 1995. Extinction selectivity and ecology of Neogene Caribbean reef corals. *Paleobiology* 21:52–73.

Johnson, K. G., and T. McCormick. 1998. STATPOD: Statistical analysis of palaeontological occurrence data. Computer program distributed by Division of Earth Sciences, University of Glasgow (http://www.earthsci.gla.ac.uk/palaeo/statpod).

———. 1999. The quantitative description of faunal change using palaeontological databases. In *Statistical palaeontology,* ed. D. Harper, 227–47. Chichester: John Wiley and Sons.

McKinney, M. L., and W. D. Allmon. 1995. Metapopulations and disturbance: From patch dynamics to biodiversity dynamics. In *New approaches to speciation in the fossil record,* ed. D. H. Erwin and R. L. Anstey, 123–83. New York: Columbia University Press.

McKinney, M. L., J. L. Lockwood, and D. R. Frederick. 1996. Does ecosystem and evolutionary stability include rare species? *Palaeogeography, Palaeoclimatology, Palaeoecology* 127:191–207.

McNeill, D. F., A. F. Budd, and P. F. Borne. 1997. Earlier (late Pliocene) first appearance of the Caribbean reef-building coral *Acropora palmata:* Stratigraphic and evolutionary implications. *Geology* 25:891–94.

Miller, A. I. 1997. "Coordinated stasis" or coincident relative stability? *Paleobiology* 23:155–64.

Patzkowsky, M. E., and S. M. Holland. 1997. Patterns of turnover in Middle and Upper Ordovician brachiopods of the eastern United States: A test of coordinated stasis. *Paleobiology* 23:420–43.

Pimm, S. L. 1991. *The balance of nature?* Chicago: University of Chicago Press.

Prothero, D. R. 1994. The late Eocene-Oligocene extinctions. *Annual Review of Earth and Planetary Science* 22:145–65.

Prothero, D. R., and T. H. Heaton. 1996. Faunal stability during the early Oligocene climatic crash. *Palaeogeography, Palaeoclimatology, Palaeoecology* 127:257–83.

Saunders, J. B., P. Jung, and B. Biju-Duval. 1986. Neogene paleontology in the northern Dominican Republic: 1, Field surveys, lithology, environment, and age. *Bulletins of American Paleontology* 89:1–79.

Stanley, S. M. 1986. Population size, extinction, and speciation: The fission effect in Neogene Bivalvia. *Paleobiology* 12:89–110.

———. 1990. The general correlation between rate of speciation and rate of extinction: Fortuitous causal linkages. In *Causes of evolution,* ed. R. Ross and W. D. Allmon, 103–27. Chicago: University of Chicago Press.

Vrba, E. S. 1985a. African Bovidae: Evolutionary events since the Miocene. *South African Journal of Science* 81:263–66.

———. 1985b. Environment and evolution: Alternative causes of the temporal distribution of evolutionary events. *South African Journal of Science,* 81:229–36.

Wells, J. W., and J. C. Lang. 1973. Systematic list of Jamaican shallow-water Scleractinia. *Bulletins of Marine Science* 23:55–58.

11 Asexual Propagation in Cheilostome Bryozoa

Evolutionary Trends in a Major Group of Colonial Animals

ECKART HÅKANSSON AND ERIK THOMSEN

Introduction

Colonial organisms by way of their organization possess a high potential for regeneration. Breakage of colonies therefore becomes a potential means of reproduction, which essentially occurs in all colonial groups, although with highly varying frequency (Boschma 1925; Tuncliffe 1981; Highsmith 1982; Hughes 1989; Wulff 1991; Hunter 1993; Benzie et al. 1995; Fong and Lirman 1995; Thomsen and Håkansson 1995; Riegl and Riegl 1996; Lirman and Fong 1997; Tsurumi and Reisweig 1997). In corals and bryozoans the frequency is, among other things, correlated with colony design; thus massive colonies are less prone to breakage than fragile, branching colonies, and, in more sophisticated examples, breakage may be directly facilitated through budding or various constructional features in the colony (Ayre and Dufty 1994; Chen et al. 1995; Thomsen and Håkansson 1995; Kramarsky-Winter and Loya 1996). One interesting aspect of this potential is that, in species employing asexual reproduction by fragmentation or fission, growth in itself becomes a reproductive effort.

Unlike most other colonial groups the cheilostome bryozoans frequently possess skeletally distinct brood chambers (ovicells), which provide a measure for sexual reproductive effort in living and fossil species. Furthermore, in most species skeletal morphology in the initial part of the colony allows a very clear distinction between sexual and asexual recruits. Hence, in population-based studies of many cheilostome species, the importance of asexual reproduction may be assessed directly in relation to the energy invested in sexual processes.

In a previous study we investigated such relations in a total of 25 species from the Maastrichtian and Danian of Denmark (fig. 11.1; Thomsen and Håkansson 1995). This study revealed strong correlations between the proportion of sexually recruited colonies, the investment in sexual reproduction, and growth form (fig. 11.2). Thus, in encrusting species an average of 18% of the zooids is brooding as opposed to only 3% in rigidly

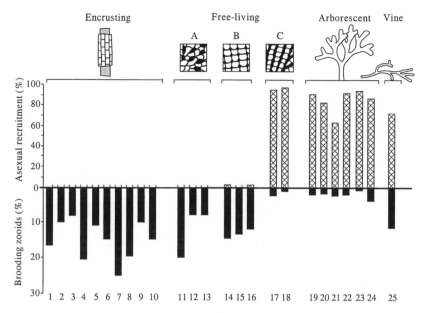

Fig. 11.1. Relative frequency of brooding zooids and asexual recruits in 25 species of cheilostome bryozoans from the Maastrichtian and Danian of Denmark. Note the very strong correlation between reproductive characteristics and colony architecture in all but a vine-growing species; free-living, lunulitid architectural designs *A–C* relate to the distribution pattern of polymorphs *(black)*. 1, *"Membranipora" johnstrupi*; 2, *Callopora* sp. 1; 3, *Ellisina brittanica*; 4, *E. reticulata*; 5, *Floridina* sp. 1; 6, *Gargantua parvicella*; 7, *Pliophloea subcornatu*; 8, *Diacanthopora* sp. 1; 9, *Monoceratopora quadrisulcata*; 10, *Balantiostoma hians*; 11, *Lunulites goldfussi*; 12, *L. spiralis*; 13, *L. mitra*; 14, *L.* n. sp. 1; 15, *L. pseudocretacea*; 16, *L.* n. sp. 2; 17, *L. semilunaris*; 18, *L. patelliformis*; 19, *Pithodella sincta*; 20, *"Membranipora" declivis*; 21, *Onychocella columella*; 22, *Floridina gothica*; 23, *Coscinopleura angusta*; 24, *Porina salebrosa*; 25, *Columnotheca cribrosa*. Data from Thomsen and Håkansson 1995.

arborescent species. All encrusting species investigated propagate exclusively through sexually produced larvae, whereas the arborescent species by and large propagate asexually through fragmentation. Similar correlations between the proportion of brooding zooids and the rate of sexual recruitment are found among the investigated free-living species. However, in this group some species reproduce mainly sexually, others mainly asexually. All species were investigated through a relatively short time span (some 2–3 million years), and except for a single species we were not able to detect any significant change through time in these reproductive parameters (fig. 11.2). In the single species thus deviating from the general pattern, the reproductive strategy was apparently linked to variation in the environment (Thomsen and Håkansson 1995).

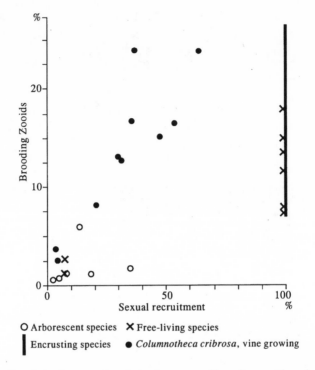

Fig. 11.2. Correlation between the relative frequency of brooding zooids and asexual recruitment in the three basic growth forms in the 24 stable species of cheilostome bryozoans
from the Maastrichtian and Danian of Denmark, together with ten conspecific populations
of the vine-growing species *Columnotheca cribrosa*. Data from Thomsen and Håkansson
1995.

The stability of reproductive strategy found for the majority of species
investigated in our previous work suggests that the relative importance
of sexual versus asexual propagation is genetically controlled and therefore may be subject to long-term evolutionary changes. In this study we
have selected two groups of cheilostome bryozoans for investigation of
possible long-term trends in reproductive strategy: the well-constrained
family Coscinopleuridae and the phylogenetically less constrained free-
living forms. The two groups represent each of the major growth forms
that were found frequently to employ asexual reproduction in our previous study (Thomsen and Håkansson 1995), and they have a fossil record
covering substantial portions of cheilostome history (fig. 11.3). The earliest known cheilostome bryozoans are all encrusting, and we therefore
regard our previous data pertaining to this growth form as an indication
of the ancestral mode of reproduction within the brooding clades of
cheilostome bryozoans.

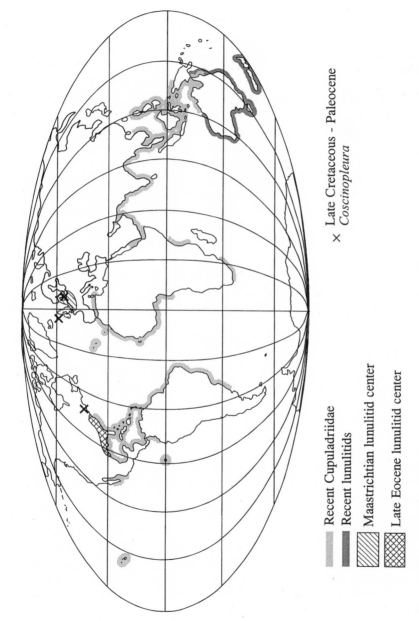

Recent Cupuladriidae

Recent lunulitids

Maastrichtian lunulitid center

Late Eocene lunulitid center

× Late Cretaceous – Paleocene
Coscinopleura

Fig. 11.3. Biogeography and stratigraphic distribution of the Coscinopleuridae and the free-living cheilostomes.

Coscinopleuridae

The Coscinopleuridae (figs. 11.4, 11.5) constitutes a wholly arborescent clade ranging from the Santonian through to the very latest Paleocene (ca. 85–55 million years ago [Ma]; Voigt 1956). They have a fairly restricted biogeographic distribution (fig. 11.3) but occur in a wide range of lithologies ranging from coccolith ooze (chalk) through bryozoan limestone to siliciclastic-rich skeletal sands. The populations investigated comprise virtually nothing but colony fragments, whether as a result of postdepositional compaction or transport. The overall colony design remains constant throughout the history of the group, and in the ordinary process of growth dichotomous branching is nearly universal. Branches are initially flattened bilamellar, but with age they become more rounded (fig. 11.4C). This subsequent thickening occurs in the proximal part of the colony where the frontal wall of individual zooids thickened and eventually coalesced to form a virtual stem where feeding no longer took place. Bases of sexually recruited colonies are small encrusting holdfasts covered entirely by kenozooids (fig. 11.4D). Zooidal morphology shows little variation through the history of the family, as does the morphology of the ovicells. Polymorphs are located along the edge of the branches, and variation in this arrangement constitutes one of the central characters in taxon differentiation within the family. Ovicells typically are arranged in clusters along the midline of the branches (figs. 11.4A and B), and commonly these clusters are paired, with adjacent clusters on both sides of the branch.

Breakage and repair is widespread in the Coscinopleuridae. However, not all cases of repair represent a successful event of asexual recruitment. In figure 11.5 four characteristic examples of recognizable asexual recruits are illustrated in addition to a single specimen displaying successful repair of no relevance to the distinction between asexual and sexual recruitment. In two fragments (figs. 11.5A and D) the repair of the proximal fracture by growth of kenozooids leaves no doubt that these fragments continued life subsequent to separation from the parent colony. One of these fragments has preserved also a distal fracture from which distally oriented growth of an ordinary budding pattern was resumed (fig. 11.5D). This particular fragment has clearly retained its original polarity after separation. Alternatively, in broken-off fragments lodged upside down, growth could be resumed with an ordinary budding pattern of opposite polarity (fig. 11.5C). Similarly, from more or less flat-lying fragments, growth may be resumed through the development of secondary branches at odd angles to the colony surface (fig. 11.5E). The direction of regenerative growth therefore appears largely independent of the original polarity of the fragment. This inference is supported by observations

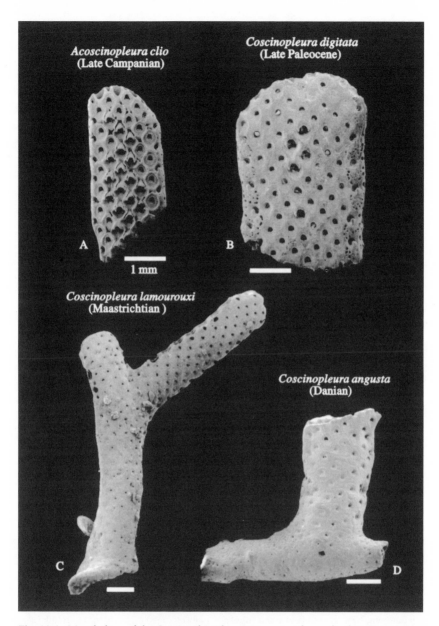

Fig. 11.4. Morphology of the Coscinopleuridae. *A, Acoscinopleura clio* (late Campanian, Vigny, France), growing tip with extensive ovicell cluster. *B, Coscinopleura digitata* (late Paleocene, Vincentown, New Jersey, USA), growing tip with a small ovicell cluster plus an isolated ovicell. *C, Coscinopleura lamourouxi* (late Maastrichtian, Maastricht, Netherlands), basal part of the colony with secondarily thickened frontal walls in most of the fragment, and functional feeding zooids restricted to the lateral branch. *D, Coscinopleura angusta* (early Danian, Voxlev, Denmark), basal part of sexually recruited colony with secondarily thickened frontal walls and well-developed cenozooidal "holdfast" encrusting a thin bryozoan branch. *Scale bars* = 1 mm.

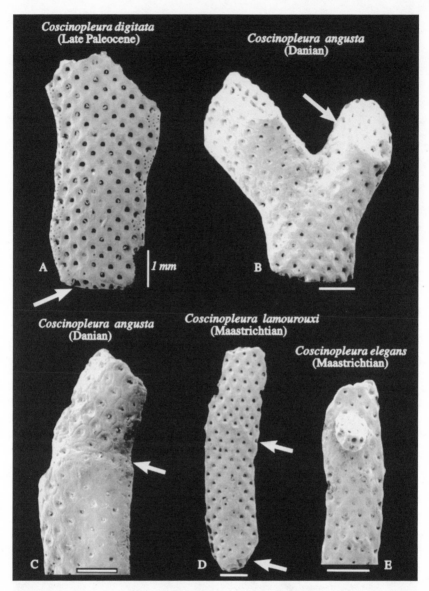

Fig. 11.5. Morphology of the Coscinopleuridae. *A, Coscinopleura digitata* (late Paleocene, Vincentown, New Jersey, USA), fragment of asexually recruited colony with cenozooidal closure of the basal fracture *(arrow)*. *Coscinopleura angusta* (late Danian, Klintholm, Denmark). *B,* Fragment with resumed zooidal growth from distal fracture *(arrow)* recognizable through abrupt change in frontal wall thickness; note that this example of regenerative repair has no bearing on the type of recruitment. *C,* Fragment of asexually recruited colony with resumed zooidal growth from proximal fracture *(arrow)*; note change of polarity in growth direction. *D, Coscinopleura lamourouxi* (late Maastrichtian, Maastricht, Netherlands), nearly complete, asexually recruited colony with cenozooidal closure of the basal fracture *(arrow)* and resumed zooidal growth from distal fracture *(arrow)*. *E, Coscinopleura elegans* (early Maastrichtian, Rügen, Germany), with young lateral, secondary branch. *Scale bars* = 1 mm.

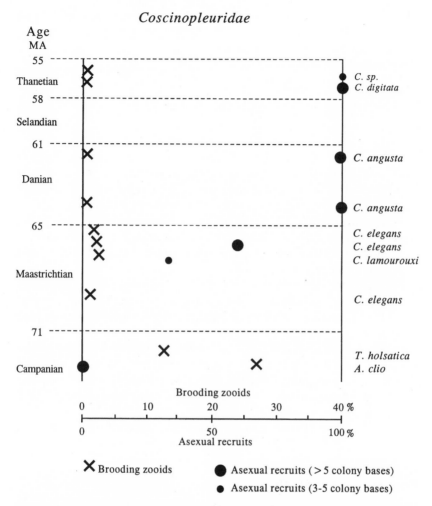

Fig. 11.6. Relative frequency of brooding zooids and asexual recruitment in ten populations of coscinopleurid species representing most of the stratigraphic distribution of the family. Note the very near total reversal in relative frequency of asexual recruits over time, and the associated negative correlation between the proportion of asexual recruits and brooding zooids (ovicells). Populations (from below): *Acoscinopleura clio,* Vigny, France; *Temnocoscinopleura holsatica,* Lägerdorf, Germany; *Coscinopleura elegans,* Rügen, Germany; *C. lamourouxi,* Maastricht, Netherlands; *C. elegans,* Curfs, Netherlands; *C. elegans,* Nye Kløv, Denmark; *C. angusta,* Voxlev, Denmark; *C. angusta,* Klintholm, Denmark; *C. digitata,* Vincentown, New Jersey, USA; *C. sp.,* Rockall Bank, North Atlantic.

Table 11.1. Comparison of brooding zooids in Campanian to Paleocene Coscinopleuridae

Age (Species)	Brooding Zooids (%)	Cluster-Size Mean (Range)	Bilateral Clusters (%)	Number of Zooids (and Fragments)
Late Paleocene (*Coscinopleura digitata*)	0.06	3.1 (1–6)	40	83,200 (402)
Maastrichtian (*Coscinopleura lamourouxi*)	6.33	9.0 (1–22)	86	3,600 (36)
Campanian (*Acoscinopleura clio*)	26.6	>15 (2–31)	100	2,500 (73)

of live colonies of branching bifoliate cheilostomes from the Adriatic Sea (Ken McKinney, pers. comm. 1999), where regenerative growth of broken-off branches lodged in the parent colony is by and large oriented toward the colony periphery, away from the crowded center.

Ten populations of seven species-level taxa representing most of the history of the Coscinopleuridae have been available for study in addition to a range of smaller collections. The investigated populations reveal a clear trend over time from wholly sexual recruitment in the Campanian to essentially asexual recruitment in the Paleocene (fig. 11.6). The investigation further corroborates the strong negative correlation between proportion of brooding zooids (= investment in sexual reproduction) and asexual recruitment observed by Thomsen and Håkansson (1995). Thus, the proportion of brooding zooids decreases from about 25% in the Campanian to 0.06% in the late Paleocene (fig. 11.6; table 11.1). Parallel with this trend we have noted that the size of ovicell clusters decreases over time, and the proportion of unilateral clusters increases (table 11.1). It should be stressed that investigations of several smaller collections, representing also the earliest history of the family, are consistent with these results.

Free-Living Cheilostomes

Free-living cheilostome bryozoans (figs. 11.7, 11.8) constitute a tightly knit ecological group conspicuous in numerous soft-bottom communities from the middle part of the Late Cretaceous until today (fig. 11.3). The free-living cheilostomes are probably highly heterogeneous phylogenetically (Håkansson, unpublished), comprising the well-constrained family Cupuladriidae and the more suspect family complex associated with the Lunulitidae, in this chapter termed "lunulitids." The unifying morphological characteristics of the group as a whole are therefore all more or less directly related to the particular way of life of its members: free-living directly on the soft, particulate seafloor, supported by stiff marginal setae emanating from regularly dispersed polymorphs (e.g., Cook 1963; Hå-kansson 1975). Lunulitid history dates back to the Coniacian or possibly Turonian (ca. 90 Ma; Håkansson and Voigt 1996), but they are nowhere common until the later part of the Campanian, when they evolved rapidly in northwest Europe (fig. 11.3; Håkansson, unpublished). Subsequent evolutionary bursts among the lunulitids were localized in North America (Eocene) and Australasia (Neogene). The Cupuladriidae, on the other hand, constitute a comparatively young group with rapid Neo-gene diversification subsequent to a sporadic record dating back to the Paleocene (Gorodiski and Balavoine 1961; Cook and Chimonides 1983). Thus, the present global, circumtropical distribution of the free-living cheilostomes is largely due to the Neogene success of the Cupuladriidae (fig. 11.3).

Free-living cheilostomes typically have radially budded, disk- to cone-shaped unilamellar colonies with feeding zooids and setate polymorphs distributed in predictable patterns. Cupuladriids invariably have poly-morphs situated distally to each feeding zooid, whereas in the lunulitids three fundamentally different budding patterns exist (see fig. 11.1). Ovi-cells in lunulitids are usually recognizable as more or less projecting, hoodlike structures at the distal margin of brooding zooids (fig. 11.7B). In a number of species the distinguishing characters are so vague, how-ever, that we have found that the number of ovicells may not be assessed with sufficient certainty to be considered in this context. In the cupuladri-ids, on the other hand, brooding takes place in zooids that are indistin-guishable from nonbrooding feeding zooids.

Throughout their history free-living cheilostome bryozoans have maintained highly predictable (albeit specifically distinct) budding pat-terns in the ancestral region of sexually recruited colonies (figs. 11.7B and F, 11.8A and F). Deviations from these patterns (figs. 11.7C–E and 11.8C–E, 11.8H and I) almost invariably indicate asexual recruitment, particularly when the thickness of the secondary skeleton in the basal wall is markedly different (figs. 11.7C and 11.8H). Two fundamentally different methods of asexual recruitment are known: fragmentation and colonial budding. Fragmentation is widespread through the history of both free-living groups investigated, whereas colonial budding is known only from the Neogene onward, most particularly among the cupu-ladriids. While fragments of colonies, regardless of species, have no pre-dictable shape (e.g., figs. 11.7E and 11.8H), colonial buds of at least the cupuladriids are recognizable as such through a distinctive early, fan-shaped budding pattern (Marcus and Marcus 1962). In contrast to sexu-ally recruited cupuladriids with a very pronounced radial symmetry cen-tered around an ancestral triad (figs. 11.8A and F; Håkansson 1973), the fan-shaped budding pattern of budded colonies in the more advanced

A-B, *Lunulites pseudocretaceae*
(Maastrichtian)

0.5 mm

Lunulites subsemilunaris
(Maastrichtian)

Lunulites semilunaris
(Maastrichtian)

E-F, *Trochopora bouei*
(Late Eocene)

species is initiated with a broken, half ancestral zooid (figs. 11.8C and D), which subsequently may develop a dual polarity due to regeneration (fig. 11.8E; Marcus and Marcus 1962). It has previously been noted that asexual recruits of free-living bryozoans will attempt to restore the ordinary, circular outline of the colony as fast as possible, maintaining the fundamental budding pattern characteristic of the species (Thomsen and Håkansson 1995). This restoration of colonial properties takes place regardless of the process of asexual recruitment (figs. 11.7C and 11.8I).

Evolutionary trends within the free-living cheilostomes are elucidated through investigation of a total of 35 populations of 22 species-level taxa representing four time levels: Late Cretaceous, Paleogene, Neogene, and Recent.

The first time level is represented by three mid–early Maastrichtian samples from the White Chalk of Denmark (age ca. 70 Ma). Eight free-living species are present in sufficient numbers to be investigated for mode of recruitment. Six of these species relied exclusively or dominantly on sexual recruitment, whereas two were heavily dominated by asexual reproduction (fig. 11.9). When compared with ovicell abundance, these data highlight our previous conclusion that the proportion of brooding zooids (i.e., investment in sexual reproduction) is negatively correlated with the importance of asexual recruitment. It also appears that at this early stage of evolution the preference for asexual reproduction through breakage or fission was of limited importance. Nevertheless, asexual reproduction appears to be correlated to the skeletal architecture of the colony, since it was essentially restricted to species where the polymorphs formed confluent rows between autozooidal rows (colonial architecture type C). Such species may have had a built-in predisposition for mechanical separation (figs. 11.8C–E; Thomsen and Håkansson 1995).

◁ **Fig. 11.7.** Morphology of free-living cheilostomes—lunulitids. *A–B, Lunulites pseudo-cretaceae* (late Maastrichtian, Ellidshøj, Denmark) displaying the pronounced radial symmetry characteristic of sexually recruited lunulitid colonies both in frontal *(B)* and basal *(A)* views; brooding zooids (ovicells) along the margin of the colony recognizable through their enlarged opesia; note the discontinuous rows of polymorphs characterizing architectural type B (cf. fig. 11.1). *C, Lunulites subsemilunaris* (late Maastrichtian, Rørdal, Denmark), basal view of asexually recruited colony regenerated from a wedge-shaped segment (with thick basal wall); note the rapid restoration of the circular outline typical of sexually recruited colonies. *D, Lunulites semilunaris* (early Maastrichtian, Møns Klint, Denmark), frontal view of asexually recruited colony in the early stage of regeneration from a single row of zooids. *E–F, Trochopora bouei* (late Eocene, Jackson, Mississippi, USA), frontal view of two colonies, one displaying the radial budding typical of sexually recruited lunulitids *(F)*, the other the aberrant budding of a colony recruited asexually through fragmentation *(E)*; note the confluent rows of polymorphs characterizing architectural type C (cf. fig. 11.1). *Scale bars* = 0.5 mm.

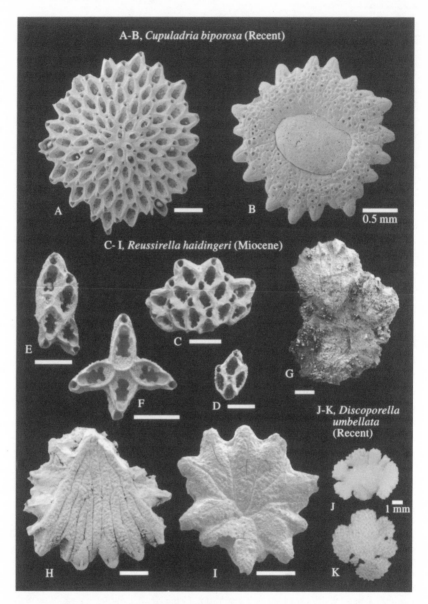

Fig. 11.8. Morphology of free-living cheilostomes—Cupuladriidae. *A–B, Cupuladria biporosa* (Recent, Florida, USA; from Håkansson 1973) displaying the pronounced radial symmetry characteristic of sexually recruited colonies in both frontal *(A)* and basal *(B)* views; note the characteristic cupuladriid budding pattern with a polymorph distal to each zooid and the partially incorporated substrate in basal view. *C–I, Reussirella haidingeri* (late Miocene, Gram, Denmark): *C,* asexually produced colony bud; note the "half" zooid at the base of the fan-shaped budding pattern, which provided the only link to the mother

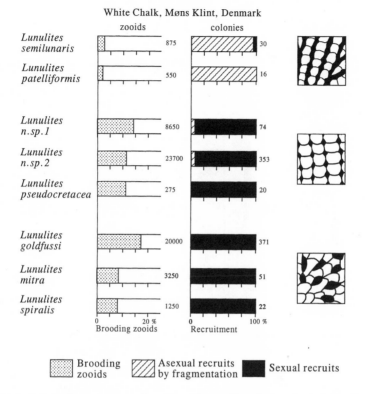

EARLY MAASTRICHTIAN
Lunulites

White Chalk, Møns Klint, Denmark

Fig. 11.9. Relation between proportion of brooding zooids and asexual recruitment in eight lunulitid species from the early Maastrichtian of Denmark. Note the fairly low overall proportion of asexual recruitment, plus correlation between colony architecture (cf. fig. 11.1) and dominant type of recruitment. *Numbers* at right end of each bar indicate number of zooids and colonies investigated.

◁ colony; *D*, comparable colony bud of only four zooids; *E*, asexually recruited juvenile colony produced by budding; note that the proximal zooid in the original bud (comprising only three zooids) has regenerated with a reversed polarity as it carries a distal polymorph in both ends; *F*, sexually recruited juvenile colony recognizable through its triple ancestrular complex (compare the central region of *A*); *G*, basal view of sexually recruited colony carrying two colonial buds near the upper margin preserved in their life position as a result of early diagenetic pyrite impregnation; *H*, basal view of asexually recruited colony produced through fragmentation; *I*, basal view of asexually recruited colony produced through colonial budding. *J–K*, *Discoporella umbellata* (Recent, Venezuela), frontal *(K)* and basal *(J)* views of colonies with abundant, broadly lobate colonial buds along the margin. Note that in contrast to *Reussirella haidingeri* this species has retained a skeletal connection to the buds. *Scale bars* = 0.5 mm *(A–I)* and 1 mm *(J–K)*.

LATE EOCENE
lunulitids
Gulf Coast Plain, USA

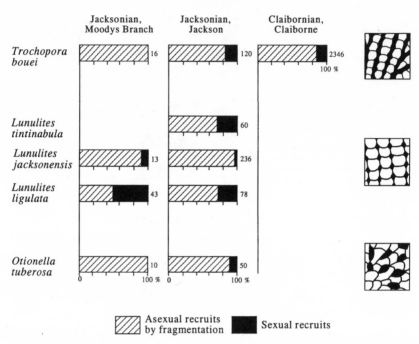

Fig. 11.10. Relation between proportion of brooding zooids and asexual recruitment in five lunulitid species from the late Eocene of the Gulf Coast Plain, USA. Note the dominance of asexual recruitment, the overall lack of correlation between colony architecture (cf. fig. 11.1) and type of recruitment, plus the comparatively high specific stability across localities. *Numbers* at right end of each bar indicate number of colonies investigated.

The Paleogene time level is represented by three samples from the late Eocene of the Gulf Coast Plain, United States (ca. 35 Ma), representing a range in soft-bottom environments from mud to fine sand. A total of five species are present in sufficient numbers to be investigated for mode of recruitment (fig. 11.10). None of these species possesses unambiguous ovicells; hence information on investment in sexual reproduction is not available. Reproduction is mainly asexual regardless of colonial architecture, and within-species variation is modest in spite of clear environmental variation (fig. 11.10).

Samples from two horizons in the late Miocene Gram Formation, Denmark (age ca. 10 Ma), representing largely similar shallow marine,

LATE MIOCENE
cupuladriids & lunulitids
Gram Formation, Gram, Denmark

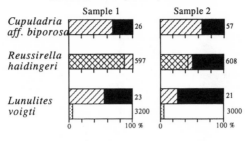

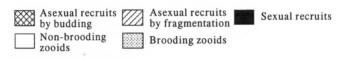

Fig. 11.11. Relation between proportion of brooding zooids and asexual recruitment in three species of free-living cheilostomes (two cupuladriid species and one lunulitid) from the late Miocene of Denmark. Note the dominance of asexual recruitment, plus the comparatively low specific stability across even very moderate environmental variation. *Numbers* at right end of each bar indicate number of colonies and zooids (*L. voigti* only) investigated.

muddy environments, are included to illustrate the Neogene time level. Three free-living species are present in sufficient numbers to be included in the analysis, one of which possesses unambiguous ovicells. The overall impression is a continuous dominance of asexual reproduction. It is noteworthy, however, that two of the species display markedly different reproductive strategy in the two populations even though their environments appear largely similar (fig. 11.11). At this point in the history of free-living cheilostomes a complete novelty in reproductive strategy is introduced, as colonial budding plays a major role in the reproduction in the cupuladriid species *Reussirella haidingeri* (figs. 11.8 and 11.11).

As judged by the four Recent samples included in this study, the situation in the present-day seas seems more complex than ever before in the history of the free-living cheilostomes. Two sets of populations have been investigated, comprising two samples of cupuladriids from a fairly shallow sandy shelf setting in West Africa and two samples of lunulitids from intermediate to fairly deep sandy mud shelf settings in New Zealand. Both Cupuladriidae and lunulitids exhibit a wide variety of strategies in co-occurring populations, and among the cupuladriids in particular,

RECENT

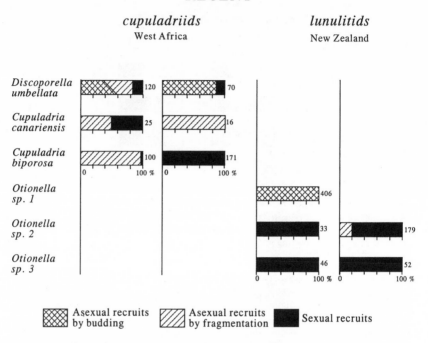

Fig. 11.12. Relation between proportion of brooding zooids and asexual recruitment in six species of free-living cheilostomes from the Recent shelves of West Africa (Ghana) and New Zealand (Otago Shelf). Note the dominance of asexual recruitment plus the near total lack of specific stability on the West African shelf. *Numbers* at right end of each bar indicate number of colonies investigated.

within-species variation is dramatic even in largely comparable environments (fig. 11.12).

Discussion and Conclusions

A number of conclusions and inferences seem warranted by this survey of asexual propagation in two groups of cheilostome bryozoans through time.

Throughout cheilostome history, species that rely heavily on asexual reproduction invest comparatively less energy in production of larvae as compared to species that utilize mainly sexual reproduction. Whether a similar correlation exists when comparing conspecific populations is less clear. Due to the lack of recognizable ovicells in most of the free-living species investigated, such a relationship has thus far been verified only for the ecologically highly flexible, vine-growing species *Columnotheca cribrosa* from the Danian of Denmark (fig. 11.2; cf. Thomsen and Håkansson 1995). It should be noted that the proportion of brooding zooids in species with extensive asexual propagation is considerably smaller than about 20%, which could be regarded as "ancestral" to the ovicellate cheilostomes.

Both the arborescent Coscinopleuridae and the free-living forms exhibit an increase in proportion of asexual reproduction through time. In the Coscinopleuridae the trend is fairly clear, whereas among the free-living cheilostomes it is somewhat blurred by intraspecific variation judged to be controlled by environmental factors, particularly in the younger part of the group history. In both groups, near total asexual recruitment is recorded regularly in single populations, but for a few species this situation seems to be the rule in most or all populations. The Danian arborescent species *Coscinopleura angusta* may provide an example. Throughout its existence this species maintained a conspicuous and constant presence in all reaches of the Danish Basin regardless of facies. Seven populations have been investigated in detail, representing most of the stratigraphic and environmental range of the species, and *C. angusta* consistently produced very few ovicells and sexual recruits were present only occasionally (fig. 11.13).

Among the free-living cheilostomes the importance of asexual reproduction is strongly correlated with small variations in colony design (architectural types A–C; cf. fig. 11.1) in the early history of the group, whereas in their later history this correlation is no longer evident. Among the Coscinopleuridae, skeletal features enhancing fragmentation are not apparent, as colony design remains essentially uniform throughout the history of the clade.

Colonial budding constitutes a novel method in producing asexual offspring appearing late in the evolution of the free-living cheilostomes. The process of colonial budding in cheilostome bryozoans was first described from Recent cupuladriids by Marcus and Marcus (1962). It is presently known to occur within a number of free-living clades but has not previously been reported from fossil populations. Colonial budding in free-living cheilostomes is found in a number of versions, mostly associated with the margin of the maternal colony (figs. 11.8G, J, and K). It has reached the highest degree of perfection within the Cupuladriidae, where the budding pattern of the colony bud is usually very stable. In several cupuladriid species the connection between the maternal colony and the colony bud is reduced to a single zooid, which is calcified only in its distal half prior to separation from the maternal colony (figs. 11.8E and F; Marcus and Marcus 1962). This design, with an uncalcified connection between the maternal and daughter colony, may be considered also as an extreme example of architectural predisposition to breakage, and in

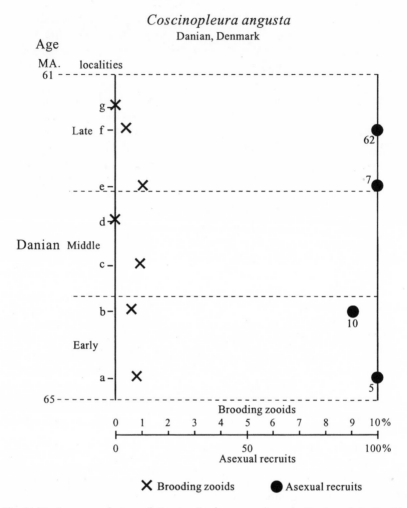

Coscinopleura angusta
Danian, Denmark

X Brooding zooids   ● Asexual recruits

Fig. 11.13. Seven populations of *Coscinopleura angusta* from the Danian of the Danish Basin representing the entire stratigraphic and geographic range of the species. Note the near total asexual recruitment throughout the history of the taxon irrespective of stratigraphic level and depositional environment. *a*, Voxlev (bryozoan mound); *b*, Skillingbro (bryozoan mound); *c*, Hanstholm ("chalk"); *d*, Faxe (bryozoan mound); *e*, Voldum ("chalk"); *f*, Klintholm (bryozoan sand); *g*, Herfølge (bryozoan sand). *Numbers* indicate colony bases available from each population.

this way it parallels the uncalcified segments characteristic of the "jumper" strategists *Membranipora triangularis* described by Håkansson and Winston (1985) and Winston and Håkansson (1986). In other cupuladriids and lunulitids employing colonial budding, however, the budding pattern of colonial buds is less regular and the means of separation from the maternal colony is less sophisticated, involving breakage without un-calcified connections (figs. 11.8J and K). In a few lunulitid species colonial buds may simply take the form of broad, irregular lobes emanating from the basal part of the maternal colony rather than from the margin (Cook and Chimonides 1986).

Few previous studies of bryozoans are directly comparable with the investigation presented here (cf. also Thomsen and Håkansson 1995), although the cheilostome literature regularly includes comments and ob-servations concerning regenerative repair, and occasionally the existence of asexual reproduction is referred to. Thus, for example, Dartevelle (1935) and Marcus and Marcus (1962) mention the existence of near total asexual populations of free-living species from the Oligocene and Recent, respectively, and Winston (1983) suggested that the ability to grow from fragments was the key to the success of some erect, jointed bryozoans around Antarctica today. Multiplication by fragmentation has even been demonstrated for the Carboniferous bryozoan *Archimedes* (McKinney 1983).

Asexual propagation by fragmentation has also been reported from various cnidarians, and it seems in particular to be important in erect branching groups of corals and octocorals (e.g., Tunicliffe 1981; High-smith 1982; Meesters and Bak 1995; Lirman and Fong 1996, 1997; Riegl and Riegl 1996). Fungiid corals and various octocorals, soft corals, and sea anemones may fissure or produce buds and asexual ramets (Boschma 1925; Chen et al. 1995; Kramarsky-Winter and Loya 1996; Dahan and Benayahu 1997; McFadden 1997) in ways that in many respects may be paralleled to the buds of the free-living cheilostomes.

In conclusion it seems that asexual reproduction through fragmenta-tion and fission occurs in all architecturally suitable groups of bryozoans. In some groups it may result in novel skeletal features facilitating fission, and in many environments it may become the dominant mode of dis-persal. In a few species sexually recruited colonies are virtually absent in all environments, as in *Coscinopleura angusta* from the Danian of Den-mark; in this species it can therefore be argued that individual genets may have survived for more than 4 million years and were potentially immortal.

Acknowledgments

Much of the material included in the present investigation has been made available from the bryozoan collections at the Natural History Museum in London (formerly the British Museum, Natural History); the Natural History Museum, Smithsonian Institution, Washington, DC; and the Voigt Collection, Hamburg. We thank P. L. Cook, P. Taylor, M. Spencer-

Jones (London), A. H. Cheetham (Washington, DC), and E. Voigt (Hamburg) for their help in finding the right material. P. Taylor and Ken McKinney (Appalachian State University) provided helpful comments on an early version of the manuscript, and P. Taylor took care of the skilled SEM operations necessary to illustrate the pyrite-impregnated budding colony of *Reussirella haidingeri* preserved in silicone oil.

References

Ayre, D. J., and S. Dufty. 1994. Evidence for restricted gene flow in the viviparous coral *Seriatopora hystrix* on Australia's Great Barrier Reef. *Evolution* 48: 1183–1201.

Benzie, J. A. H., A. Haskell, and H. Lehman. 1995. Variation in the genetic composition of coral *(Pocillopora damicornis* and *Acropora palifera)* populations from different reef habitats. *Marine Biology* 121:731–39.

Boschma, H. 1925. Papers from Dr. Th. Mortensen's Pacific Expedition 1914–16: 28, Madreporaria I: Fungiidae. *Videnskabelige Meddelelser, Naturhistorisk Forening København* 79:185–259.

Chen, C. L. A., C. P. Chen, and I. M. Chen. 1995. Sexual and asexual reproduction of the tropical corallimorpharian *Rhodactis (= Discosoma) indosinensis* (Cnidaria: Corallimorpharia) in Taiwan. *Zoological Studies* 34:29–40.

Cook, P. L. 1963. Observations on live lunulitiform zoaria of Polyzoa. *Cahiers de Biologie Marine* 4:407–13.

Cook, P. L., and P. J. Chimonides. 1983. A short history of the Lunulite Bryozoa. *Bulletin of Marine Science* 33:566–81.

———. 1986. Recent and fossil Lunulitidae (Bryozoa, Cheilostomata): 6, *Lunulites sensu lato* and the genus *Lunularia* from Australia. *Journal of Natural History* 20:681–705.

Dahan, M., and Y. Benayahu. 1997. Clonal propagation by the azooxanthellate Octocoral *Dendronephthya hemprichi. Coral Reefs* 16:5–12.

Dartevelle, E. 1935. Zoarial regeneration of free polyzoa. *Annals and Magazine of Natural History Series 10* 15:191–95.

Fong, P., and D. Lirman. 1995. Hurricanes cause population expansion of the ranching coral *Acropora palmata* (Scleractinia): Wound healing and growth pattern of asexual recruits. *Marine Ecology* 16:317–35.

Gorodiski, A., and P. Balavoine. 1961. Bryozoaires crétacés et éocènes du Sénégal. *Bulletin du Bureau du Recherches Géologiques et Minières Paris* 4:1–15.

Håkansson, E. 1973. Mode of growth of the Cupuladriidae (Bryozoa, Cheilostomata). *Living and fossil Bryozoa,* ed. G. P. Larwood, 287–98. London: Academic Press.

———. 1975. Population structure of colonial organisms: A palaeoecological study of some free-living Cretaceous bryozoans. *Documents des Laboratoires de Géologie de la Faculté des Sciences de Lyon,* hors série 3/2:385–99.

Håkansson, E., and E. Voigt. 1996. New free-living bryozoans from the northwest European chalk. *Bulletin of the Geological Society, Denmark* 42:187–207.

Håkansson, E., and J. E. Winston. 1985. Interstitial bryozoans: Unexpected life forms in a high energy environment. In *Bryozoa: Ordovician to Recent,* ed. C. Nielsen and G. P. Larwood, 125–34. Fredensborg, Denmark: Olsen and Olsen.

Highsmith, R. C. 1982. Reproduction by fragmentation in corals. *Marine Ecology Progress Series* 7:207–26.

Hughes, R. N. 1989. *A functional biology of clonal animals.* London: Chapman and Hall.

Hunter, C. L. 1993. Genotypic variation and clonal structure in coral populations with different disturbance histories. *Evolution* 47:1213–28.

Kramarsky-Winter, E., and Y. Loya. 1996. Regeneration versus budding in fungiid corals: A trade off. *Marine Ecology Progress Series* 134:179–85.

Lirman, D., and P. Fong. 1996. Sequential storms cause zone-specific damage on a reef in the northern Florida reef tract: Evidence from Hurricane Andrew and the 1993 storm of the century. *Florida Scientist* 59:50–64.

———. 1997. Patterns of damage to the branching coral *Acropora palmata* following Hurricane Andrew: Damage and survivorship of hurricane generated recruits. *Journal of Coastal Research* 13:67–72.

Marcus, E., and E. Marcus. 1962. On some lunulitiform bryozoa. *Boletin da Faculdade de Filosofia Ciêncas e Letras Universidade de São Paulo,* 261 Seria Zoologia, 24:281–324.

McFadden, C. S. 1997. Contributions of sexual and asexual reproduction to population structure in the clonal soft coral, *Alcyonium rudyi. Evolution* 51:112–26.

McKinney, F. K. 1983. Asexual colony multiplication by fragmentation: An important mode of genet longevity in the Carboniferous bryozoan *Archimedes. Paleobiology* 9:35–43.

Meesters, E. H., and R. P. M. Bak. 1995. Age-related deterioration of a physiological function in the branching coral *Acropora palmata. Marine Ecology Progress Series* 121:203–9.

Riegl, B., and A. Riegl. 1996. How episodic coral breakage can determine community structure: A South African coral reef example. *Marine Ecology* 17: 399–410.

Thomsen, E., and E. Håkansson. 1995. Sexual versus asexual dispersal in clonal animals: Examples from cheilostome bryozoans. *Paleobiology* 21:496–508.

Tsurumi, M., and H. M. Reiswig. 1997. Sexual versus asexual reproduction in an oviparous rope-form sponge, *Aplysina cauliformis* (Porifera: Verongida). *Invertebrate Reproduction and Development* 32:1–9.

Tunicliffe, V. 1981. Breakage and propagation of the stony coral *Acropora cervicornis. Proceedings of the National Academy of Sciences, USA* 78:2427–31.

Voigt, E. 1956. Untersuchungen über *Coscinopleura* Marss. (Bryoz. foss.) und verwandte Gattungen. *Mitteilungen aus der Geologisches Staatsinstitut Hamburg* 25:26–75.

Winston, J. E. 1983. Patterns of growth, reproduction, and mortality in bryozoans from the Ross Sea, Antarctica. *Bulletin of Marine Science* 33:688–702.

Winston, J. E., and E. Håkansson. 1986. The interstitial bryozoan fauna from Capron Shoal, Florida. *American Museum Novitates* 2865:1–50.

Wulff, J. L. 1991. Asexual fragmentation, genotype success, and population dynamics of erect branching sponges. *Journal of Experimental Marine Biology and Ecology* 149:227–47.

12 Macroevolutionary Trends

Perception Depends on the Measure Used

Frank K. McKinney, Scott Lidgard,
and Paul D. Taylor

Introduction

"Species lists do not alone encompass ecological communities" (Jackson 1994, 1412). This statement was accompanied by a plea for paleoecologists to supplement their work with information on relative abundance of fossil species to gain a deeper understanding of the historical basis for community structure. A parallel concern is whether temporal data on the waxing and waning of taxonomic diversity is sufficient by itself to portray macroevolution of ecological systems. Would information on relative abundance of taxa simply track the secular patterns that are based on global or local diversity? If patterns of relative abundance through time differ from patterns based on counts of taxa, then new perspectives on the history of life may be obtained (McKinney et al. 1998; Lupia et al. 1999).

An artificial dichotomy exists between methods for studying patterns over "evolutionary time" and methods employed in more localized, "ecological time" studies. Macroevolutionary patterns in the fossil record typically are demonstrated by counting numbers of taxa or by measuring morphological characteristics and plotting them against time. During the past quarter century, numerous quantitative studies based on counts of taxa have established a large array of patterns ranging from long-term trends in diversity (review in Signor 1990; Sepkoski 1993, 1995, 1997) and within-clade morphology (Cheetham et al. 1981; Cheetham 1986; McGhee 1980; Vermeij 1987; Foote 1993, 1995), demonstration of major extinction events (Raup and Sepkoski 1982) that may be periodic (Raup and Sepkoski 1984; but see Stigler and Wagner 1987), and recognizable patterns of ecosystem recovery from extinction events (Collins 1989; Kauffman and Erwin 1995; Hart 1996).

In contrast to most macroevolutionary studies, and despite Jackson's valid criticism (1994) of the dearth of abundance data in paleoecology, paleoecological studies often supplement presence/absence of taxa with indications of abundance, typically counts of numbers of individuals (e.g., Kitchell et al. 1986; Krause 1986; D'Hondt and Keller 1991; Aberhan 1994; Fowell et al. 1994; Berglund et al. 1996; Lupia et al. 1999). Counts, in some cases, are supplemented by estimates of relative biomass among taxa or biomass per unit of sediment (e.g., Phillips and Peppers 1984; M. L. McKinney 1986; Wing et al. 1993; Radenbaugh and McKinney 1998). Almost all studies that attempt to estimate biomass are based on counts of individuals, in some instances multiplied by their estimated average size. This procedure does not work well for modular organisms (Ausich 1981), which usually are fragmented, not only confounding attempts to count number of "individuals" but also producing such a broad size range of fragments that error of estimate of average size is high. Point-counted sections through rock, or picking and weighing individual fossils from disaggregated rock or poorly consolidated sediment can accommodate, within the same method of quantification, both fossils of individual organisms and skeletal fragments of modular organisms. While recognizing its limitations, counts of bryozoan colony fragments were used by Cheetham (1963) as a surrogate for volume in his study of zoogeography of Eocene bryozoans of the eastern Gulf Coast of the United States. Other studies have also used abundance of skeletal remains as the closest available proxy for biomass of preservable organisms within or on a sedimentary bed or within a facies that was produced over some interval of time (Ausich 1981; Staff et al. 1985).

In an exemplary, detailed study of the paleoecology of cheilostome bryozoans in a Danian mound in Sweden, Alan Cheetham (1971) used several measures in an attempt to understand the interactions among the organisms present and between the organisms and the environment. From multiple samples distributed across the core and flanks of the mound, he identified and tallied all of the cheilostome species present, and for each species he determined the zooidal construction, growth habit(s), and weight percentage per sample. Among Cheetham's findings was that cyclostomes were consistently more abundant than cheilostomes across the mound, making up approximately 75% of the total bryozoan skeletal mass. These results were consistent with earlier biovolume comparisons of cyclostomes and cheilostomes in Danian carbonates (Berthelsen 1962) but would not be predictable from the Danian taxonomic diversity data showing (1) a much higher number of cheilostome than cyclostome families (Lidgard et al. 1993); (2) a roughly equal number of genera present in the two clades (Viskova 1995; McKinney et al. 1998); and (3) a slightly higher within-assemblage species richness of cheilostomes over cyclostomes during the Danian (Lidgard et al. 1993).

In this chapter, we present data on the within-fauna skeletal masses of cheilostome bryozoans relative to that of cyclostome bryozoans, from the Lower Cretaceous to Holocene. We use skeletal mass as a proxy for abundance, one of several measures that have been used to infer paleoecological dominance at a given place and time (Lupia et al. 1999). We then compare the resultant temporal trend with patterns of within-fauna species richness and with global diversity of families and genera for the two clades, in order to see how the measures of numbers of taxa within the clades compare with their changing dominance, as inferred from skeletal mass. An unexpected lack of correspondence in the patterns is particularly conspicuous immediately after the Cretaceous-Tertiary (K-T) extinction event and suggests ecological opportunism in cyclostomes that is not apparent from counts of numbers of taxa.

Materials and Methods

This study is based on bulk sediment samples taken from 78 bryozoan-bearing localities distributed throughout the Cretaceous to Holocene marine sedimentary record (appendix 12.1), including data from Berthelsen 1962 and Cheetham 1971. The majority of the localities are in Europe and eastern North America, although one collection is from the Upper Cretaceous of South Africa, and some Cenozoic and Holocene collections are from New Zealand and Australia. Most contain abundant and diverse bryozoan faunas. All are from temperate to subtropical, inner- to outer-shelf deposits and include various carbonate and siliciclastic lithologies (appendix 12.1).

Sediment sample size ranged from 0.3 to 3.0 kg. Depending on the type and degree of lithification, samples were washed directly through 2-mm and 0.5-mm sieves or were first treated with a dilute solution of Quaternary-O, multiple freeze-thaw sequences, or freeze-thaw in a saturated solution of sodium sulfate decahydrate before sieving. Bryozoans were picked from the sieved residues and sorted into cyclostome and cheilostome fractions. Where free-living, cap-shaped cheilostomes ("lunulites") constituted more than a trivial proportion of the sample, they were sorted into a third group. Depending on the abundance of bryozoans within a sample, either all were picked from the sieved residues, or a weighed subsample was picked and multiplied proportionally to estimate the mass of skeleton contained within the entire bulk sample. For most of the samples, all of the >2-mm material was sorted. However, only a weighed subsample was picked for the 0.5–2-mm fraction. For bryozoan-rich sediments such as some Maastrichtian and Danian chalks, 5-g aliquots of the 0.5–2-mm fraction were sufficient (Cheetham 1971). Where sediments were less rich in bryozoans, at least 10-g aliquots were used. Representation of cyclostomes and cheilostomes may vary considerably in the <0.5-mm fraction from the proportion seen in larger fractions

(Håkansson, pers. comm.). Taxa from the small fraction are nearly always ignored in the taxonomic literature, however, and constitute such a small proportion of overall mass per sample that we excluded them from our analysis. Cyclostomes and cheilostomes were not separated by smaller taxonomic units or by growth forms but were weighed as sum of the picked cyclostomes and sum of the picked cheilostomes (omitting the free-living colonies—a newly evolved guild—where they constituted more than a trivial part of the cheilostomes).

To determine how much variation in the results was due to uncemented sediment adhering to the screened and sorted samples, a set of six samples from varying lithologies and stratigraphic horizons, each of which had been previously weighed and recorded, was cleaned ultrasonically and separated from any loosened sediment. The average change in ratio of cyclostome mass to total bryozoan mass generated by ultrasonic cleaning was 0.348%, and the coefficient of variation in the change was 0.111%. Because the change was so slight, no other samples were cleaned ultrasonically.

Thin sections were point-counted in order to determine whether there were any taxonomic and temporal differences in the proportion of mass in samples contributed by the skeletons of the bryozoans compared with nonbryozoan matrix. Adherent sediment and cement, both within the zooidal chambers and on the skeletal surface, contributes to the mass of each fossil sample. A total of 3,753 point counts of skeleton and adherent material were made from 68 bryozoans in samples from four localities of differing age. For each of the samples, the ratio of skeleton to adherent material was determined separately for cyclostomes and for cheilostomes.

The age of each sample was determined to the finest stratigraphic resolution available, generally stage or biozone within stage. Age in millions of years (My) for each sample was estimated as the midpoint of the stage or as the midpoint of the proportion of the biozone, based on the arbitrary assignment of equal proportions of the stage's duration to each of the successive biozones recognized within it. Absolute beginning and ending points of each stage are based on the estimates given in Harland et al. 1990.

Almost all data on relative biovolume in this paper were generated from the primary collections highlighted herein, but they were supplemented by some information published previously by Berthelsen (1962) and by Cheetham (1971) for Danian deposits of Scandinavia. Berthelsen and Cheetham typically sorted, weighed, and recorded values individually for bryozoans from multiple samples per locality. Their data are averaged here by locality or local facies, so that only one record per site or local facies studied by Berthelsen and by Cheetham is included herein.

Data on taxonomic diversity of bryozoan families are from Taylor

1993, modified with updated information from McKinney et al. 1998. Genus diversity data are from McKinney et al. 1998, modified by Sepkoski (pers. comm.). Within-fauna species richness data are reported for 592 fossil assemblages, modified from Lidgard et al. 1993 and McKinney et al. 1998.

Results

Post-Paleozoic Diversity History of the Cyclostomata

The only order of stenolaemate bryozoans unquestionably to have survived the combination of end-Permian and Triassic extinction events was the Cyclostomata, which—in contrast with the extinguished fenestrates, cryptostomes, trepostomes, and cystoporates—had been of minor importance during the Paleozoic (Taylor and Larwood 1988, 1990; Taylor and Wilson 1996; but see Boardman 1984 and Viskova and Morozova 1993 for argument that some noncyclostome stenolaemates survived at least until later in the Mesozoic). The oldest known post-Paleozoic cyclostomes are Late Triassic (Carnian) *Stomatopora* and other tubuliporines (Bizzarini and Braga 1981, 1985), but cyclostomes remained rare and had low diversity until they experienced a diversity radiation during the Early Jurassic and again through the Cretaceous, so that by Maastrichtian time global diversity had risen to at least 176 genera (fig. 12.1A; Taylor and Larwood 1990; Viskova 1995; McKinney et al. 1998). This Mesozoic radiation at the genus level was paralleled by a Jurassic through Late Cretaceous radiation in cyclostome families, which totaled 20 by the Maastrichtian but dropped off to the high teens at the K-T boundary, where it has remained through the Cenozoic (fig. 12.1C; Lidgard et al. 1993; Taylor 1993). The number of cyclostome genera decreased more precipitously at the K-T boundary, dropping to about 80 by the end of the Paleocene and remaining fairly stable at 70–80 thereafter (Viskova 1995; McKinney et al. 1998). Data on the extinction of cyclostome species at the K-T boundary are few. A single monographic study (Brood 1972) of cyclostomes across the boundary shows that, among 70 Danian cyclostome species in Scandinavia, 54–55 (77–78%) were holdovers from the Cretaceous.

Origin and Diversity History of the Cheilostomata

The earliest known cheilostomes are from Late Jurassic deposits (Pohowsky 1973; Taylor 1994). The early cheilostomes and their closest living relatives have features that indicate an origin from the Ctenostomata (Banta 1975; Taylor 1990; Banta et al. 1995; Todd 2000). Family-level diversification of cheilostomes was delayed until the Late Cretaceous (Cenomanian-Santonian), followed by a second interval of cladogenesis from latest Late Cretaceous (Maastrichtian) through late Oligocene (Chattian) (fig. 12.1D; Cheetham 1971; Lidgard et al. 1993; Taylor 1993). Moreover, as noted by Lidgard et al. (1993, 359), "There is no indication from this absolute diversity trend of an end-Cretaceous cheilostome extinction event."

Global genus diversity of cheilostomes blossomed during the entire Late Cretaceous, reaching 178 genera in the Maastrichtian (fig. 12.1B; Viskova 1995; McKinney et al. 1998). The number of cheilostome genera declined to approximately 110 by the end of the Paleocene, in parallel with the number of cyclostome genera, but increased rapidly in the Eocene to exceed the Maastrichtian peak; it continued to increase more slowly thereafter (Viskova 1995; McKinney et al. 1998).

Many specialized, highly derived features evolved in cheilostomes during the Campanian and Maastrichtian. Voigt (1985) noted specialized avicularia and vibracula, fenestration of ovicells, alterations in zooecial walls, frontal budding, zoarial mobility, and an expansion of flexible zoarial forms among Campanian and Maastrichtian innovations. Many genera characterized by these innovations carried through into the Tertiary. In the absence of further morphological innovation and taxonomic diversification until the Eocene (possibly late Paleocene), Voigt (1985)

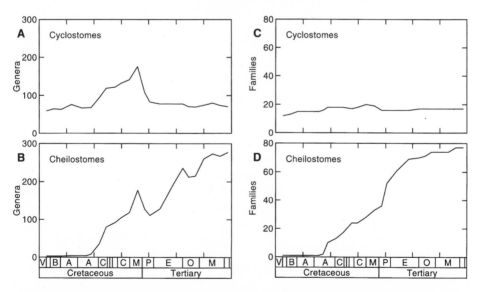

Fig. 12.1. Global taxonomic diversity of cyclostomes and cheilostomes from Early Cretaceous to Holocene, plotted at midpoints of geologic stages (Harland et al. 1990). Genus diversity of cyclostomes *(A)* and cheilostomes *(B)* based on data from McKinney et al. 1998 with modifications by J. J. Sepkoski, Jr. Family diversity of cyclostomes *(C)* and cheilostomes *(D)* modified from data in McKinney et al. 1998 and Taylor 1993.

considered the depauperate Danian cheilostome fauna to be more closely allied taxonomically and ecologically to the Maastrichtian than to later Tertiary faunas.

As with cyclostomes, data on cheilostome species across the K-T boundary are limited. However, based on a single monographic study (Berthelsen 1962) of 117 cheilostome species in the Danian of east Denmark, 31–33 (26–28%) were holdovers from the Cretaceous. Therefore, the proportion of holdover species among Danian cheilostomes was almost a third less than among Danian cyclostome species based on these two studies in Scandinavia (Berthelsen 1962; Brood 1972), implying greater cheilostome extinction at the K-T boundary and greater post-extinction radiation of cheilostomes (see Sepkoski et al. 2000).

Within-Fauna Species Richness

The number of species present in local faunas must be examined very differently from global diversity patterns. For one thing, biogeographic constraints and conditions of the local environments and microenvironments will interact with the total number of species in existence at any given time to determine the number of species potentially inhabiting an area. Local assemblage diversity as reported in the literature may be subject both to discovery biases (species occurrence) and recognition biases (species differentiation) (Lidgard and Crane 1990; Lupia et al. 1999) that could alter the total reported number of cyclostomes, cheilostomes, or both.

Most species are rare. It is therefore unlikely that all species occurring in a once-living fauna were preserved. In addition, sampling effort varies widely, and species may not be differentiated adequately as reported in the literature. In fact, details of sampling effort and tests of sampling adequacy (Hayek and Buzas 1997) are seldom reported in paleofaunistic or -floristic studies. For example, species with very small colonies or colony fragments may be overlooked, which would generally result in under-reporting the actual number of preserved species at the site. Additionally, the standard taxonomic practices used in virtually all records of fossil bryozoan species diversity may consistently underestimate the number of "basal" species that might be revealed by detailed but time-consuming morphometric methods (Jackson and Cheetham 1990; Lidgard and Buckley 1994). However, there is no a priori expectation that reporting or preservational biases have any directional trend through time. Thus, although these problems may be a source of noise that potentially could obscure biological trends, they would not be expected to generate spurious temporal trends.

From deposits spanning the past 200 million years, Lidgard et al. (1993) tallied species from 676 assemblages that included the entire bryozoan fauna (i.e., that were not intentionally restricted either to cyclostomes or to cheilostomes). They found that, among these two calcified clades of post-Paleozoic bryozoans, cheilostomes accounted for a trivial proportion of within-fauna bryozoan species richness from Late Jurassic until late Early Cretaceous, when they began a near-linear increase in percentage that culminated at about 80% in the mid-Eocene. Following that, they dropped to a median of about 60% of within-fauna species during the late Miocene but rebounded to equal the mid-Eocene median by the end of the Pliocene. Figure 12.2D plots the percentage of cyclostome species within local faunas, which is the reciprocal of that of cheilostomes. Although there have been a few post-Campanian local faunas with more cyclostome than cheilostome species, the median percentage of cyclostomes within faunas has not exceeded 50% since the Campanian.

The actual number of species of cyclostomes and cheilostomes within faunas exhibits a different time-series pattern than the corresponding

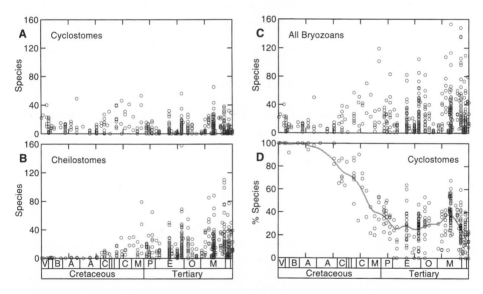

Fig. 12.2. Within-fauna actual species richness of cyclostomes *(A)*, cheilostomes *(B)*, and all bryozoans *(C)*, Early Cretaceous to Holocene (592 faunas). Based on an expansion of the data set used in Lidgard et al. 1993 and McKinney et al. 1998. Relative within-fauna species richness of cyclostomes (*D;* 351 faunas) plotted as the percentage of total bryozoan species richness consisting of cyclostomes. Each point represents the percentage of cyclostomes for a single faunal assemblage with ten or more bryozoan species; studies that explicitly excluded either cyclostomes or cheilostomes were omitted here. The *trend line* was fitted using a distance-weighted least squares method (Wilkinson 1990). The algorithm permits different levels of sensitivity to the variations seen in the scatter of data. All fitted trend lines in figures are shown at the default sensitivity (0.5). Other smoothing methods for bivariate data produced similar results.

within-faunal percentages (fig. 12.2A and B; Lidgard et al. 1993). The median number of species within faunas slowly increased from fewer than 5 during the Late Triassic to fewer than 15 by mid-Cretaceous but then rapidly increased to a Maastrichtian high of about 25 (fig. 12.2C). Through the Cenozoic, this number has been less than the Late Cretaceous high but has fluctuated and nearly reached the Late Cretaceous value in the Miocene. Throughout the 200-million-year record, variance in number of species has increased, and during the Cenozoic some local faunas have contained well over 100 bryozoan species (see Lidgard et al. 1997).

Analysis of a subset of the 676 faunas that contained at least ten species demonstrated that median cyclostome species richness increased significantly ($p < 0.001$) through the Jurassic and Cretaceous, reaching a maximum during the Campanian, after which their species richness declined (fig. 12.2A; Lidgard et al. 1993). Although there is a graphically conspicuous drop-off in the trend line of cyclostome species richness across the K-T boundary, the latest Cretaceous through Tertiary decline was not statistically significant. From single species of cheilostomes within local faunas during the Late Jurassic and usually only single species locally until the end of Early Cretaceous, cheilostomes diversified rapidly within local faunas to a median of about 15 to 20 species by the Campanian, and, with the exception of a brief decrease at the K-T boundary that almost equaled that of cyclostomes, their within-fauna species richness— while fluctuating somewhat—has remained high through the Cenozoic (fig. 12.2B).

Relative Skeletal Mass

No cheilostome bryozoans were found in the Late Jurassic and Early Cretaceous faunas for which bryozoan skeletal mass was determined, but they were present in all Late Cretaceous and Cenozoic faunas (fig. 12.3A). Two clearly demarcated intervals of decrease in skeletal mass of cyclostomes relative to cheilostomes are present in the data. Figure 12.3A shows declines from the Cenomanian to the latest Maastrichtian and from the early Eocene to the Holocene. Figure 12.3B shows the standardized residual values from the trend line fitted to these data. Despite the visible inflexion across the K-T boundary interval, the observed cyclostome percentages immediately before and after the boundary are clearly separate. We therefore partitioned the data at the K-T boundary for further analysis.

From the earliest (Cenomanian) to the latest (Maastrichtian) Late Cretaceous, the percentage of bryozoan skeletal mass consisting of cyclostomes generally decreased to an average of 21% in Maastrichtian faunas (SD = 12.4%, N = 11). Cyclostomes dominated earliest Cenozoic

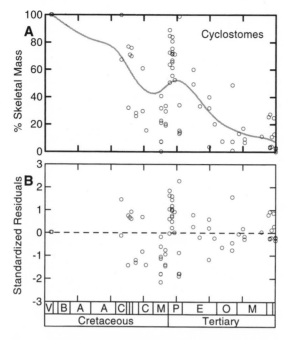

Fig. 12.3. *A,* Relative within-fauna skeletal mass of cyclostomes from Early Cretaceous to Holocene. See figure 12.2 for plotting methods. *B,* Standardized residual values for trend line fitted in *A* shows a disjunct pattern across the K-T boundary interval.

(Danian) faunas, however, constituting on average 75% (SD = 14.5%, N = 5) of the bryozoan skeletons in the early Danian faunas examined. Later Danian faunas (coccolith zones 3–8 of Thomsen) also have a high proportion of cyclostome skeletal mass (mean = 72%, SD = 9.1%, N = 12). With minor fluctuations, there has been a decrease in the cyclostome proportion of bryozoan skeletal mass from the Paleocene to the Holocene.

Results from point-counting skeletal walls versus sediment plus cement in thin sections are summarized in table 12.1. On average, a greater proportion of the cheilostome than cyclostome specimens in all four studied faunas comprised skeletal walls, although variance was high. There is no obvious temporal trend. In particular, a comparison of the relative cheilostome and cyclostome proportions for the Maastrichtian and Danian samples shows that the striking change from cheilostome to cyclostome dominance by weight at the K-T boundary cannot be explained as an artifact resulting from a change in the cement and/or sediment adhering to the skeletons of members of the two clades at this time. Summing all points counted in thin sections, the percentage area of cheilostome

Table 12.1. Proportion of bryozoan skeleton versus cement plus sediment in cheilostomes and cyclostomes

	Cheilostome Skeletal Proportion	Cyclostome Skeletal Proportion	Cheilostomes/ Cyclostomes
Upper Pliocene, Castell'Arquato	Mean = 0.74 $N = 9$, SD = 0.064, R = 0.66–0.87	Mean = 0.61 $N = 9$, SD = 0.142, R = 0.45–0.87	1.21
Lower Pliocene, Crag Farm	Mean = 0.51 $N = 7$, SD = 0.099, R = 0.38–0.65	Mean = 0.43 $N = 10$, SD = 0.148, R = 0.20–0.70	1.19
Danian, Karlby Klint	Mean = 0.61 $N = 9$, SD = 0.088, R = 0.50–0.68	Mean = 0.58 $N = 6$, SD = 0.150, R = 0.38–0.76	1.05
Maastrichtian, Nye Kløv	Mean = 0.61 $N = 8$, SD = 0.191, R = 0.24–0.84	Mean = 0.48 $N = 10$, SD = 0.073, R = 0.36–0.58	1.27
All points counted	0.625	0.497	1.26

Notes: Proportions were determined by point-counting thin sections of resin-embedded bryozoans from four faunas (N is the number of individual specimens; SD is standard deviation; R is observed range of values). Higher mean values indicate greater areas of bryozoan skeletal wall relative to cement and sediment infilling chambers or adhering to specimen surfaces. On average, cheilostome specimens are found to have more skeletal wall area than cyclostomes (ratio given in right-hand column), but there is no detectable trend relative to the age of the fauna.

specimens comprising skeletal wall was 62.5%, compared with 49.7% for cyclostomes, which therefore contained correspondingly more diagenetic cement and sediment than did cheilostomes. A scaling factor of 1.26 (62.5/49.7) based on these differences was applied to all pre-Holocene cheilostome raw weight data to give a more accurate estimate of skeletal biomass for comparison with cyclostome skeletal biomass. Visual inspection suggests that the generally thicker skeletal walls of cheilostomes are responsible for the difference rather than a propensity for cyclostomes to contain more cement and/or sediment.

Discussion

The history of relative skeletal mass of cyclostome and cheilostome bryozoans within individual faunas falls somewhere within a spectrum bounded by a radical short-term excursion from a long-term trend (i.e., "typical" disaster fauna), and a profound resetting of bryozoan ecology—and species-level evolution—at the K-T boundary that required over 10 million years for cheilostomes to return to ecological dominance. Regardless of the duration of the interval of cyclostome skeletal mass following the end-Cretaceous, it stands in conspicuous contrast with patterns of global diversity and within-fauna species richness that have been documented previously. It is not what one would expect if ecological dominance is reflected accurately and sufficiently in counts of taxa.

We have too few upper Paleocene and lower Eocene data at present to discriminate between the short-term excursion versus the profound resetting possibilities. If the Danian values are omitted, there is a continuous decrease in proportion of cyclostome skeletal volume from Early Cretaceous to the present, suggesting that the Danian excursion to high cyclostome skeletal volumes may have been an aberration in an otherwise monotonic, long-term trend on which the end-Cretaceous extinction had no effect beyond a temporary perturbation.

If the Danian values represent a resetting of the relative ecological dominance of cyclostomes and cheilostomes at local scales, as suggested by the abrupt, large-scale change at the K-T boundary, the second pattern is suggested. If two separate trend lines are fitted to the data (fig. 12.4A and C), the trend in within-fauna cyclostome dominance through the Late Cretaceous declines steeply. The Cenozoic trend line likewise declines steeply but begins well above the end-Cretaceous values. Moreover, the

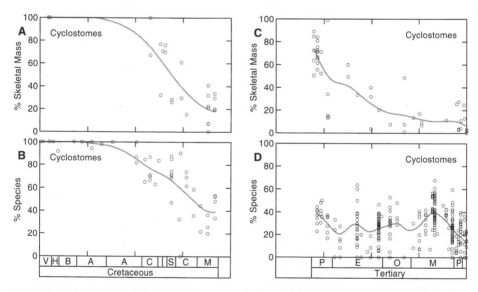

Fig. 12.4. Relative within-fauna skeletal mass of cyclostomes, plotted as percentage of total bryozoan skeletal mass consisting of cyclostomes for the Cretaceous (A; 27 faunas) and the Cenozoic (C; 53 faunas). Each point represents the percentage of cyclostomes in a sample(s) from a single faunal assemblage as given in appendix 12.1. Relative within-fauna species richness of cyclostomes, plotted as a percentage of total bryozoan species richness for the Cretaceous (B; 106 faunas) and the Cenozoic (D; 484 faunas). See figure 12.2 for plotting methods.

trend for the Cenozoic approximates an exponential decay curve. More Thanetian through Bartonian data are needed in order to decide between the two possibilities (or variants of them).

Geographic Correspondence of Data Sets

Time of diversification and rate of recovery from major perturbations may vary with geographic region (Miller 1997; Jablonski 1998), which could potentially generate a spurious decoupling of taxonomic diversity and abundance if the taxonomic data and abundance data were based on different or only partially overlapping geographic regions. In addition, the longer-term trends reported here could be unrelated to one another if based on compilations for different geographic regions. The patterns, even the global patterns, described in this paper are based on information from essentially the same geographic bases.

Mesozoic and Cenozoic bryozoan faunas are best known from Europe and North America, as is clearly reflected in the faunal compilations of Sepkoski (1982 and subsequent unpublished updates). The endpoints of ranges of the vast majority of families and genera are based on these European and North American faunas (Taylor, pers. observation). Therefore, family and genus diversities largely reflect what is known from these regions.

Data on species richness and skeletal masses used in this chapter are also based overwhelmingly on European and North American collections (table 12.2). Only the Miocene data for species-richness proportions have

Table 12.2. Geographic distribution of faunas used for species richness and skeletal mass determinations

Stage	Species Richness (%)		Skeletal Mass (%)	
	Europe	N. America	Europe	N. America
Pleistocene	0	67	0	50
Pliocene	52	42	100	0
Miocene	58	14	100	0
Oligocene	44	56	40	60
Eocene	68	25	13	50
Paleocene	54	31	55	27
Maastrichtian	83	17	55	36
Campanian	100	0	100	0
Santonian	88	12	100	0
Turonian	100	0	100	0
Cenomanian	92	8	100	0
Hauterivian	100	0	100	0
Valanginian	100	0	100	0

Note: Where percentages for Europe and North America do not total 100% for any stage, the residual represents faunas from other regions.

a high proportion of collections from elsewhere, due to the work of El Hajjaji (1992) and Moissette (1988) in North Africa. These North African localities contain bryozoans that lived in the Mediterranean Sea and are closely allied with the even more extensively studied Mediterranean-related bryozoans in southern Europe.

The proportions of European versus North American faunas on which relative species richness and skeletal mass were determined for this study are strongly correlated through the stratigraphic section (Kendall tau = 0.589, r = 0.005, N = 13). Most important, the proportions are closely similar for Maastrichtian and Paleocene (table 12.2), the two units that define the abrupt decoupling of diversity and skeletal mass at the K-T boundary. At a finer geographic scale, data on relative species richness and skeletal mass are drawn from the same local regions within continents. For example, all European Maastrichtian and Danian data are from the Danish Basin and adjacent areas, and equivalent North American data are all from the Atlantic and Gulf coastal plain. Consequently, differences in the diversity and abundance patterns compared in this paper do not appear to be due to using differing geographic bases for different data sets.

Taphonomy Generated Scatter, Not Trends

Cyclostome bryozoans are well skeletonized, and their skeletons are constructed entirely of low- to mid-magnesium calcite (Smith et al. 1992; Smith and Nelson 1993). Most are small relative to the size of cheilostome colonies (McKinney et al. 1998; McKinney 2000).

Cheilostome skeletons also are all mineralized, but they range from low degree of mineralization in some erect, continuously flexible forms (notably cellurines) to rigid or articulated skeletons that are highly mineralized. Over half the well-mineralized cheilostome species that have been studied have skeletons entirely of calcite, about one-third have mixed calcitic and aragonitic skeletons, and fewer than 10% have 100% aragonitic skeletons (Smith et al. 1992; Smith and Nelson 1993). The calcite in cheilostome skeletons contains a greater variation in proportion of magnesium and typically is slightly higher in weight percentage of magnesium than is calcite in cyclostome skeletons (Smith and Nelson 1993).

The mineral composition of cyclostomes makes them on average less likely to dissolve than cheilostomes. In contrast, the smaller typical size of cyclostomes would seem to put them more at risk of dissolving than the typically larger cheilostomes. However, dissolution rates of bryozoan skeletons result from a complex interaction of mineralogy and colony construction, growth habit, and, for erect forms, size of branches; morphology appears to be as important in skeletal durability as is mineralogy (Smith et al. 1992; Smith and Nelson 1993, 1994).

Cyclostome and cheilostome bryozoans have roughly similar growth habits (McKinney and Jackson 1989). Different colony growth habits of bryozoans abrade at different rates in highly kinetic environments (Smith and Nelson 1996), yielding a different suite of most durable growth habits than does dissolution.

Different taphonomic responses of cyclostome and cheilostome bryozoans in different kinetic and chemical environments likely have added some unknown amount of noise to the data presented here (Kidwell and Flessa 1995). Because of the differences discussed above, cyclostomes and cheilostomes may have on average somewhat different rates of degradation within any given environment. However, we are at present unaware of mid-Mesozoic to Holocene changes in mineralogy or growth forms that would have generated a trend in relative skeletal abundance through time, although on average cheilostomes may be consistently underrepresented because of their skeletal mineralogy. If cheilostomes are consistently and proportionally underrepresented in relative skeletal mass, then temporal trends in increase in cheilostome skeletons reflects an even stronger temporal trend in the relative abundance of cheilostomes in living faunas.

Breadth of Ecologies in Cyclostomes and Cheilostomes

The phylum-wide *Bauplan* of bryozoans restricts the diversity of colony morphologies and ecological contexts that individual species have been able to adopt (McKinney and Jackson 1989; McKinney 1991; Hageman et al. 1998). Both cyclostomes and cheilostomes have evolved diverse encrusting and erect colony habits (McKinney 1986a, 1986b, 1991), and, by the Cenomanian, cheilostomes had developed the full range of basic colony habits and encompassed the entire range of environments and microenvironments inhabited by cyclostomes (Jablonski et al. 1997). During the Late Cretaceous, some cheilostomes developed free-living and rooted colony habits that have not been developed by cyclostomes (McKinney and Jackson 1989). Cheilostomes have also invaded less saline waters (<18 ppt) than have cyclostomes (Winston 1977; Mitchell et al. 1997). Despite these ecological innovations of cheilostomes, there is a strong positive correlation between within-fauna species richness of cyclostomes and of cheilostomes through time ($r = 0.77$, based on 315 faunas; Lidgard et al. 1993), which is one indication of the broadly equivalent ecological ranges of the two groups.

Although we have not performed rigorous tests, it appears to us that evolution of cheilostomes into a broader range of ecologies does not significantly affect the data on relative skeletal mass used in this chapter. All faunas for which skeletal mass was determined are normal-salinity shelf deposits, so cyclostomes are not excluded because of reduced salinity. The vast majority of shelf bryozoan species cement themselves to a substratum (McKinney and Jackson 1989), and most of the "rooted" forms, such as the cheilostome genus, *Cellaria,* are attached by rhizoids to solid substrata rather than inhabiting loose sediment as occurs more frequently in deeper waters. Only a trivial proportion of sediment-rooted skeletons were seen, and they were present in only a few faunas included here. Free-living (lunulitiform) cheilostomes were much more widespread within the faunas. If they accounted for only a very small proportion of the cheilostomes (<5%), they were included. If they were relatively more abundant, they were kept separate and not included in the skeletal mass recorded.

Cyclostomes overall may show a greater preference for cryptic environments than do cheilostomes in Holocene seas (e.g., Harmelin 1977; McKinney 2000). Nevertheless, the two clades have occupied the same range of shelf environments and microenvironments from the Cenomanian to the present (except for the addition of free-living and sediment-rooted cheilostomes as noted above). There is even evidence that bryozoan-favorable habitats on shelves result in occupancy by cyclostomes and cheilostomes in comparable proportions across biofacies. In a detailed study of weight percentage of skeletal elements across the core and two different slope conditions of a Danian mound at Limhamn, Sweden, Cheetham (1971) found that cyclostomes and cheilostomes formed the most tightly knit group among all lithic components in cluster analyses, with abundance correlation between the two clades at 0.91 (fig. 12.5).

Global Environmental Changes

Global environmental changes that occurred over the interval of cyclostome to cheilostome transition could conceivably affect taxonomic diversity or skeletal mass differentially in the two bryozoan groups. Several indicators of global paleotemperature show greatly elevated world temperatures during the mid-Cretaceous, with low thermal gradients from equator to poles (fig. 12.6A; Crowley 1991; Spicer and Corfield 1992; Huber et al. 1995; Compton and Mallinson 1996). This interval is also characterized by increased global sea levels and major transgressions, development of broad carbonate platforms, increased ocean-crust formation with volcanism and outgassing from mantle upwellings, and elevated global levels of atmospheric carbon dioxide (Larson 1991; Scott 1995; Compton and Mallinson 1996). These mid-Cretaceous environmental patterns correspond with the initial phase of cheilostome diversification.

Global paleotemperatures exhibit an overall downward trend from a mid-Cretaceous peak to the Recent, punctuated by shorter excursions of rapid temperature decline and increase. While oxygen isotope ratios (Huber et al. 1995; Richards 1997), atmospheric carbon dioxide levels,

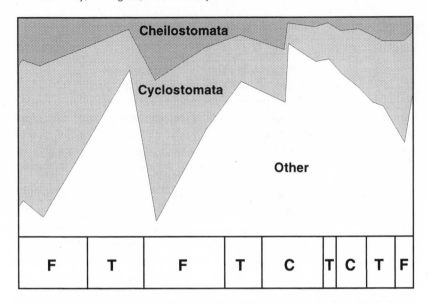

Fig. 12.5. Relative skeletal mass of cyclostomes, cheilostomes, and other (largely biotic) constituents distributed in a quarry wall across a Danian carbonate mound at Limhamn, Sweden, based on the >500-μm fraction. A complex of three facies occurred within the mound, core *(C)*, transition *(T)*, and flank *(F)*. Modified from Cheetham 1971, fig. 16.

and inferred paleotemperature all appear to track the direction of the transition from cyclostome prevalence in taxonomic diversity and dominance in abundance to cheilostomes, more detailed correspondence is unclear. During one interval of higher resolution, however, the rapid paleotemperature decline near the K-T boundary and the interval of exceptionally low organic productivity in the world oceans above the boundary (D'Hondt et al. 1996, 1998) appear to correspond with a rapid increase in relative abundance of cyclostomes (see below).

Hardie (1996) and Stanley and Hardie (1998) have proposed a model linking secular changes in seawater chemistry with trends in mineralogies of marine limestones and of a number of skeletonized marine invertebrates, including bryozoans. The model predicts that, during the interval from the mid-Cretaceous to the Recent, the magnesium/calcium mole ratio of seawater underwent a gradual shift from values favoring the secretion of calcite by "hypercalcifying" organisms that "produced massive skeletons, large reefs, or voluminous bodies of sediment" (Stanley and Hardie 1998, 3) to secretion of high-magnesium calcite and aragonite (fig. 12.6B). Concurrently, calcium concentration in seawater rose to a plateau in early to mid-Cretaceous, followed by gradual decline from latest Cretaceous onward (fig. 12.6C).

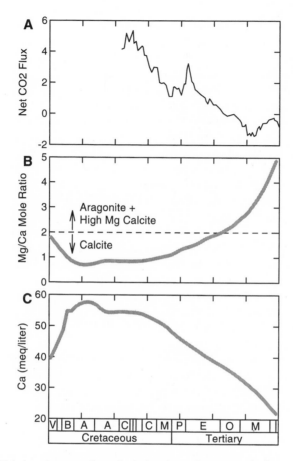

Fig. 12.6. Global environmental trends during the cyclostome-to-cheilostome transition. *A*, Carbon dioxide imbalance between weathering and organic carbon burial rates (sinks) and outgassing from mantle upwelling and oceanic crust formation (sources) over the past 100 million years. Warmer global paleotemperatures are assumed to be related to intervals of increased carbon dioxide imbalance shown here. The overall trend accords reasonably well with paleotemperature estimates derived independently from oxygen isotope ratios derived from Tertiary benthic foraminiferans (e.g., Richards 1997). Figure modified from Compton and Mallinson 1996. *B*, Global trends in magnesium/calcium mole ratio of seawater over the past 140 million years, based on the model presented by Hardie 1996. The *horizontal line* is inferred to represent a division between nucleation fields of high-magnesium calcite and aragonite relative to calcite (see text for discussion). *C*, Global trends in overall calcium concentration in seawater (milliequivalents/liter) over the past 140 million years, based on the model presented by Hardie 1996. Figure C modified from Stanley and Hardie 1998, fig. 1.

The relevance of this model for the present chapter lies in the skeletal mineralogy of cyclostomes versus cheilostomes. Cyclostomes are known to secrete skeletons of low- to moderate-magnesium calcite, whereas those of cheilostomes are moderate-magnesium calcite or aragonite (Rucker and Carver 1969; Sandberg 1971, 1983; Poluzzi and Sartori 1975; Smith and Nelson 1993), predominantly 6 to 9 mole percent magnesium carbonate (Sandberg 1983; Smith and Nelson 1993), which is at the low end of the moderate-magnesium calcite range. The key questions are, (1) can diversity or skeletal mass of bryozoan taxa that secrete skeletons of different mineralogical compositions be enhanced over long spans of geologic time by uptake of calcium and magnesium in varying proportions, by overall calcium concentration, and by paleotemperature, and (2) to what extent are skeletal mineralogies constrained within cyclostomes, cheilostomes, and various subclades of both? The elevated paleotemperatures and calcium concentration in seawater during the mid- to Late Cretaceous and parts of the early Paleogene would seemingly reduce metabolic costs of calcification during the early to mid-Cretaceous cyclostome diversification and the two major phases of cheilostome diversification. However, the decline in relative within-fauna abundance of cyclostomes during the Late Cretaceous, and the cyclostome resurgence in abundance immediately after the end-Cretaceous extinction, are difficult to reconcile with predictions of this model.

The diversification of aragonitic cheilostomes appears to have begun in the Eocene (see discussion above) and continued through the Neogene. Today, aragonitic cheilostomes are relatively more common in the warm-water tropics, and most living cheilostomes studied thus far secrete moderate-magnesium calcite. Relative within-fauna diversities of cyclostomes appear to have declined sharply in the latter part of the Neogene. These observations regarding diversity suggest a possible linkage to the model's prediction of increasing high-magnesium calcite/aragonite skeletal secretion through the Neogene. However, if the conditions predicted by the Hardie model affected the history of post-Paleozoic bryozoans, the effects should be most clearly seen in relative skeletal abundance of cyclostomes and cheilostomes. For the most part, the observed relative abundance trends in the Late Cretaceous and early Paleogene appear at variance with the model.

At present we have insufficient data to evaluate the Hardie model's explanatory role in these bryozoan trends. Skeletal mineralogies have been reported for fewer than 200 bryozoan species (Smith and Nelson 1993), on the order of 2–3% of recognized living species. We know of no actualistic or experimental studies of the mineralogical composition of Recent bryozoan skeletons as a function of ambient seawater temperature, magnesium/calcium ratio, or calcium concentration. Moreover, given the unresolved phylogenetic relationships of subgroups within cyclostomes and cheilostomes, we lack an adequate basis for assessing possible phylogenetic constraints on skeletal mineralogy. This last point is amplified by the occurrence of moderate-magnesium calcite and aragonitic skeletons in two conspecific colonies of the cheilostome *Celleporaria vagans* (Rucker and Carver 1969), and high levels of variance in magnesium carbonate reported within many species (Smith and Nelson 1993). Thus, we conclude that, for the history of post-Paleozoic bryozoans, the relevance of these secular environmental trends in seawater chemistry, while tantalizing, is uncertain.

Ambiguous Indicators of Relative "Success"

Cheilostome bryozoans are a more energy-intensive group of organisms than are cyclostomes. Cheilostomes generate higher feeding current velocities, can feed on larger particles, have larger zooids, grow faster, and appear to invest more energy in reproduction relative to cyclostomes (McKinney 1993). Competitive overgrowth success of encrusting organisms has been demonstrated to correspond with ability to deprive competitors of food (Buss 1980; Okamura et al., this volume), higher growth rate (Buss 1981; Winston and Jackson 1984), and ability to overtop the competitors (Buss 1981; Walters and Wethey 1986).

Colonies of typical cheilostome species that encounter colonies of typical cyclostome species would therefore be predicted to be competitively superior and to win a majority of the overgrowth encounters. This has been demonstrated to be true for encounters between live cheilostomes and cyclostomes that encrust shell substrata in the northern Adriatic Sea, where cheilostomes were found to overgrow cyclostomes in 78% of the encounters, whereas cyclostomes overgrew cheilostomes in only 8% of the encounters (the remaining 14% were stalemates; McKinney 1992). From the mid-Cretaceous through the Cenozoic, the fossil record documents consistently prevalent overgrowth of cyclostomes by cheilostomes at a remarkably constant rate consistent with that seen in the northern Adriatic (fig. 12.7; McKinney 1995). Therefore, one might predict an increase in cheilostomes relative to cyclostomes through time, if direct interactions between the two groups were the only mediating force, isolated from all other biotic interactions during all life-history stages (especially larval) and from abiotic environmental factors that affect one group more than another.

The direct evidence for competition between cyclostome and cheilostome bryozoans is based on interactions between encrusting species, but the diversities and skeletal masses used in this chapter include both encrusting and erect bryozoans. Erect bryozoans constituted about half the species during the Cretaceous, although that proportion has declined to

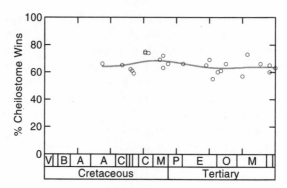

Fig. 12.7. Percentage of cheilostome wins as measured by overlap of colonies where encrusting cyclostome and cheilostome colonies are in contact on hard substrates, mid-Cretaceous to Holocene. The *trend line* was fitted using a distance-weighted least squares method (Wilkinson 1990). Figure modified from McKinney 1995, fig. 2.

about 25% in the Recent (McKinney and Jackson 1989; Jackson and McKinney 1990). Little is known about competition among erect or between erect and encrusting bryozoans. Upstream colonies of an erect cheilostome have been demonstrated to diminish feeding success of downstream conspecifics, and smaller colonies of the species show greater reduction of feeding success in flowing water than do larger colonies (Okamura 1984). Erect cyclostomes are generally smaller and have lower pumping velocities than do cheilostomes (McKinney 1993); consequently we have a working hypothesis that, where competition occurs between erect cyclostomes and cheilostomes, it would normally be more deleterious to the cyclostomes. At present, however, there are no direct data for interactions between erect cyclostomes and cheilostomes, and our inference of persistent competitive disadvantage of cyclostomes is based on data derived solely from encrusting forms.

Measured by diversification, cheilostomes have been much more successful than have cyclostomes. Tracking global diversity of families through time, cyclostomes gradually diversified from Late Triassic through Late Cretaceous, with a relatively brief plateau during the Late Jurassic, but the number of families declined slightly at the K-T boundary and has remained relatively stable to the present (fig. 12.1C). The slight Late Cretaceous to early Cenozoic decline in cyclostome family diversity occurred during an extended interval of vigorous cladogenesis in cheilostomes at the family level, which continued without interruption across the K-T boundary and began to accelerate during the early Paleogene (fig. 12.1D).

In addition, the modest decline in within-fauna species richness of cyclostomes since their mid–Late Cretaceous peak (fig. 12.2A) might in part be attributed to an increase in within-fauna species richness of the competitively superior, ecologically similar cheilostomes, which increased from less than 5 in the Cenomanian to a mean value of over 20 by the Maastrichtian (fig. 12.2B; Lidgard et al. 1993; McKinney et al. 1998). The drop in species richness of cyclostomes, which have never recovered appreciably, began and then reached its nadir during the period of rapid increase in within-fauna species richness of cheilostomes. This "replacement" of cyclostomes by cheilostomes had stabilized by the Oligocene, with cheilostomes constituting nearly 80% of the species richness in the vast majority of Oligocene to Holocene bryozoan faunas; the K-T boundary is not even seen as an inflection in the steady increase in proportion of within-fauna cheilostome species from late Early Cretaceous to Oligocene (figs. 12.2D, 12.3C and D.)

The K-T boundary is, however, reflected globally at the genus level in both cyclostomes and cheilostomes (fig. 12.1A and B; Viskova 1995; McKinney et al. 1998). The conspicuous decrease in number of cyclostome genera into the Paleocene is perceptibly reflected in the very slightly decreased number of cyclostome families. But in the cheilostomes, decrease in number of genera across the K-T boundary must have been more than compensated for by rapid morphological divergence in cheilostomes that generated new families on both sides of the K-T boundary. The lack of correspondence between global diversity of genera and families for cyclostomes and especially for cheilostomes across the K-T boundary stands in conspicuous contrast with the expectation of close correspondence of diversity patterns of families and genera from the much larger sample size of many phyla combined in the overall marine fauna (Sepkoski 1997). Throughout the Phanerozoic, this combined marine family diversity mirrors genus diversity, although scaled differences in amplitude for periods of diversification versus extinction are greater for genera than for families. Sepkoski (1997) explains the similar generic and family diversity, but more accentuated generic curve, as a consequence of the hierarchical nesting of genera within families so that a family is present within an interval of time regardless of whether it is represented by a single or multiple genera at that time. Thus, this general pattern of conformity for cheilostomes, but not for cyclostomes, across the K-T boundary illustrates that the statistical expectation based on a large number of marine families is not necessarily met by all marine taxa when they are examined in isolation.

The Paleocene decrease in number of global cyclostome taxa is probably better attributed to the end-Cretaceous extinction than to direct competition from cheilostomes. However, the lack of recovery of genus-level diversity in cyclostomes following the end-Cretaceous extinction is consistent with a coupled logistic model of clade interaction (Sepkoski et al.

2000) and may be the result of a switch in preemptive ecological roles with cheilostomes. Niches or suites of niches that cyclostomes held before the end-Cretaceous extinction may have been eliminated by the extinction event, and similar niches may then have been filled by the rapidly diverging cheilostomes. Cheilostomes seemingly became the ecological incumbents (Rosenzweig and McCord 1991) that have prevented a resumption of the stately radiation of cyclostomes.

Implications of Relative Skeletal Abundance

Assessment of relative skeletal masses of cyclostomes and cheilostomes presents a radically different record of the effects of the end-Cretaceous extinction. The abrupt increase from an average of 21% of the bryozoan skeletal mass contributed by cyclostomes in the Maastrichtian to about 70% in the Danian clearly shows that the end-Cretaceous extinction had a profound effect on the relative ecological dominance of the two bryozoan clades. Cyclostomes abruptly became the dominant bryozoans in biomass at the K-T boundary, even though (1) the number of species within faunas relative to cheilostomes continued to decrease, (2) the global generic diversity of the two clades declined in parallel, and (3) the number of cyclostome families dropped slightly while the number of cheilostome families increased substantially. Long-term ecological success measured in biomass is in this instance abruptly decoupled from evolutionary success measured in numbers of taxa.

The two types of measures—skeletal mass and taxic diversity (in particular, within-fauna species richness)—are not always as independent of one another as they were for bryozoans across the K-T boundary. During the Cretaceous, the two measures changed in the same direction. Note, however, that from near-zero cheilostomes during Early Cretaceous the curve for relative skeletal mass was steeper (ending at ~20% at the K-T boundary) through the Late Cretaceous than was the curve for relative within-fauna species richness, which ended at about 40% at the K-T boundary (fig. 12.4A and B). In this case, the two types of measures could be used as approximate surrogates for one another; both indicate an increase in cheilostomes relative to cyclostomes, although skeletal mass was increasing at a greater rate. The discrepancy in the rate of increase in numbers of species relative to skeletal mass must have been due to greater number of colonies per species of cheilostomes, to larger colony size for cheilostomes, or to an interplay of the two. The discrepancy in rate does not in itself indicate that the cheilostomes were ecologically displacing the cyclostomes, because the cheilostomes could have been (and were, at least until late in the Cretaceous) adding on to overall bryozoan diversity and skeletal abundance within habitats (Lidgard et al. 1993).

During the Cenozoic, as the cheilostomes have rebounded in relative skeletal mass from the Danian low (inverse of fig. 12.4C), that pattern has proceeded at a much steeper rate than increase in within-fauna species richness of cheilostomes relative to cyclostomes (fig. 12.4D). Slowly, cheilostomes recovered from their Danian low in ecological abundance following the end-Cretaceous extinction and have risen to overwhelming dominance over cyclostomes in most late Cenozoic and Holocene faunas. Danian bryozoan faunas had on average about 1.5 times as many cheilostome species as cyclostomes, yet cheilostomes constituted only 30% of the skeletal mass. Thus, on average, each cheilostome species contributed only one-fourth the skeletal mass of a typical cyclostome species. In Holocene faunas, there are about four times as many strongly calcified cheilostome species as cyclostomes, and they contribute about 97% of the skeletal mass, so, at present, each cheilostome species contributes an average of eight times the skeletal mass of a cyclostome species. If one were to choose skeletal mass as an exclusive measure of dominance, on average a cheilostome species is now 32 times more "important" ecologically with respect to the average cyclostome species than during the Danian!

The proliferation in abundance of cyclostomes following the K-T extinction suggests that they responded ecologically to the extinction event in a very different way than did the cheilostomes, which—while surviving well in taxonomic diversity—became relatively less abundant. An abrupt decrease in nannofossil production at the K-T boundary initiated a 2-million-year interval characterized by low nannofossil production (D'Hondt et al. 1996). In addition, excursions in $^{13}C/^{12}C$ ratios between planktic and benthic foraminifera, and between infaunal and epifaunal benthic foraminifera, began at the K-T boundary and lasted at least 500,000 years (Stott and Kennett 1990; D'Hondt et al. 1996). These measures, plus the preferential survival of deposit-feeding macrobenthos over suspension-feeding benthos at the K-T boundary (Sheehan and Hansen 1986; Rhodes and Thayer 1991; Sheehan et al. 1996), indicate that there was an abrupt decline in primary productivity (i.e., phytoplankton) at the K-T boundary. Recovery from this decline was unexpectedly slow, and normal open ocean carbon flux may have been disrupted for up to 3 million years (D'Hondt et al. 1998).

The change in primary productivity is one possible cause for the differing responses of cyclostomes and cheilostomes to the end-Cretaceous extinction event. Rhodes and Thayer (1991) argued that, for mass extinctions caused by reduced primary productivity, extinctions should be concentrated among animals with starvation-susceptible feeding, greater activity, and high energy budgets. This was based on an evaluation of articulate brachiopods, which have low energy budgets relative to suspension-feeding bivalves. Brachiopods had comparatively higher extinction rates than did suspension-feeding bivalves during the Permian-Triassic extinction and during the Jurassic, yet during the K-T extinction event, rates of extinction of articulates and suspension-feeding bivalves were equal. Rhodes and Thayer (1991) attributed the K-T pattern to the advantage that articulates have for survival when food is scarce, and organisms with greater energy requirements are more susceptible to extinction.

As noted earlier, cheilostome bryozoans have a substantially higher energy budget than do cyclostomes (McKinney 1993). If the Rhodes-Thayer hypothesis is applicable here, then cheilostomes should have been more adversely affected than cyclostomes during the end-Cretaceous extinction. There is no convincing evidence that a larger proportion of cheilostome than cyclostome species became extinct at or immediately following the K-T boundary. Study of bryozoans below and above the K-T boundary in the Nye Kløv section found that, of 115 cheilostome species found within the exposure, only 11 occurred both above and below the boundary beds (Håkansson and Thomsen 1979). Håkansson and Thomsen's study (1979) constitutes a detailed documentation of the sequence of environments of deposition and of cheilostome life habits across the K-T boundary at a single locality, and some additional Maastrichtian cheilostome species are known from other Danian localities in Scandinavia (MacLeod et al. 1997). To get a dependable approximation of extinction of Scandinavian cheilostome species at the K-T boundary, many more sections above and below the boundary in other regions require study to even out local environmental fluctuations that so strongly affect occurrences of benthic species. But differential extinction at the species level is not supported on a more global level, because there is no increase in relative number of cyclostome species in local faunas above the K-T boundary (fig. 12.2D).

Relative skeletal abundance shows a much more profound negative effect on cheilostomes than on cyclostomes. Reversal from cheilostome to cyclostome dominance at the K-T boundary may well have resulted from a greater stress on cheilostomes than on cyclostomes during the ensuing period of low primary productivity. Cheilostomes did not become proportionally less species-rich in local communities than did cyclostomes across the K-T boundary, but their population densities were devastated. One would predict that such severe reduction in abundance would correspond with a decrease in species richness, but it does not.

We had not expected the decoupling of taxonomic and ecological response of cheilostomes and cyclostomes so that they did not respond with parallel patterns at the K-T boundary. Other taxa, including some ferns, stromatolites, the bivalve *Claraia,* and some planktic foraminiferans, are found in abrupt but short-lived high abundance immediately after mass extinctions (Tschudy et al. 1984; Wolfe and Upchurch 1986; Schubert and Bottjer 1992; Fowell and Olsen 1993; Fowell et al. 1994) and have been tagged as "disaster taxa" or—for the stromatolites—"disaster forms."

The abrupt increase in relative abundance of cyclostomes above the K-T boundary suggests their possible proliferation as disaster taxa. However, disaster taxa described elsewhere have low diversity and are abundant for relatively short periods of time. One would predict that following the interval of low primary productivity (maximum of 2 million years, according to D'Hondt et al. 1996), cheilostome population sizes would recover quickly. This did not happen, based on our middle and late Danian data, and cyclostome species-level diversities in our lower Danian samples are not noticeably different from earlier and later samples.

Other ecological disturbances following mass extinctions, from the blooming of disaster taxa to the disruption of predator-prey relationships (Kelley and Hansen 1996), lasted at most about 4 million years. The pattern of abrupt increase in cyclostome skeletal mass relative to cheilostome mass is either similar to the most long-lived of the other mass extinction–induced ecological disturbances, or it represents a resetting of the trend of increased cheilostome ecological dominance over cyclostomes back to mid-Cretaceous levels, with a recovery that spanned tens of millions of years. In either case, the ecological disruption can be seen only by quantitative study of relative abundance of skeletal remains, and it is completely masked where only taxonomic information is considered.

Conclusions

The macroevolutionary history of post-Paleozoic bryozoans cannot be characterized fully with only one class of data (McKinney et al. 1998) as with some other groups (Roy 1996; Lupia et al. 1999). In this chapter, we have compared four different databases for post-Paleozoic bryozoans: global family diversity, global genus diversity, within-fauna species richness, and within-fauna relative skeletal abundance of the two major clades. The only patterns that correspond consistently with one another are global family and global genus diversity of cyclostomes. With this one exception, several patterns exhibited by individual databases were not shared by others.

Species richness and global diversity patterns carry only a portion of evolutionary patterns, and they may not correspond with other important aspects of the ecological history of the biosphere. While counts of fossil taxa are useful for demonstrating many types of trends, they are not always adequate proxies for abundance data. The profound ecological devastation of cheilostome bryozoans and resurgence of cyclostomes at the K-T extinction event cannot be seen in any of the three taxon-based databases (although data available are too imprecise to test for high global extinction rate of cheilostome *species* at the K-T boundary).

Jackson's criticism (1994) of the dependence on species lists in many studies of ecological communities also applies to the interface between macroevolution and evolutionary macroecology. We think that, while taxonomic lists convey incisive facts, have been the basis for important advances in the understanding of the history of life, and should certainly be retained as valuable analytical tools, they should be supplemented with various kinds of information, particularly relative abundance. We will hold a limited and sometimes misleading view of the history of life if our concept is based only on taxonomic diversity and not on measures of abundance as well, which give an indication of prevalence of the studied taxa within evolving ecosystems.

Acknowledgments

For information, guidance, and assistance in making field collections, we thank Giampietro Braga, Jon Bryan, Rossella Capozzi, Stanislav Cech, David T. Dockery, III, Andy Gale, Eckart Håkansson, Elizabeth Harper, William B. Harris, J. W. M. Jagt, Andrej Jaklin, Jiří Kříž, Marjorie McKinney, Timothy J. Palmer, Angelo Poluzzi, Simone Pouyet, Erik Thomsen, Norbert Vavra, Bernard Walter, and Fred Webb, Jr. We thank Jon Bryan, Roger Cuffey, Tatyana Favorskaya, and Steven J. Hageman for providing supplementary bulk samples. We benefited from discussions with S. J. Hageman, R. Lupia, and J. J. Sepkoski, Jr. A. I. Miller's and R. K. Bambach's readings of the manuscript resulted in substantial improvements. FKM's research was supported by the National Science Foundation (NSF EAR 9117289), a U.S.-U.K. Fulbright research scholarship, a visiting fellowship at Wolfson College, University of Cambridge, and the National Geographic Society. SL's research was also supported by the NSF (DEB 9306729). PDT's contribution is part of the Global Change and the Biosphere Programme of the Natural History Museum and University College, London.

Appendix 12.1: Locality Register

Bathonian
1. Caillasse de Blainville (basal upper Bathonian, aspidoides zone), approximately 100 m south of pedestrian bridge, west side of highway D515, Blainville, Normandy, France. 2. Caillasse de la Basse-Ecarde

(upper Bathonian, distans zone), cliffs along beach extending west from center of Saint Aubin-sur-Mer, Normandy, France.

Valanginian

1. Marnes à bryozoaires (uppermost Valanginian), outcrop along highway D119, just west of bridge, 0.75 km due west of Nozeroy, northern Jura, France.

Hauterivian

1. Marne d'Hauterive (lower Hauterivian), penultimate switchback below plateau on road between Confort and Mentieres, near Bellgarde, Jura, France.

Cenomanian

1. Approximately 150 m south of Port petrolier du Havre-Antifer at Saint Jouin Bruneval, Normandy, France. 2. Sandy facies ("Wilmington Sands") of Beer Head Limestone, White Hart Sandpit, Wilmington, Devon, England (ST208999).

Turonian

1. Lower Turonian carbonate sands and conglomerates in the third of several small contiguous quarries at Kank, by Kutná Hora, Czech Republic. 2. Upper Turonian chalk, approximately 200 m north of access, beach at Tilleul, near Etretat, Normandy, France.

Turonian-Coniacian

1. White chalk at approximate Turonian-Coniacian boundary, beach at Etretat, Normandy coast, France.

Coniacian

1. Hope Gap, near Cuckmere Haven, Seaford, Sussex, England (TV 509937). 2. Beach at Vaucottes, southwest of Fécamp, Normandy, France.

Santonian

1. Chalk at beach, Sotteville-sur-Mer, east of Saint Valery-en-Caux, Normandy, France. 2. Chalk at beach, Vasterival, west of Varengeville-sur-Mer, Normandy, France.

Campanian

1. Carbonate grainstone (upper lower Campanian), Svenske Kalk Quarry, 1.5–3.5 km southeast of Nya Ingaberge, west side of highway E21, Scania, Sweden. 2. Carbonate packstone (upper Lower Campanian), Ivö Klack, north end of Ivö island, east of Kristianstad, Sweden. 3. Carbonate grainstone midway along northwest side of Pointe de Vallières, Royan, France.

Maastrichtian

1. Meerssen Chalk, approximately 2.5 m above Caster Horizon, ENCI Quarry, approximately 4 km south of center of Maastricht, Netherlands. 2. Meerssen Chalk, NEKAMI Quarry, 't Rooth-Margraten, Bemelen, Netherlands. 3. Upper Maastrichtian, approximately 30 cm

below boundary clay, Nye Kløv, northwestern Jutland, Denmark.
4. Northeast end of Rørdal Ålborg Portland Cement Quarry, Ålborg,
Denmark. 5. Southwest wall of Rørdal Ålborg Portland Cement
Quarry, Ålborg, Denmark. 6. Ellidshøj Kalk Quarry, south of Ålborg,
Denmark. 7. Peedee Formation, west bank of northeast Cape Fear
River, approximately 0.3 km east of NC highway 53, near Burgaw,
North Carolina. 8. Peedee Formation, west bank of Peedee River at
Burches Ferry, east end of Mill Branch Road, east of SC highway 57,
southeast of Florence, South Carolina. 9. Peedee Formation, bluff on
right bank of Neuse River, 35°20'01" N, 77°26'31" W, Ayden Quadrangle, Lenoir County, North Carolina (USGS Sohl locality 32192).
10. Prairie Bluff Chalk, Lake Ridge Apartments, NW1/4 NE1/4 sec.
28, R2W, T19N, Livingston Quadrangle, Sumter County, Alabama.
11. Igoda Formation, Needs Camp, approximately midway between
Kingwilliamstown and East London, South Africa.

Danian

1. Bryozoan chalk (Thomsen coccolith zone 2), approximately 1.3 m
above K-T boundary clay, Dania Quarry, north of Assens, on south
side of Mariager fjord, Denmark. 2. Bryozoan chalk (Thomsen coccolith zone 2), approximately 2 m above K-T boundary clay, Dania
Quarry, north of Assens, on south side of Mariager fjord, Denmark.
3. Bryozoan chalk (Thomsen coccolith zone 2), second promontory
north from south end of sea cliff, Karlby Klint, Denmark. 4. Chalk
(Thomsen coccolith zone 3), Hammelev Quarry near Sangstrup, Denmark. 5. Chalk (Thomsen coccolith zone 4), Hammelev Quarry near
Sangstrup, Denmark. 6. Chalk (Thomsen coccolith zone 8), 0–1.5 m
below Selandian clays, wall of small quarry within waste dump complex, Klintholm, south of Nyborg, Denmark. 7. Clayton Formation,
roadcut on west side of AL highway 263, 8 km south of intersection
with AL highway 21, Lowndes County, Alabama.

Thanetian

1. Vincentown Formation, 150 m west of Snuff Mill Road, across from
New Egypt Middle School, New Egypt, New Jersey. 2. Vincentown
Formation, embankment on north side of South Branch of Rancocas
Creek, approximately 1 km north of center of Vincentown, New Jer-
sey (bulk sample provided by Roger Cuffey). 3. Sullukapy Formation
(= Sullukapinskaya Formation), Sullukapy, Mangyshlak, Kazakhstan
(bulk sample provided by Tatyana Favorskaya). 4. Sullukapy For-
mation (= Sullukapinskaya Formation), Chakyrgan, Mangyshlak,
Kazakhstan (bulk sample provided by Tatyana Favorskaya). 5. Salt
Mountain Limestone, Salt Mountain, county highway 15, sec. 33,
T6N, R2E, Clarke County, Alabama (lithified sample provided by Jon
Bryan).

Ypresian-Lutetian

1. Matanginui Limestone, Kekerione Group, quarry near Lake Mara-kapai, Chatham Island, New Zealand. 2. Matanginui Limestone, Kekerione Group, Rocky Side (689226), Tarawhenua Peninsula, Pitt Island, Chatham Islands, New Zealand.

Lutetian

1. Castle Hayne Limestone, sequence 1 of Zullo and Harris 1987, Fussell Quarry, Rose Hill, Duplin County, North Carolina.

Bartonian

1. Moodys Branch Formation, Jackson Group, Fossil Gulch, Le Fleurs Bluff State Park, Jackson, Mississippi. 2. Castle Hayne Limestone, sequence 2 of Zullo and Harris 1987, quarry on north side of NC highway 53, approximately 1.5 km east of intersection with NC highway 50 near Maple Hill, Pender County, North Carolina. 3. Castle Hayne Limestone, sequence 2 of Zullo and Harris 1987, Fussell Quarry, Rose Hill, Duplin County, North Carolina.

Priabonnian

1. Slope, approximately 50 m above Museum of Priabonna, Priabonna, Italy.

Rupelian

1. Echinoid Limestone, entrance to Chateau Pavie caves, Saint Emilion, France.

Chattian

1. Mint Spring Formation, deeply incised creek on east side of MS highway 3, 0.6 km south of entrance to International Paper Company plant, NE1/4 NW1/4 sec. 26, T18N, R4E, Floweree 7.5′ quadrangle, Warren County, Mississippi. 2. Road into International Paper Company plant, 0.9 km from entrance off MS highway 3, approximately 60 km south of intersection with MS highway 49, NE1/4 SE1/4 sec. 23, T18N, R4E, Floweree 7.5′ quadrangle, Warren County, Mississippi. 3. Byram Formation, excavation into bank at Keyes Iron and Metal, US 61 Business, 2 km south of intersection with US 61 Bypass, north side of Vicksburg, Mississippi. 4. Col del'Asse outcrops, Possagno, Treviso province, Venetia, Italy.

Aquitanian

1. Approximately 3-m-high exposure along road between Sigmund-sherberg and Brugg, near Eggenburg, Austria. 2. *Ostrea aquitanica* bank, 50 m SSW of church, Sainte-Croix-du-Mont, near Cadillac, France.

Lower Miocene

1. Shale at base of Calcarenite di Castelcucco, roadcut immediately north of gap at Bocca di Serra, near Possagno, 45°50′56″ N, 11°55′32″ E, Italy.

Burdigalian

1. Approximately 20 m below top of Burdigalian limestone, along road approximately 1 km north of Baux, France.

Tortonian

1. Savigneyan facies, preserved quarry 0.5 km south of Channay-sur-Lathan, Touraine, France.

Lower Pliocene (Zanclean)

1. "Spungone" (*Globorotalia puncticulata* zone), quarry near hilltop on Strada per Bagnolo, Castrocaro, Italy.

Pliocene

1. Coralline Crag, Broom Pit near Orford, Suffolk, England (TM467499), biofacies 1. 2. Coralline Crag, Broom Pit near Orford, Suffolk, England (TM467499), biofacies 2. 3. Coralline Crag, Crag Farm Pit, northeast of Orford, Suffolk, England (TM428523). 4. Yorktown Formation, Jacks Creek, Washington, North Carolina.

Upper Pliocene (Piacenzian)

1. Clay-rich sands, Parco Fluviale di Stirone, along riverbank near S. Sicomede Church near Fidenza, Italy. 2. Mudstone along right bank of Arda River, 50 m downstream from bridge at Castell'Arquato, Italy.

Pleistocene

1. Petane Limestone, Devil's Hollow, on highway 2 approximately 15 km north of Napier, New Zealand (V20 458079). 2. Waccamaw Formation, abandoned quarry at Old Dock, Columbus County, North Carolina.

Holocene

1. 1–2-m depth, Tar Landing bay, Bogue Sound between Morehead City and Bogue Banks, North Carolina. 2. 35-m depth, at base of escarpment on west side of Banjole Island, Rovinj, Croatia. 3. 44–61-m depth, 0.3 km NNW-NW of Little San Giovanni Island, Rovinj, Croatia. 4. 21-m depth, approximately 1 km north of Figarole Island, Rovinj, Croatia. 5. 225-m depth, Lacepede Shelf edge, approximately 137° E, 37° S, south coast of Australia (bulk sample provided by Steven J. Hageman).

References

Aberhan, M. 1994. Guild-structure and evolution of Mesozoic benthic shelf communities. *Palaios* 9:516–45.

Ausich, W. I. 1981. Biovolume revisited: A relative diversity index for paleoecological analyses. *Ohio Journal of Science* 81:268–74.

Banta, W. C. 1975. Origin and early evolution of cheilostome Bryozoa. *Documents des Laboratoires de Géologie de la Faculté des Sciences de Lyon, hors série* 3:565–82.

Banta, W. C., F. M. Perez, and S. Santagata. 1995. A setigerous collar in *Membranipora chesapeakensis* n. sp. (Bryozoa). *Invertebrate Biology* 114:83–88.

Berglund, B. E., H. J. B. Birks, M. Ralska-Jasiewiczowa, and H. E. Wright, eds. 1996. *Palaeoecological events during the last 15,000 years.* Chichester: Wiley.

Berthelsen, O. 1962. Cheilostome Bryozoa in the Danian deposits of east Denmark. *Danmarks Geologiske Undersøgelse* 83:1–290.

Bizzarini, F., and G. Braga. 1981. Prima segnalazione del genere *Stomatopora* (Bryozoa: Cyclostomata) new Trias Superiore delle Dolomite Orientali (Italia). *Società Veneziana di Scienze Naturali Lavori* 6:135–44.

———. 1985. *Braiesopora voigti* n. gen. n. sp. (cyclostome bryozoan) in the S. Cassiano Formation in the eastern Alps (Italy). In *Bryozoa: Ordovician to Recent,* ed. C. Nielsen and G. P. Larwood, 25–33. Fredensborg: Olsen and Olsen.

Boardman, R. S. 1984. Origin of the post-Triassic Stenolaemata (Bryozoa): A taxonomic oversight. *Journal of Paleontology* 58:19–39.

Brood, K. 1972. Cyclostomatous Bryozoa from the Upper Cretaceous and Danian in Scandinavia. *Stockholm Contributions in Geology* 26:1–464.

Buss, L. W. 1980. Bryozoan overgrowth interactions: The interdependence of competition for space and food. *Nature* 281:475–77.

———. 1981. Mechanisms of competition between *Onychocella alula* (Hastings) and *Antropora tincta* (Hastings) on an eastern Pacific rocky shoreline. In *Recent and fossil Bryozoa,* ed. G. P. Larwood and C. Nielsen, 39–49. Fredensborg: Olsen and Olsen.

Cheetham, A. H. 1963. *Late Eocene zoogeography of the eastern Gulf Coast region. Geological Society of America, Memoir* 91. New York: Geological Society of America.

———. 1971. *Functional morphology and biofacies distribution of cheilostome Bryozoa in the Danian stage (Paleocene) of southern Scandinavia.* Smithsonian Contributions to Paleobiology no. 6. Washington, DC: Smithsonian Institution Press.

———. 1986. Branching, biomechanics, and bryozoan evolution. *Proceedings of the Royal Society, London,* series B, 228:151–71.

Cheetham, A. H., L. C. Hayek, and E. Thomsen. 1981. Growth models in fossil arborescent cheilostome bryozoans. *Paleobiology* 7:68–86.

Collins, L. S. 1989. Evolutionary rates of a rapid radiation: The Paleogene planktic Foraminifera. *Palaios* 4:251–63.

Compton, J. S., and D. J. Mallinson. 1996. Geochemical consequences of increased late Cenozoic weathering rates and the global CO_2 balance since 100 Ma. *Paleoceanography* 11:431–46.

Crowley, T. J. 1991. Past CO_2 changes and tropical sea surface temperatures. *Paleoceanography* 6:387–94.

Foote, M. 1993. Contribution of individual taxa to overall morphological disparity. *Paleobiology* 19:403–19.

———. 1995. Morphological diversification of Paleozoic crinoids. *Paleobiology* 21:273–99.

Fowell, S. J., B. Cornet, and P. E. Olsen. 1994. Geologically rapid Late Triassic extinctions: Palynological evidence from the Newark Supergroup. *Geological Society of America Special Paper* 288:197–206.

Fowell, S. J., and P. E. Olsen. 1993. Time calibration of Triassic/Jurassic micro-floral turnover, eastern North America. *Tectonophysics* 222:361–69.

Hageman, S. J., P. E. Bock, Y. Bone, and B. McGowran. 1998. Bryozoan growth habits: Classification and analysis. *Journal of Paleontology* 72:418–36.

El Hajjaji, K. 1992. Les bryozoaires du miocène, supérieur du Maroc nord-oriental. *Documents des Laboratoires de Géologie Lyon* 123:1–355.

Håkansson, E., and E. Thomsen. 1979. Distribution and types of bryozoan communities at the boundary in Denmark. In *Cretaceous-Tertiary boundary events: 1, The Maastrichtian and Danian of Denmark,* ed. T. Birkelund and R. G. Bromley, 78–91. Copenhagen: University of Copenhagen.

Hardie, L. A. 1996. Secular variation in seawater chemistry: An explanation for the coupled secular variation in the mineralogies of marine limestones and potash evaporites over the past 600 m.y. *Geology* 24:279–83.

Harland, W. B., R. L. Armstrong, A. V. Cox, L. E. Craig, A. G. Smith, and D. G. Smith. 1990. *A geologic time scale 1989.* Cambridge: Cambridge University Press.

Harmelin, J.-G. 1977. Bryozoaires des Iles d'Hyères: Cryptofaune bryozo-ologique des valves vides de *Pinna nobilis* rencontrées dans les herbiers de posidonies. *Travaux Scientifiques du Parc national de Port-Cros* 3:143–57.

Hart, M. B., ed. 1996. Biotic recovery from mass extinction events. *Geological Society Special Publication* 102:1–392.

Hayek, L. C., and M. A. Buzas. 1997. *Surveying natural populations.* New York: Columbia University Press.

D'Hondt, S., P. Donaghay, J. C. Zachos, D. Luttenberg, and M. Lindinger. 1998. Organic carbon fluxes and ecological recovery from the Cretaceous-Tertiary mass extinction. *Science* 282:276–79.

D'Hondt, S., T. D. Herbert, J. King, and C. Gibson. 1996. Planktic foraminifera, asteroids, and marine production: Death and recovery at the Cretaceous-Tertiary boundary. *Geological Society of America Special Paper* 307:303–17.

D'Hondt, S., and G. Keller. 1991. Some patterns of planktic foraminiferal assemblage turnover at the Cretaceous-Tertiary boundary. *Marine Micropaleontology* 17:77–118.

Huber, B. T., D. A. Hodell, and C. P. Hamilton. 1995. Middle-Late Cretaceous climates of the southern-high latitudes: Stable isotopic evidence for minimal equator to pole thermal gradients. *Bulletin of the Geological Society of America* 107:1164–91.

Jablonski, D. 1998. Geographic variation in the molluscan recovery from the end-Cretaceous extinction. *Science* 279:1327–30.

Jablonski, D., S. Lidgard, and P. D. Taylor. 1997. Comparative ecology of bryo-zoan radiations: Origin of novelties in cyclostomes and cheilostomes. *Palaios* 12:505–23.

Jackson, J. B. C. 1994. Community unity? *Science* 264:1412–13.

Jackson, J. B. C., and A. H. Cheetham. 1990. Evolutionary significance of morphospecies: A test with cheilostome Bryozoa. *Science* 248:579–82.

Jackson, J. B. C., and F. K. McKinney. 1990. Ecological processes and progressive macroevolution of marine clonal benthos. In *Causes of evolution,* ed. R. M. Ross and W. D. Allmon, 173–209. Chicago: University of Chicago Press.

Kauffman, E. G., and D. H. Erwin. 1995. Surviving mass extinctions. *Geotimes* 40 (3): 14–17.

Kelley, P. H., and T. H. Hansen. 1996. Recovery of the naticid gastropod predator-prey system from the Cretaceous-Tertiary and Eocene-Oligocene extinctions. *Geological Society Special Publication* 102:373–86.

Kidwell, S., and K. Flessa. 1995. The quality of the fossil record: Populations, species, and communities. *Palaios* 26:269–99.

Kitchell, J. A., D. L. Clark, and A. M. Gombos, Jr. 1986. Biological selectivity of extinction: A link between background and mass extinction. *Palaios* 1:504–11.

Krause, D. W. 1986. Competitive exclusion and taxonomic displacement in the fossil record: The case of rodents and multituberculates in North America. *Contributions to Geology, University of Wyoming, Special Paper* 3:95–117.

Larson, R. L. 1991. Geological consequences of superplumes. *Geology* 19:963–66.

Lidgard, S., and G. A. Buckley. 1994. Toward a morphological species concept in cheilostomates: Phenotypic variation in *Adeonellopsis varraensis* (Waters). In *Biology and palaeobiology of bryozoans*, ed. P. J. Hayward, J. S. Ryland, and P. D. Taylor, 101–5. Fredensborg: Olsen and Olsen.

Lidgard, S., and P. R. Crane. 1990. Angiosperm diversification and Cretaceous floristic trends: A comparison of palynofloras and leaf macrofloras. *Paleobiology* 16:77–93.

Lidgard, S., F. K. McKinney, and S. J. Hageman. 1997. Species-rich bryozoan gardens of the post-Paleozoic. *Geological Society of America Abstracts with Programs* 29 (6): A167.

Lidgard, S., F. K. McKinney, and P. D. Taylor. 1993. Competition, clade replacement, and a history of cyclostome and cheilostome bryozoan diversity. *Paleobiology* 19:352–71.

Lupia, R., S. Lidgard, and P. R. Crane. 1999. Comparing palynological abundance and diversity: Implications for biotic replacement during the Cretaceous angiosperm radiation. *Paleobiology* 25:305–40.

MacLeod, N., P. F. Rawson, P. L. Forey, F. T. Banner, M. K. Boudagher-Fadel, P. R. Bown, J. A. Burnett, P. Chambers, S. Carver, S. E. Evans, C. Jeffery, M. A. Kaminski, A. R. Lord, A. C. Milner, A. R. Milner, N. Morris, E. Owen, R. R. Rosen, A. B. Smith, P. D. Taylor, E. Urquhart, and J. R. Young. 1997. The Cretaceous-Tertiary biotic transition. *Journal of the Geological Society, London* 154:265–92.

McGhee, G. R. 1980. Shell form in the biconvex articulate Brachiopoda: A geometric analysis. *Paleobiology* 6:57–76.

McKinney, F. K. 1986a. Evolution of erect marine bryozoan faunas: Repeated success of unilaminate species. *American Naturalist* 128:795–809.

———. 1986b. Historical record of erect bryozoan growth forms. *Proceedings of the Royal Society, London,* series B, 228:133–49.

———. 1991. How phylogeny limits function: The example of *Exidmonea. National Geographic Research and Exploration* 7:432–41.

———. 1992. Competitive interactions between related clades: Evolutionary implications of overgrowth interactions between encrusting cyclostome and cheilostome bryozoans. *Marine Biology* 114:645–52.

————. 1993. A faster-paced world? Contrasts in biovolume and life-process rates in cyclostome (class Stenolaemata) and cheilostome (class Gymnolaemata) bryozoans. *Paleobiology* 19:335–51.

————. 1995. One hundred million years of competitive interactions between bryozoan clades: Asymmetrical but not escalating. *Biological Journal of the Linnean Society* 56:465–81.

————. 2000. Colony sizes and occurrence patterns among Bryozoa encrusting disarticulated bivalves in the northeastern Adriatic Sea. In *Proceedings of the Eleventh International Bryozoology Association Conference*, ed. A. Herrera Cubilla and J. B. C. Jackson, 282–90. Balboa: Smithsonian Tropical Research Institute.

McKinney, F. K., and J. B. C. Jackson. 1989. *Bryozoan evolution.* Boston: Unwin Hyman.

McKinney, F. K., S. Lidgard, J. J. Sepkoski, Jr., and P. D. Taylor. 1998. Decoupled temporal patterns of evolution and ecology in two post-Paleozoic clades. *Science* 281:807–9.

McKinney, M. L. 1986. Estimating volumetric fossil abundance from cross-sections: A stereological approach. *Palaios* 1:79–84.

Miller, A. I. 1997. Dissecting global diversity patterns: Examples from the Ordovician radiation. *Annual Review of Ecology and Systematics* 27:85–104.

Mitchell, J. S., R. J. Cuffey, J. M. Masters, and J. A. Devera. 1997. Nodular cheilostome bryozoans from the basal Paleocene, southernmost Illinois. *Geological Society of America Abstracts with Programs* 29 (4): 61–62.

Moissette, P. 1988. Faunes de bryozoaires du messinien d'Algerie occidentale. *Documents des Laboratoires de Géologie Lyon* 102:1–351.

Okamura, B. 1984. The effects of ambient flow velocity, colony size, and upstream colonies on the feeding success of Bryozoa: I, *Bugula stolonifera* Ryland, an arborescent species. *Journal of Experimental Marine Biology and Ecology* 83:179–93.

Phillips, T. L., and R. A. Peppers. 1984. Changing patterns of Pennsylvanian coal-swamp vegetation and implications for climatic control on coal occurrences. *Journal of Coal Geology* 3:205–55.

Pohowsky, R. A. 1973. A Jurassic cheilostome from England. In *Living and fossil Bryozoa*, ed. G. P. Larwood, 447–61. London: Academic Press.

Poluzzi, A., and R. Sartori. 1975. Report on the carbonate mineralogy of Bryozoa. *Documents des Laboratoires de Géologie de la Faculté des Sciences de Lyon, hors série* 3:193–210.

Radenbaugh, T. A., and F. K. McKinney. 1998. Comparison of the structure of a Mississippian and a Holocene pen shell assemblage. *Palaios* 13:52–69.

Raup, D. M., and J. J. Sepkoski, Jr. 1982. Mass extinctions in the marine fossil record. *Science* 215:1501–3.

————. 1984. Periodicity of extinctions in the geologic past. *Proceedings of the National Academy of Sciences, USA* 81:801–5.

Rhodes, M. C., and C. W. Thayer. 1991. Mass extinctions: Ecological selectivity and primary production. *Geology* 19:877–80.

Richards, G. R. 1997. Interpolating a palaeoclimatic time series: Nonlinearity, persistence, and trending in the Cenozoic. *Palaeogeography, Palaeoclimatology, Palaeoecology* 128:12–27.

Rosenzweig, M. L., and R. D. McCord. 1991. Incumbent replacement: Evidence for long-term evolutionary progress. *Paleobiology* 17:202–13.

Roy, K. 1996. The roles of mass extinction and biotic interaction in large-scale replacements: A reexamination using the fossil record of stromboidean gastropods. *Paleobiology* 22:436–52.

Rucker, J. B., and R. E. Carver. 1969. A survey of the carbonate mineralogy of cheilostome Bryozoa. *Journal of Paleontology* 43:791–99.

Sandberg, P. A. 1971. Scanning electron microscopy of cheilostome bryozoan skeletons: Techniques and preliminary observations. *Micropaleontology* 17: 129–51.

———. 1983. Ultrastructure and skeletal development in cheilostomate Bryozoa. In *Treatise on invertebrate paleontology: G, Bryozoa, revised,* ed. R. A. Robison, 1:238–86. Boulder: Geological Society of America; Lawrence: University of Kansas.

Schubert, J. K., and D. J. Bottjer. 1992. Early Triassic stromatolites as post–mass extinction disaster forms. *Geology* 20:883–86.

Scott, R. W. 1995. Global environmental controls on Cretaceous reefal ecosystems. *Palaeogeography, Palaeoclimatology, Palaeoecology* 119:187–99.

Sepkoski, J. J., Jr. 1982. A compendium of fossil marine families. *Milwaukee Public Museum Contributions in Biology and Geology* 51:1–125.

———. 1993. Ten years in the library: New data confirm paleontological patterns. *Paleobiology* 19:43–51.

———. 1995. Patterns of Phanerozoic extinction: A perspective from global data bases. In *Global events and event stratigraphy in the Phanerozoic,* ed. O. H. Walliser, 35–51. Berlin: Springer-Verlag.

———. 1997. Biodiversity: Past, present, and future. *Journal of Paleontology* 71: 533–39.

Sepkoski, J. J., Jr., F. K. McKinney, and S. Lidgard. 2000. Competitive displacement among post-Paleozoic cyclostome and cheilostome bryozoans. *Paleobiology* 26:7–18.

Sheehan, P. M., P. J. Coorough, and D. E. Fastovsky. 1996. Biotic selectivity during the K/T and late Ordovician extinction events. *Geological Society of America Special Paper* 307:477–89.

Sheehan, P. M., and T. Hansen. 1986. Detrital feeding as a buffer to extinction at the end of the Cretaceous. *Geology* 14:868–70.

Signor, P. W. 1990. The geological history of diversity. *Annual Review of Ecology and Systematics* 21:509–39.

Smith, A. M., and C. S. Nelson. 1993. Mineralogical, carbonate geochemical, and diagenetic data for modern New Zealand bryozoans. Occasional Report 17. Hamilton, New Zealand: Department of Earth Sciences, University of Waikato.

———. 1994. Selectivity in sea-floor processes: Taphonomy of bryozoans. In *Biology and palaeobiology of bryozoans,* ed. P. J. Hayward, J. S. Ryland, and P. D. Taylor, 177–80. Fredensborg: Olsen and Olsen.

———. 1996. Differential abrasion of bryozoan skeletons: Taphonomic implications for paleoenvironmental interpretation. In *Bryozoans in space and time,* ed. D. P. Gordon, A. M. Smith, and J. A. Grant-Mackie, 305–13. Wellington, New Zealand: National Institute of Water and Atmospheric Research.

Smith, A. M., C. S. Nelson, and P. J. Danaher. 1992. Dissolution behaviour of bryozoan sediments: Taphonomic implications for nontropical shelf carbonates. *Palaeogeography, Palaeoclimatology, Palaeoecology* 93:213–26.

Spicer, R. A., and R. M. Corfield. 1992. Review of terrestrial and marine climates in the Cretaceous with implications for modelling the "Greenhouse Earth." *Geological Magazine* 129:169–80.

Staff, G., E. N. Powell, R. J. Stanton, Jr., and H. Cummins. 1985. Biomass: Is it a useful tool in paleocommunity reconstruction? *Lethaia* 18:209–32.

Stanley, S. M., and L. A. Hardie. 1998. Secular oscillations in the carbonate mineralogy of reef-building and sediment-producing organisms driven by tectonically forced shifts in seawater chemistry. *Palaeogeography, Palaeoclimatology, Palaeoecology* 144:3–19.

Stigler, S. M., and M. J. Wagner. 1987. A substantial bias in non-parametric tests for periodicity in geophysical data. *Science* 238:940–45.

Stott, L. D., and J. P. Kennett. 1990. The paleoceanographic and paleoclimatic signature of the Cretaceous/Paleogene boundary in the Antarctic: Stable isotopic results from ODP Leg 113. *Proceedings of the Ocean Drilling Program, Scientific Results* 113:829–48.

Taylor, P. D. 1990. Bioimmured ctenostomes from the Jurassic and the origin of the cheilostome Bryozoa. *Palaeontology* 33:19–34.

———. 1993. Bryozoa. In *The fossil record 2,* ed. M. J. Benton, 465–89. London: Chapman and Hall.

———. 1994. An early cheilostome bryozoan from the Upper Jurassic of Yemen. *Neues Jahrbuch für Geologie und Paläontologie Abhandlungen* 191:331–44.

Taylor, P. D., and G. P. Larwood. 1988. Mass extinctions and the pattern of bryozoan evolution. In *Extinction and survival in the fossil record,* ed. G. P. Larwood, 99–119. Oxford: Clarendon Press.

———. 1990. Major evolutionary radiations in the Bryozoa. In *Major evolutionary radiations,* ed. P. D. Taylor and G. P. Larwood, 209–33. Oxford: Clarendon Press.

Taylor, P. D., and M. A. Wilson. 1996. *Cuffeyella,* a new bryozoan genus from the Late Ordovician of North America, and its bearing on the origin of the post-Paleozoic cyclostomates. In *Bryozoans in space and time,* ed. D. P. Gordon, A. M. Smith, and J. A. Grant-Mackie, 351–60. Wellington, New Zealand: National Institute of Water and Atmospheric Research.

Todd, J. A. 2000. The central role of cheilostomes in bryozoan phylogeny. In *Proceedings of the 11th International Bryozoology Association Conference,* ed. A. Herrera Cubilla and J. B. C. Jackson, 104–35. Balboa: Smithsonian Tropical Research Institute.

Tschudy, R. H., C. L. Pillmore, C. J. Orth, J. S. Gilmore, and J. D. Knight. 1984. Disruption of the terrestrial plant ecosystem at the Cretaceous-Tertiary boundary, Western Interior. *Science* 225:1030–32.

Vermeij, G. J. 1987. *Evolution and escalation.* Princeton: Princeton University Press.

Viskova, L. A. 1995. Dinamika rodovogo raznoobraziya mshanok (klassy Steno-laemata i Eurystomata) v intervale trias-nyne. *Ekosistemiye peresmroyki i evo-lyutziya biosfery* 2:80–83.

Viskova, L. A., and I. P. Morozova. 1993. Evolyutzionnye preobrazovaniya mor-skikh mshanok i krizisnye situatzii fanerozoya. *Paleontologicheskii Zhurnal* 1993 (3): 49–55.

Voigt, E. 1985. The Bryozoa of the Cretaceous-Tertiary boundary. In *Bryozoa: Ordovician to Recent,* ed. C. Nielsen and G. P. Larwood, 329–42. Fredens-borg: Olsen and Olsen.

Walters, L. J., and D. S. Wethey. 1986. Surface topography influences competitive hierarchies on marine hard substrate: A field experiment. *Biological Bulletin* 170:441–49.

Wilkinson, L. 1990. *SYSTAT: The system for statistics.* Evanston, IL: Systat.

Wing, S. L., L. J. Hickey, and C. C. Swisher. 1993. Implications of an exceptional fossil flora for late Cretaceous vegetation. *Science* 363:342–44.

Winston, J. E. 1977. Distribution and ecology of estuarine ectoprocts: A critical review. *Chesapeake Science* 18:34–57.

Winston, J. E., and J. B. C. Jackson. 1984. Ecology of cryptic coral reef communi-ties: 6, Community development and life histories of encrusting cheilostome Bryozoa. *Journal of Experimental Marine Biology and Ecology* 76:1–21.

Wolfe, J. A., and G. R. Upchurch. 1986. Vegetation, climatic, and floral changes at the Cretaceous-Tertiary boundary. *Nature* 324:148–52.

Zullo, V. A., and W. B. Harris. 1987. Sequence stratigraphy, biostratigraphy, and correlation of Eocene through Lower Miocene strata in North Carolina. *Cushman Foundation for Foraminiferal Research Special Publication* 24:197–214.

Contributors

Ann F. Budd
Department of Geology
University of Iowa
Iowa City, IA 52242

Efstathia Bura
Department of Statistics
George Washington University
Washington, DC 20052

Leo W. Buss
Department of Ecology and
    Evolutionary Biology
Department of Geology and
    Geophysics
Yale University
New Haven, CT 06520

Mike Foote
Department of the Geophysical
    Sciences
University of Chicago
5734 South Ellis Avenue
Chicago, IL 60637

Jörn Geister
Geologisches Institut
Universität Bern
Baltzerstrasse 1
CH-3012 Bern
Switzerland

Stephen Jay Gould
Museum of Comparative Zoology
Harvard University
26 Oxford Street
Cambridge, MA 02138

Eckart Håkansson
Geological Institute
University of Copenhagen
Øster Voldgade 10
DK-1350 Copenhagen K
Denmark

Jean-Georges Harmelin
Centre d'Océanologie de Marseille
CNRS-UMR 6540
Station Marine d'Endoume
F-13007
Marseille
France

Lee-Ann C. Hayek
Smithsonian Institution
NHB MRC 136
Washington, DC 20560

Jeremy B. C. Jackson
Scripps Institution of
    Oceanography
University of California, San
    Diego
La Jolla, CA 92093
and
Smithsonian Tropical Research
    Institute
Apartado 2072
Balboa
Republic of Panama

Kenneth G. Johnson
Scripps Institution of
    Oceanography
University of California, San
    Diego
La Jolla, CA 92093

Nancy Knowlton
Marine Biology Research Division
Scripps Institution of
    Oceanography
University of California, San
    Diego
La Jolla, CA 92093-0202
and
Smithsonian Tropical Research
    Institute
Apartado 2072
Balboa
Republic of Panama

Scott Lidgard
Department of Geology
Field Museum of Natural History
1400 South Lake Shore Drive
Chicago, IL 60605

Frank K. McKinney
Department of Geology
Appalachian State University
Boone, NC 28608

Daniel W. McShea
Department of Biology
104 Biological Sciences Building
Duke University, Box 90338
Durham, NC 27708

Ross H. Nehm
Department of Integrative Biology
Museum of Paleontology
University of California
Berkeley, CA 94720

Beth Okamura
School of Animal and Microbial
    Sciences
University of Reading
Reading RG6 6AJ
United Kingdom

John M. Pandolfi
Department of Paleobiology,
    MRC-121
National Museum of Natural
    History
Smithsonian Institution
Washington, DC 20560-0121

Paul D. Taylor
Department of Paleontology
Natural History Museum
Cromwell Road
London SW7 5BD
United Kingdom

Erik Thomsen
Geological Institute
Aarhus University
Universitetparken
DK-8000 Aarhus C
Denmark

Index